CELL APOPTOTIC SIGNALLING PATHWAYS

CELL APOPTOTIC SIGNALLING PATHWAYS

CHARLES O. PICKENS
EDITOR

Nova Biomedical Books
New York

For permission to use material from this book please contact us:
Telephone 631-231-7269; Fax 631-231-8175
Web Site: http://www.novapublishers.com

LIBRARY OF CONGRESS CATALOGING-IN-PUBLICATION DATA

Cell apoptotic signaling pathways / Charles O. Pickens (editor).
p. ; cm.
Includes bibliographical references and index.
ISBN-13: 978-1-60021-507-0 (hardcover)
ISBN-10: 1-60021-507-6 (hardcover)
1. Apoptosis. 2. Cellular signal transduction. I. Pickens, Charles O.
[DNLM: 1. Apoptosis--physiology. 2. Signal Transduction--physiology. QU 375 C3923 2007]
QH671.C387 2007
571.9'36--dc22

2006037847

Published by Nova Science Publishers, Inc. ✢ New York

CONTENTS

PREFACE

Apoptosis is the regulated form of cell death. It is a complex process defined by a set of characteristic morphological and biochemical features that involves the active participation of affected cells in a self-destruction cascade. This programmed cell death plays a critical role in physiological functions such as cell deletion during embryonic development, balancing cell number in continuously renewing tissues and immune system development. Additionally, a dysregulation of apoptosis is underlying in numerous pathological situations such as Parkinson, Alzheimer´s disease and cancer. A number of studies have pointed out an association between consumption of fruits and vegetables, and certain beverages such as tea and wine, which are rich in polyphenols, with reduced risk of chronic diseases, including cancer. Apoptosis is also the regulatory mechanism involved in the removal of unnecessary cells during development and in tissue homeostasis in a wide range of organisms from insects to mammals. This new book focuses on cell apoptotic signalling pathways.

Chapter I - Sepsis and septic shock represent the systemic immunologic and pathophysiologic response to overwhelming infection. Approximately 800,000 cases of sepsis are admitted every year to hospitals in North America and despite aggressive antibiotics and supportive care, over 200,000 patients per year succumb to this disorder. Septic deaths are typically due to early refractory cardiovascular failure (hypotension or shock) or later multiple organ failure. Shock and organ failure may occur and progress to death despite the fact that complete eradication of the invading organism can usually be achieved with antibiotic therapy. Sepsis and septic shock are able to initiate a genetic/metabolic cascade in the host that progresses to organ injury, organ system dysfunction and death despite elimination of the initial trigger. The host response to septic shock includes regulation of the systemic inflammatory cytokine response, cell-adhesion molecule expression, inducible nitric oxide synthase (iNOS) induction and apoptosis. Here, we highlight the intra- and inter-cellular signal transduction pathways that modulate the host response to sepsis and we present data which suggests that the intrinsic and extrinsic apoptotic pathways are activated during sepsis. Identification and characterization of key signal transduction pathways that are modulated during sepsis will lead to a better understanding of the disease and to the development of therapeutics.

Chapter II - Apoptosis is a cell death program essential for development and tissue homeostasis in metazoan organisms. There are two major intracellular apoptotic signaling pathways defined as the extrinsic pathway or the death receptor pathway and the intrinsic pathway or the mitochondrial pathway. In this chapter we will focus on the key role played by mitochondria during apoptosis. The mitochondrial apoptotic pathway is linked to a change in the permeability of the outer mitochondrial membrane (OMM), which eventually leads to the release of intermembrane space apoptogenic factors such as cytochrome c, Smac/DIABLO, AIF, endoG, HtraA2/OMI and DDP/TIMM8a. Redistribution of these factors to the cytoplasm activates the apoptotic program, this however, is not the result of a simultaneous massive release of all these proteins but as some reports have pointed out, to a scheduled timing of their release. The permeabilization of the OMM is under the control of the Bcl-2 family of proteins and we will discuss the interactions of this family of proteins and the OMM through putative receptors (lipids or proteins). Since important alterations in the mitochondrial morphology and network are observed in the early stages of apoptotic cell death resulting in fragmentation of mitochondrial reticulum and remodeling of mitochondrial cristae, the mechanisms involving the fission/fusion machinery of mitochondria will also be discussed. Moreover some members of the Bcl-2 family of proteins such as Bax, Bak, Bid and Bcl-2 appear to have the capacity to modulate morphological remodeling of mitochondria suggesting a relationship between the apoptotic molecular events and the dynamics of mitochondrial morphology.

Chapter III - Intracellular pH plays an important role in the maintenance of normal cell function and has therefore to be tightly controlled. This control is largely based upon the activity of membrane transporters, among which the ubiquitous Na^{+}/H^{+} exchanger isoform 1 (NHE1). Besides its role as a pH and volume regulator, more recent works have suggested NHE1 as a multifaceted regulator, notably via its scaffolding and cytoskeleton anchoring properties. Such properties appear to be important notably for activation of survival pathways. Indeed, it has been recently shown that NHE1 forms a signalling complex through binding to ezrin/radixin/moiesin (ERM) proteins, which allows activation of the PI3K/Akt pathway and hence survival. Besides this role, the exchanger can be modulated by diverse extracellular or intracellular stimuli, leading to pH_i changes that have been implicated in cell proliferation and migration, as well as in cell death. Regarding proliferation, it has thus been shown that NHE1 expression is often up-regulated in cancer cells, possibly resulting from reactive oxygen species (ROS) disturbances. Due to this up-regulation, inhibition of NHE1 has been proposed as an alternative for cancer therapy. Indeed, intracellular acidification, such as that resulting from NHE1 inhibition, seems to be a common apoptotic feature helping for caspase activation. Also, a few diverse apoptotic stimuli have been reported to lead to NHE1 inhibition, notably through cleavage of the transporter by caspases. Despite several studies indicating activation of NHE1 during proliferation processes, such an effect has also been observed upon apoptotic stimuli such as toxic xenobiotics (eg. polycyclic aromatic hydrocarbons) or cytokine deprivation. However, the precise role for such an early activation in the apoptotic cascade has still to be determined and might involve mitochondrial dysfunction. The aim of this review is to present and discuss new findings implicating NHE1 as a central signalling event activated by stress conditions and modulating cell proliferation and death processes.

Chapter IV - Cells entering the route of programmed cell death are recognized and removed by professional phagocytes. The uptake of apoptotic cells prior to cell membrane disintegration, averts an inflammatory response in the surrounding tissue by preventing contact of neighboring cells with cellular constituents derived from secondary necrosis. During the last decade it became apparent that prevention/termination of inflammation not only is due to apoptotic cell clearance, but also results from apoptotic cell-derived signals, which provoke a macrophage phenotype shift. Along this phenotype switch, a proinflammatory mediator profile characterized by the formation of e.g. TNFα (tumor necrosis factor-α) or nitric oxide changes to an immunosuppressive one, dominated by the release of e.g. interleukin-10, TGF-β (transforming growth factor-β) or PGE_2 (prostaglandin E_2).

Cell clearance is tightly controlled by a variety of signaling molecules and receptors. Thus, phagocyte attraction is facilitated, among other signals by bioactive lipids secreted from apoptotic cells. At the same time, macrophages primed by contact with apoptotic cells produce antiinflammatory cytokines and upregulate antiapoptotic proteins which guarantee their survival in an otherwise hostile environment. Recognition, binding and subsequent uptake of apoptotic cells by macrophages involves so-called 'eat-me flag' on the surface of apoptotic cells but also the formation of soluble factors by dying cells ('remember me signals'). These alterations initiated during apoptosis orchestrate a process in macrophages known as 'alternative activation' to affect cell viability, cytokine production as well as NO and/or O_2^- formation. We discuss the impact of apoptotic cells on macrophage function and focus on apoptosis-derived signals, which determine apoptosis and inflammatory responses.

Chapter V - Acute pancreatitis has numerous causes, culminating in a common pathway of changes within the pancreatic acinar cell. The pancreatic acinar cell being the functional unit of the exocrine pancreas, it synthesizes, stores, and secretes digestive enzymes. Evidence has accumulated of the significance of apoptotic cell death in the systemic manifestation associated with acute pancreatitis and studies suggest that apoptotic cell death may play a considerable role in affecting mortality and morbidity in severe acute pancreatitis. Apoptosis in pancreatitis is mediated by caspase activation and mitochondrial dysfunction in acinar cells. It is suggested that caspases play an important role in apoptosis. The expression of caspases is a frequent event in chronic pancreatitis and its distribution pattern may reflect the functions of this protease in its role in the apoptotic pathway in atrophic acinar. Also acute pancreatitis modifies mRNA level of both pro and anti-apoptotic genes simultaneously. Numerous studies suggest that pancreatic bcl2, bax and p53 mRNA levels increased extensively in the mild acute pancreatitis and the levels were unchanged during severe acute pancreatitis. This article presents a brief overview of the apoptotic machinery in pancreatitis, including specific and indirect death signals and the underlying molecular events in cell.

Chapter VI - Introduction: Bone resorption and formation are closely linked during bone remodeling. The cessation of bone resorption precedes bone formation, and is characterized by osteoclast apoptosis. Although apoptosis has been studied in rodent models of osteoclast differentiation, little is known about osteoclast apoptosis in humans.

Osteoclast apoptosis: The two most common apoptotic pathways are the activation of one of the membrane death receptors, and that of the mitochondrion-activated pathway involving members of the Bcl-2 family. The induction of apoptosis by extracellular signals involves ligands related to TNF. Some death receptors have been described in detail, including CD95

(Fas) that binds the CD95 ligand (FasL), and the TRAIL receptors that bind TRAIL (TNF-Related Apoptosis-Inducing Ligand). Among others, caspase-8, -9, and -3 are situated at pivotal junctions in apoptotic pathways. Caspase-8 activation is initiated during receptor-induced cell death. Caspase-9 is activated in response to agents or insults that impair the internal Bcl-2 family members balance. Both caspase-8 and caspase-9 eventually lead to the activation of executor caspases, such as caspase-3 or caspase-7. When apoptosis occurs in osteoclasts, caspases are likely to be activated. Osteoclast apoptosis and its down-regulation by the survival factor M-CSF, may involve members of the Bcl-2 family, particularly Bcl-xL. The Fas-FasL and TRAIL-TRAILR systems may also be involved in osteoclast apoptosis; however TRAIL-/- or Fas-/- deficient mouse models suggest that these factors cannot be essential for bone biology under normal conditions. However, the two main apoptotic pathways, involving TRAIL receptors and Fas, could be implicated in human osteoclast apoptosis. The regulation of this process may be important in controlling bone homeostasis, and could be altered *in-vivo* under conditions characterized by a high level of osteoclast formation.

Conclusion: A better knowledge of the role of apoptosis as a major point of control of bone resorption, would fairly increase our understanding of osteoclast biology. This field is of considerable clinical importance, and can be expected to advance our therapeutic approach to diseases characterized by a high level of bone resorption.

Chapter VII - Apoptosis, or programmed cell death, is a physiological process that not only plays an essential role during development, morphogenesis, and tissue homeostasis but is also part of various defense mechanisms of multicellular animals. Following exposure to countless stimuli, physiological and non-physiological, apoptotic cells actively participate in their own demise through activation of an intrinsic program that consists of a series of signal cascades. Dysregulation of the mechanisms controlling these apoptotic cascades may result in various disorders, such as auto-immune or neurodegenerative diseases, tumorigenesis, and infections.

Sponges (phylum Porifera) represent the next closest relatives to the common ancestral phylum Urmetazoa, ancestor of all metazoan phyla. Thus, the discovery of core components of the poriferan apoptotic cell death machinery contributes to advancing our understanding of both molecular evolution and basic control of apoptotic mechanisms. The apoptotic molecules that have been preserved from Porifera to man include members of the Bcl-2 family (apoptotic regulators), death domain proteins (adapters), caspases (effectors), and various proteins, whose part in apoptosis has not been completely elucidated yet.

Since sponges are filter feeders, they are abundantly exposed to hazardous microorganisms. Consequently, they are provided with efficient defense mechanisms. However, breach of those defensive barriers inevitably requires the disposal of infected and of damaged cells through apoptotic processes. Here, we describe the identification of apoptotic and defensive molecules and their implication in the poriferan response to bacterial lipopolysaccharide (LPS), an endotoxic component of the outer envelope of Gram-negative bacteria. These poriferan proteins comprise the transcription factor LITAF_SUBDO (responsible for the regulated expression of apoptotic molecules), AIF_SUBDO (a mitochondrial pro-apoptotic effector), and VLP_SUBDO (mediator of the innate immune response). Characterization of the respective sponge molecules revealed both conserved as well as sponge specific features. In combination with functional and phylogenetic analyses we attempt to substantiate the overlapping pathways of poriferan apoptosis and innate immunity.

In: Cell Apoptotic Signaling Pathways
Editor: Charles O. Pickens, pp. 1-40

ISBN: 978-1-60021-507-0

Chapter I

MOLECULAR BIOLOGY OF SEPSIS: IDENTIFICATION AND CHARACTERIZATION OF KEY APOPTOTIC SIGNALLING PATHWAYS

Danielle Brabant[1], Paul Michael[1], Tyler Kirwan[1], Chilakamarti V. Ramana[2], Joseph E. Parrillo[3], Anand Kumar[3,4] and Aseem Kumar[1,*]

[1]Department of Chemistry and Biochemistry and the Biomolecular Sciences Programme, Laurentian University, Sudbury, ON, Canada, P3E 2C6;
[2]Department of Medicine, Yale University, New Haven, CT, USA, 06516;
[3]Division of Cardiovascular Disease and Critical Care Medicine, Cooper University Hospital, Robert Wood Johnson Medical School, Camden, New Jersey, USA, 08103;
[4]Section of Critical Care Medicine, University of Manitoba, MB, Canada, R3A 1R9.

ABSTRACT

Sepsis and septic shock represent the systemic immunologic and pathophysiologic response to overwhelming infection. Approximately 800,000 cases of sepsis are admitted every year to hospitals in North America and despite aggressive antibiotics and supportive care, over 200,000 patients per year succumb to this disorder. Septic deaths are typically due to early refractory cardiovascular failure (hypotension or shock) or later multiple organ failure. Shock and organ failure may occur and progress to death despite the fact that complete eradication of the invading organism can usually be achieved with antibiotic therapy. Sepsis and septic shock are able to initiate a genetic/metabolic cascade in the host that progresses to organ injury, organ system dysfunction and death despite elimination of the initial trigger. The host response to septic shock includes regulation of

[*] Correspondence concerning this article should be addressed to Aseem Kumar Ph.D. Department of Chemistry and Biochemistry, Biomolecular Sciences Programme Laurentian University, 935 Ramsey Lake Rd. Sudbury, ON, Canada, P3E 2C6. Tel: 705-675-1151 ext.2103; Fax: 705-675-4844; akumar@laurentian.ca.

the systemic inflammatory cytokine response, cell-adhesion molecule expression, inducible nitric oxide synthase (iNOS) induction and apoptosis. Here, we highlight the intra- and inter-cellular signal transduction pathways that modulate the host response to sepsis and we present data which suggests that the intrinsic and extrinsic apoptotic pathways are activated during sepsis. Identification and characterization of key signal transduction pathways that are modulated during sepsis will lead to a better understanding of the disease and to the development of therapeutics.

INTRODUCTION

Sepsis and septic shock are the leading causes of death in intensive care units in North America. Approximately 800,000 cases of sepsis are reported every year in the USA and 215,000 will succumb to the disease despite aggressive antibiotic and supportive care [1-7]. The incidence of sepsis is increasing at a rate of 9% per year even with advances in the understanding of the pathophysiology of the disease [1].

In 1992, the American College of Chest Physicians (ACCP) and the Society of Critical Care Medicine (SCCM) assigned broad definitions to describe sepsis and the different stages of the disease [8]. The concept of Systemic Inflammatory Response Syndrome (SIRS) was introduced and describes a systemic activation of the innate immune response. This response can be triggered by noninfectious stimuli, for example, SIRS may develop in patients with major trauma, burns or pancreatitis. Inclusion criteria for SIRS are: 1) temperature >38°C or <36°C; 2) heart rate >90 beats/min; 3) respiratory rate >20 breaths/min or $PaCO_2$ <32 mmHg; 4) white blood cell count >12 000 cells/μL or < 4 000 cells/μL or >10% immature band forms. In contrast, sepsis is defined as the systemic inflammatory response to an infection. More specifically, sepsis is defined as two or more of the SIRS criteria with a documented or suspected infection. Severe sepsis is sepsis with associated organ dysfunction, hypoperfusion or hypotension. Septic shock is sepsis-induced hypotension despite adequate fluid resuscitation which may be accompanied by perfusion abnormalities [8]. These broad definitions prove to be problematic for clinicians and researchers in relation to the precise characterization and staging of the disease. Recently, an international conference was held in order to revisit the common definitions as such and to modify them according to a better understanding of the pathophysiology of the disease [8].

Sepsis can be initiated with a single or localized infection in the host which is termed a nidus of infection [9]. Examples of such infections include but are not limited to abscess, pneumonia, peritonitis, pyelonephritis and cellulitis. The invading organism which include bacteria, parasites or viruses cause a local infection and may circulate in the bloodstream and release components to initiate an inflammatory response [10]. Bacteria cause the majority of sepsis cases, whereas viruses and parasites are reported in approximately 2-4% of the sepsis cases [1]. Gram-positive bacterial induced sepsis is found to be most common with an occurrence of 30-50%. Gram-negative bacterial induced sepsis, which was predominant in the 1970s and 1980s is less frequent with an incidence rate of 25-30% [1]. In addition, polymicrobial infections are identified in 25% of sepsis cases. The inflammatory response can be triggered by the presence of bacterial toxins, either endotoxins or exotoxins. For example, a component of the Gram-negative bacterial outer cell membrane, lipopolysaccharide (LPS), is a classic inducer of the inflammatory cascade [4,10,11]. Examples of bacterial exotoxins include Toxic Shock Syndrome Toxin-1 (TSST-1) and Toxin A released from *Staphylococcus* and *Pseudomonas* species, respectively. Further, structural components of the bacteria may initiate a host response including peptidoglycan and lipoteichoic acid from Gram-positive bacterial cell wall [1,6,9]. More recently, studies have demonstrated that pure bacterial DNA and RNA can stimulate a host response [9].

Initiating Molecules of the Inflammatory Response

The immunological host response to invading organisms includes innate and adaptive immunity [7,12,13]. The innate response is the body's first line of defense and is therefore responsible for the recognition of pathogens for subsequent elimination [7,12]. Immune cells including neutrophils, monocytes and macrophages are key components of the host response. These cells are capable of recognizing structural components or products from bacteria including LPS, peptidoglycan and lipoteichoic acid [12,14-18]. These products vary among different strains or species; however, all microbes possess an invariable pathogen-associated molecular pattern (PAMP) that can be recognized by pattern recognition receptors (PRRs) on innate immune cells [16,19,20]. Consequently, PRRs will specifically activate signalling pathways to mediate the inflammatory response by releasing pro-inflammatory mediators.

Gram-Negative Bacteria

Previously, the most common infections leading to the development of sepsis were initiated from Gram-negative bacteria [1]. The PAMP for this type of bacteria is LPS since it is the principal component of the bacterial outer cell membrane [6,15,21,22]. LPS is an amphiphilic macromolecule that is composed of three major elements [9,11]. These include the polysaccharide side chain, the phospholipid and the lipid A; the last component being the most toxic element of the endotoxin [9]. The lipid A moiety was shown to reproduce many clinical and metabolic manifestations observed during septic shock [9]. The presence of LPS in the body, even at low levels, activates monocytes and macrophages through recognition of the invariant pattern of the lipid A portion [15]. Initially, LPS forms a complex with a plasma protein called lipopolysaccharide binding protein (LBP) [9,15,17,22]. The LBP catalyzes the transfer of the LPS to a membrane bound receptor termed CD14 on the surface of monocytes and macrophages or to a soluble CD14 molecule found in the plasma for activation of CD14 negative cells such as endothelial cells [9,17,21]. CD14 along with Toll-like receptors (TLRs) belong to the family of PRRs. Subsequent to LPS binding to CD14, LPS is transferred to TLR 4 [9]. TLR4 belongs to a family of ten members which are characteristically identified by the presence of a leucine-rich repeat (LRR) domain in the extracellular region [12-14]. In addition, TLRs possess a cytoplasmic region that is homologous to the interleukin-1 receptor (IL-1R) referred to as the Toll/IL-1 receptor (TIR) domain [16]. Consequently, the signalling pathways between the pro-inflammatory cytokine IL-1 and the TLRs possess similarities. However, the binding of LPS to TLR4 is not sufficient to stimulate an adequate inflammatory response. Shimazu *et al.* have shown that an adaptor protein termed MD-2 interacts with TLR4 on the cell surface which is required in order to enhance the inflammatory response. Therefore, LPS recognition by immune cells necessitates the formation of a large complex whereby TLR4 will transduce the signal downstream to activate transcription factors such as Nuclear Factor-κB (NF-κB) and Activator Protein 1 (AP-1) that modulate the host response [17,23]. TLR4 is not exclusive to LPS recognition. It is also able to bind to other ligands such as Taxol, heat shock proteins (HSP 60 and 70) as well as host extracellular matrix components (fibronectin, hyaluronic acid and fibrinogen) [13].

Gram-Positive Bacteria

Gram-positive bacteria components such as peptidoglycan, a large polymer that provides most of the strength and rigidity of bacterial cell walls and teichoic acid, a diverse group of polymers found in the cell wall or cell membrane are also able to initiate the inflammatory response. The presence of these structural components will be recognized by another member of the TLR family, more specifically, TLR2 [12-14]. Similarly to LPS, peptidoglycan and lipoteichoic acid will bind to the TLR receptor in a CD14 dependent manner [18]. However, the accessory protein MD-2 is unique to TLR4 recognition and is not required for the activation of the inflammatory response through TLR2 [13]. Other ligands distinguished by TLR2 include lipoproteins, lipoarabinomannan (mycobacteria), atypical LPS and porins (*Neisseria*) [13]. In addition, researchers have shown that TLR2 is able to cooperate with other members of the TLR family to recognize other microbial components. More specifically, TLR1 and TLR6 are functionally associated with TLR2 and are able to discriminate between other ligands such as tri-acyl lipopeptides and di-acyl lipopeptides, respectively [13]. Upon ligand recognition by TLR2, a signalling pathway will be initiated to activate transcription factors such as NF-κB and AP-1 that modulate the inflammatory response.

Bacterial DNA

Another potent activator of the inflammatory response is bacterial DNA. Bacterial genomic DNA contains unmethylated CpG dinucleotides [13,24,25]. In contrast, mammalian DNA has a low level of CpG motifs and is usually found to be methylated. As such, innate immune cells are able to discriminate between their own DNA and prokaryotic DNA. Thus, CpG motifs are yet another example of a PAMP by which TLRs are able to recognize. In this case, TLR9 has been identified as the family member responsible for the stimulatory effects of the innate response in the presence of bacterial genomic DNA [13,14,16,26]. Unlike TLR1, 2 and 4 which are found to be expressed on the cell surface, TLR9 has been identified as being in the endosome [13,24]. Studies have demonstrated that CpG DNA enters the cell by endocytosis and that TLR9 is expressed on the surface of the endosome [13]. Consequently, TLR9 is able to initiate a signalling pathway to activate the AP-1 and NF-κB transcription factors.

Double-stranded RNA

The innate response has also evolved to recognize the presence of viral components such as double stranded RNA (dsRNA). dsRNA may be produced during the viral replication cycle [13]. It has been demonstrated to be a strong inducer of type I interferons such as interferon alpha (IFN-α) and interferon beta (IFN-β) [13]. Furthermore, dsRNA is able to promote the transcription of interferon inducible genes and thus stimulate the host response. A synthetic dsRNA called polyinosinic-polycytidylic acid [poly (I:C)] has been shown to exhibit similar activity to dsRNA [13,26]. The stimulatory effects of dsRNA were thought to originate from a protein termed dsRNA-dependent protein kinase (PKR). However, previous studies have indicated that other molecules are involved in viral recognition. More specifically, TLR3 is able to respond to the presence of dsRNA and elicit an immunological response through NF-κB activation [13,14,16].

Bacterial RNA

Bacterial RNA can also elicit an immunological response. More recently, the notion that TLRs can distinguish between mammalian and bacterial RNA through nucleoside modifications was investigated [27]. The majority of cellular RNA is comprised of ribosomal RNA (rRNA) and is modified in mammalian cells versus bacterial cells. For example, human rRNA possesses approximately ten-fold and twenty-five fold the amount of pseudoouridine and 2'-O-methylated nucleosides respectively when compared to bacterial rRNA [27]. Another component of cellular RNA is the transfer RNA (tRNA). Humans minimally possess 25% of tRNA modified nucleosides whereas bacterial tRNA is rarely modified. Similarly, bacterial messenger RNA (mRNA) does not contain nucleoside modifications as compared to human mRNA [27]. Typical examples of human mRNA nucleoside modifications include 5-methylcytidine (m5C), N6-methyladenosine (m6A), inosine, 2'-O-methylated nucleosides and N7-methylguanosine (m7G) [27]. Results from *in vitro* studies indicate that unmodified nucleosides are recognized by immune cells through TLR3, 7 and 8 [27].

TLR Receptor Signalling Pathways

As mentioned previously, a variety of stimuli are able to initiate a host response. These foreign antigens are specifically recognized by different members of the TLR family where they transduce the signal to downstream effectors. The most common effectors are the transcription factors AP-1, NF-κB and Interferon Regulatory Factor (IRF) [14]. Despite the fact that all the TLR members are able to activate NF-κB and Mitogen Activated Protein (MAP) kinases with the exception of TLR3, the resulting gene expression profile varies among the individual TLR [14]. A possible explanation for this specificity is the adaptor proteins being recruited by the TLRs.

MyD88-dependent Pathway

The most common intracellular signalling pathway utilized by all TLRs except for TLR3 is the activation of NF-κB and MAP kinases through the recruitment of the adaptor protein called Myeloid Differentiation Factor 88 (MyD88) [12,14]. The signalling pathway will begin in the common TIR domain on the cytoplasmic surface of each TLR. A homophilic interaction between the TIR domains of the TLR and the C-terminal portion of the adaptor protein MyD88 will occur [12,23]. In turn, the N-terminus of MyD88 contains a death domain which will recruit a death domain containing serine/threonine kinase called IL-1 receptor-associated kinase 4 (IRAK-4) [20,23]. Consequently, IRAK-4 becomes activated and phosphorylates IRAK-1. Activated IRAK-1 recruits TNF receptor associated factor 6 (TRAF6) which ultimately leads to the activation of the following kinases: 1) Inhibitor of NF-κB kinase (IKK) complex; 2) p38; and 3) Jun N-terminal kinase (JNK) [14,20]. The IKK pathway eventually leads to the activation of the transcription factor NF-κB whereas the p38 and JNK mitogen-activated protein kinase signalling pathways converge to activate the transcription factor AP-1 [12,23].

MyD88-independent Pathway

In addition, a MyD88-independent pathway has been identified. MyD88 deficient cells are able to activate NF-κB and JNK in response to LPS stimulation. However, a delayed reaction is observed [13]. In this pathway, a novel adaptor protein termed TIR domain-containing adaptor protein (TIRAP)/MyD88-adaptor-like (Mal) was discovered. TIRAP/Mal has a C-terminal TIR domain but lacks the N-terminal death domain observed in MyD88 [13-15,20]. Recruitment of IRAK-2 by TIRAP/Mal occurs and eventually leads to NF-κB and JNK activation [14,18]. Furthermore, stimulation of MyD88 deficient cells by LPS also activates IRF3 [13,15]. Consequently, IRF3 activation induces the expression of many IFN-inducible genes.

Other Adaptor Proteins

In addition to MyD88, other adaptor proteins have recently been identified that contribute to the specificity of TLRs. For instance, TLR4 and TLR3 are both able to recruit a novel protein called TIR-related adaptor protein inducing interferon (Trif) [15,26]. Trif signals through an IKK-like kinase called TBK-1 which is required for the phosphorylation and activation of the transcription factor IRF3 and IRF7 [20,26]. In addition, another adaptor protein was shown to be involved in TLR4 signalling. This protein named Trif-related adaptor molecule (Tram) has been shown to act as a bridging adaptor for the subsequent recruitment of Trif [14]. Therefore, this distinct individual TLR signalling may explain the various biological responses to different stimuli. Furthermore, PAMPs have been shown to activate transcription factors independent of the TLR family. For instance, LPS has been shown to activate another family of transcription factors, namely, the Signal Transducer and Activator of Transcription (STAT) [13]. STAT1 deficient mice were subjected to LPS and a reduction of IFN-regulated genes was observed. This data indicates a potential role of STAT1 in the response to LPS [13].

In summary, the recognition of PAMPs by innate immune cells will activate complex signalling pathways which ultimately lead to the activation of transcription factors such as NF-κB, AP-1, STAT and IRF. Accordingly, these transcription factors will induce the inflammatory response through the expression of inflammatory mediators such as cytokines, chemokines, cell-adhesion molecules and inducible nitric oxide synthase (iNOS) [19].

Transcription Factors

NF-κB

Nuclear Factor-κB (NF-κB) proteins play a significant role in the host response against pathogen invasion. Recent studies in animal models of sepsis and septic shock have identified NF-κB as having a key role in the pathophysiology of sepsis [28]. NF-κB is a family of structurally related proteins that form homodimers and heterodimers with various members of the NF-κB/Rel family [28,29]. Such members include NF-κB1 (p50/p105), NF-κB2 (p52/p100), RelA (p65), RelB and C-Rel. NF-κB1 (p50) and NF-κB2 (p52) are synthesized as precursor p105 and p100, respectively [19,28-30]. Members of this family can associate to form a variety of dimers which have distinct properties and are able to differentially regulate gene expression. The most common association is the heterodimer p50/p65 [28]. A common characteristic of this family is the presence of a 300 amino acid Rel homology domain. The Rel domain is important for dimerization, as well as DNA binding. It is also responsible for binding to a family of inhibitory proteins called Inhibitors of NF-κB (IκB). In unstimulated cells, NF-κB subunits are bound to IκB proteins. These proteins include IκBα, IκBβ, IκBγ, IκBε, IκBζ, Bcl-3, p105 and p100 [19,28]. These inhibitory proteins can interact with NF-κB proteins though their characteristic ankyrin repeat domain. This interaction masks the nuclear localization sequence on NF-κB proteins and thus, serves as a regulatory mechanism by sequestering the transcription factor to the cytoplasm [28,29]. The IκB proteins will be phosphorylated by IKKs and thus signalled for ubiquitination and degradation upon NF-κB activation by LPS or cytokines [28-31]. Therefore, the NF-κB proteins can translocate to the nucleus and bind to its cognate regulatory element to modulate gene expression.

NF-κB broadly regulates biological functions including the regulation of inflammatory mediators. In addition, it has been found to play a pivotal role in the regulation of cell proliferation and apoptosis [28,30]. Under normal conditions, NF-κB generates pro-survival signals by modulating the anti-apoptotic genes such as inhibitors of apoptosis-1 and -2 (IAP-1 and IAP-2), the X chromosome-linked inhibitor of apoptosis protein (XIAP) and the cellular Fas-associated death domain-like IL-1β-converting enzyme (FLICE)-like inhibitory protein (c-FLIP), an inhibitor of caspase 8 [28,29]. However, if the activity of NF-κB is deficient or inhibited, the cell will undergo apoptosis. In contrast, recent findings have indicated that activation of NF-κB can lead to apoptosis through the expression of pro-apoptotic genes such as Fas ligand and c-Myc. A plausible explanation for this contradiction is the fact that NF-κB may switch from one response to another depending on the time and the stimulus [28].

NF-κB also plays an important role in adaptive and innate responses. For example, NF-κB is required for hematopoiesis and for myeloid and lymphoid cell differentiation in the adaptive response [28]. Similarly, during innate immunity, NF-κB plays a pivotal role in the development and maturation of hematopoietic cells such as macrophages and neutrophils [28]. More importantly is NF-κB's role in the inflammatory response. NF-κB modulates the expression of many pro-inflammaotory mediators such as cytokines, chemokines, cell-adhesion molecules and iNOS [19,29]. An increased transcriptional level of pro-inflammatory mediators is thought to be the hallmark of the pathophysiology of sepsis [28]. Previous studies have implicated NF-κB in the pathogenesis of septic shock. Approximately

200 genes that have been implicated in the pathophysiology of sepsis and septic shock contain NF-κB regulatory elements in their promoters [28]. NF-κB regulated genes include: cytokines (TNF-α, IL-1β); chemokines (macrophage inflammatory protein (MIP)-1α, regulated on activation normal T cell expressed and secreted (RANTES); adhesion molecules (ICAM-1, VCAM-1, E-selectin); and enzymes (iNOS, phospholipase A2). Interestingly, the activity of NF-κB seems to correlate with the severity of the disease [19,28].

AP-1

The Activator Protein 1 (AP-1) is a complex of DNA binding proteins that belong to the basic region-leucine zipper (bZIP) family [32-34]. The basic leucine zipper is required for DNA binding as well as for interaction with other family members to form dimers. The formation of dimers is essential for recognition and binding to the palindrome DNA promoter sequence [32,34]. The AP-1 complexes can form homodimers with Jun family members (c-Jun, v-Jun, Jun-B and Jun-D) or heterodimers with Fos family members (c-Fos, Fos-B, Fra-1, or Fra-2) [32-34]. The properties of the basic leucine zipper of Fos family members does not allow homodimerization [32]. In addition, Jun and Fos can form heterodimers with other bZIP family members such as Activating Transcription Factor (ATF) and Maf [33]. Therefore, the subunit composition of the AP-1 complexes can vary significantly which leads to differential gene expression [34].

The heterodimer c-Jun/c-Fos is expressed in the majority of cell types [33]. However, a variety of stimuli can promote dimerization. These stimuli include cytokines (TNF-α, IL-1, IFN-γ), growth factors, serum, oncoproteins and cellular stress [32,34]. The induction of AP-1 activity upon these stimuli indicates that AP-1 is involved in growth regulation as well as inflammatory and innate immune responses [34]. Several stimuli dependent mechanisms exist that lead to AP-1 activation. For example, growth factors will induce AP-1 activity through a subfamily of the mitogen-activated protein kinases (MAPKs) called the extracellular signal-regulated kinases (ERK) [32,34]. Upon pro-inflammatory cytokine stimulation, the activation of AP-1 will most likely utilize two other MAPKs subfamilies called JNK and p38 [34]. The p38 MAPKs are involved in *fos* and c-*jun* gene induction [34]. Subsequently, JNK will bind and phosphorylate c-Jun thereby increasing the transcriptional activity. The Fos/Jun heterodimers subsequently bind to AP-1 promoter sites and regulate the gene expression. AP-1 regulates the expression of numerous cytokines and tissue-destructive enzymes such as collagenase [33]. Similar to NF-κB, AP-1 has also been shown to induce the expression of apoptotic and survival related genes. For example, AP-1 has previously been shown to induce the expression of the proapoptotic gene Fas-ligand (FasL) [34]. Cyclin D is another important gene regulated by AP-1 and may play a role in proliferation and cell death [34]. Cyclin D primarily induces cell proliferation, however, when overexpressed, cyclin D contributes to apoptosis in certain cell types [34]. An additional target gene for induction of apoptosis by AP-1 is p53. High levels of c-Jun repress p53-mediated p21 induction thereby allowing p53 activity to promote apoptosis [34]. In summary, AP-1 is a transcription factor that is activated in response to growth factors and cytokines that mediates gene expression involved in inflammatory and immune responses as well as cell survival and apoptosis.

STAT

The Signal Transducers and Activators of Transcription (STAT) are cytoplasmic proteins that belong to a family of seven members [35,36]. The STAT members are STAT1, STAT2, STAT3, STAT4, STAT5a, STAT5b and STAT6 [37-39]. A wide variety of extracellular signalling polypeptides can activate STAT transcription factors. Examples of STAT activators are the epidermal growth factor (EGF), platelet-derived growth factor (PDGF), colony-stimulating factor-1 (CSF-1), the pro-inflammatory cytokines interferon alpha (IFN-α), interferon beta (IFN-β), interferon gamma (IFN-γ) and interleukins (IL) 2 to 7, 10 to 13, and 15 [35,37,38]. During cytokine signalling, the extracellular ligand binds to its cognate cell surface receptor and activate the Janus protein tyrosine kinases (JAKs) [39,40]. The JAKs autophosphorylate and phosphorylate tyrosine residues on the cytoplasmic domain of the surface receptor [37]. Consequently, the STAT proteins are recruited to the receptor through their src-homology 2 (SH2) domains which recognize and bind to phosphorylated tyrosines [35,36]. The bound STAT proteins will in turn be tyrosine phosphorylated by JAKs. Once phosphorylated, the STAT proteins will form homodimers or heterodimers depending on the ligand and dissociate from the receptor. Once dissociated, the dimer translocates to the nucleus and regulate gene expression [35,37,40,41]. Conversely, non-cytokine activators such as EGF, PDGF and CSF-1 are able to activate the STATs directly through their receptor tyrosine kinases, and therefore, do not require the JAK activation.

Differential gene expression will occur due to the various dimer combinations. The majority of the STAT proteins form homodimers, however, heterodimers such as STAT1-STAT2 as well as STAT1-STAT3 are able to form in response to cytokines [35,37]. Subsequently, the dimers translocate to the nucleus and bind to specific promoter sequences and mediate gene expression of target genes. The dimerized STATs bind to a specific sequence called the gamma-activating sequence (GAS) [37]. The GAS is present in IFN-γ inducible genes including transcription factors (IRF1, IRF8), adhesion molecules (ICAM) and chemokines (IFN-γ-induced monokine [MIG]) [42].

However, if STAT1-STAT2 heterodimer forms a complex with a DNA binding adaptor protein called p48, also known as IRF9 and ISGF3γ, the complex will preferentially bind to an interferon alpha-stimulated gene response element (ISRE) [37]. The ISRE is present in a variety of genes including; transcription factors (IRF1, IRF2); cytokines (IL-6, IL-2); enzymes (2'-5' oligoadenylate synthetase, iNOS, PKR), chemokines (chemokine ligand 5 [CCL5]) and adhesion molecules (ICAM-1, VCAM-1) [42]. In summary, the STAT family members broadly regulate cell cycle, cell adhesion, apoptotic and iNOS genes which are found to be implicated in the pathogenesis of sepsis.

IRF

Interferon Regulatory Factor (IRF) is a family of transcription factors comprised of nine members; IRF1, IRF2, IRF3, IRF4, IRF5, IRF6, IRF7, IRF8 and IRF9/p48/ISGF3γ [39,43]. All members contain a characteristic DNA binding motif in the N-terminal region to which they bind to ISRE, interferon consensus sequence (ICS) or interferon regulatory factor element (IRF-E) also known as positive regulatory domain (PRD) I and III [44-47]. In general, the expression of IRF members is constitutive and is activated by post translational modifications such as phosphorylation or acetylation. Certain IRF members can be induced in response to cytokines such as IFNs, TNF, IL-1, IL-6, dsRNA and viral infection [39]. The IRF family members differentially regulate the transcription of many genes including: cytokines (IFN-α, IFN-β, IL-18), chemokines (RANTES, IP-10) and enzymes (PKR, iNOS, 2'-5'OAS) [42,44]. These genes are directly involved in the innate immune response against viral infections [39]. Furthermore, IRFs are involved in regulating processes such as cell growth inhibition, hematopoietic differentiation, cytokine signalling and apoptosis [43,44].

In summary, transcription factors such as NF-κB, AP-1, STAT and IRF are activated in response to pathogen invasion. Consequently, these proteins will modulate the expression of hundreds of genes involved in biological processes such as inflammatory and innate responses, cell cycle and apoptosis.

Inflammatory Mediators

The inflammatory response is designed to eradicate the microorganism and allow for potential tissue repair. However, an excessive pro-inflammatory response may lead to tissue injury and organ dysfunction [7]. Therefore, the host response should be tightly regulated in order to maintain a balance between a protective inflammatory response and one that is detrimental to the individual. During sepsis, this balance is shifted towards an overwhelming pro-inflammatory response and thus leads to cellular dysfunction, tissue damage, organ dysfunction and potentially, death [7]. This next section will describe important pro-inflammatory mediators which have been shown to play a role in the pathogenesis of sepsis. Cytokines are important mediators of the inflammatory response and also serve to amplify the host response. Of these, tumor-necrosis factor alpha (TNF-α), interleukin-1 beta (IL-1β), and interferon gamma (IFN-γ) have been shown to be elevated in serum from septic patients [48,49].

Tumor Necrosis Factor-alpha

Tumor necrosis factor-alpha (TNF-α) is a potent cytokine synthesized by a variety of cell types which include macrophages, monocytes and lymphocytes in response to inflammation or injury [49-51]. TNF-α plays a pivotal role in the immune response by modulating the expression of other cytokines, chemokines and cell-adhesion molecules in order to promote the extravasation and attraction of leukocytes to the site of infection [51]. However, systemic and chronic production of TNF-α which occurs during sepsis, can have harmful effects on various tissues and organs possibly resulting in death [50]. For example, TNF-α induces fever (via IL-1β) and stimulates the acute-phase response (via IL-6) [49,50]. Further, TNF-α activates the coagulation cascade and induces myocardial depression as well as vascular leakage when present at high concentrations [50]. This finding was confirmed through the administration of TNF-α to animals in a variety of sepsis experiments. The physiological response of the animals to TNF-α closely mimicked the clinical manifestations observed during septic shock including cardiovascular collapse, hypotension, intravascular coagulopathy and myocardial suppression [50]. Elevated levels of circulating TNF-α have been reported in patients with sepsis and septic shock [49]. In addition, a correlation was established between elevated levels of TNF-α and the severity or poor outcome of septic patients [49].

TNF-α exerts its cellular effects when the soluble trimer binds to its cell-surface receptor [49]. There are two distinct cell surface receptors that have been identified for TNF-α: type I or TNFR1 and type II or TNFR2 which preferentially bind to soluble and membrane bound TNF-α, respectively [49,51-54]. The majority of cell types constitutively express TNFR1 whereas TNFR2 is tightly regulated. The majority of the biological actions of TNF-α will be initiated through TNFR1 [49,50,52,53]. Binding of TNF-α to its receptors will ultimately lead to the activation of the transcription factors AP-1 and NF-κB. TNF-α can also initiate a caspase signalling cascade leading to apoptosis [51,52]. Recent findings of the formation of two distinct signalling complexes have provided a plausible explanation in the selection of either the pro-survival or apoptosis processes [55].

The TNF-α homotrimer will bind to TNFR1 which will initiate receptor trimerization. This conformational change will result in the recruitment of the adaptor protein called TNFR1-associated death domain protein (TRADD) through TNFR1's cytoplasmic death domain (DD) [55]. Once TRADD is recruited to TNFR1, it will serve as a platform to recruit subsequent proteins. These secondary proteins are receptor-interacting protein 1 (RIP1), TNF-receptor-associated factor 2 (TRAF2) and a member of the Inhibitors of Apoptosis family called cIAP1 [55]. Hence, TNFR1, TRADD, RIP1, TRAF2 and cIAP1 will form a complex termed Complex I which is formed early after receptor activation. Consequently, RIP1 will recruit a macromolecule called the "signalosome" [55]. The signalosome is composed of three major proteins; inhibitor of NF-κB kinase 1 (IKK1), inhibitor of NF-κB kinase 2 (IKK2) and inhibitor of NF-κB kinase γ (IKKγ) also called NEMO [56]. These kinases are able to phosphorylate the inhibitor proteins of NF-κB (IκB). NF-κB is tightly regulated by residing in the cytoplasm when associated with IκB. Once phosphorylated, IκB will be ubiquitinated and degraded by the proteosome [56]. Once IκB is degraded, the liberated NF-κB monomers dimerize and translocate to the nucleus where they bind to their cognate regulatory elements and modulate the expression of target genes.

In contrast, a few hours after TNFR1 activation or possibly after receptor internalization, the TRADD-based complex can dissociate from the receptor [55]. Once dissociated, this allows for the formation of Complex II through the recruitment of Fas-associated death domain protein (FADD) and Caspase 8 [55]. This complex will promote cell death which will be discussed in further details in the following sections.

In summary, TNF-α is a primary mediator of the immune response through the activation of transcription factors such as NF-κB and AP-1. Hence, TNF-α is able to amplify the host response by modulating the expression of inflammatory mediators.

Interleukin-1 Beta

Interleukin-1β (IL-1β) is a pro-inflammatory cytokine that is produced mainly by mononuclear phagocytes and neutrophils when triggered by bacterial products such as LPS or by cytokines such as TNF-α [7,49]. IL-1β is synthesized as a 31 kDa precursor and is proteolytically cleaved by interleukin-1 converting enzyme (ICE), also known as caspase 1, into an active 17 kDa fragment [57]. IL-1β plays a pivotal role in the regulation of innate immunity by stimulating the expression of genes such as chemokines, cytokines and cell surface receptors. Similar to TNF-α, the biological effects of IL-1β are dependent on its concentration. At low levels, IL-1β will primarily act on endothelial cells to increase surface molecules for leukocyte recruitment as well as promote a coagulation response [7]. In high quantities, IL-1β will exert systemic effects such as fever, acute phase plasma protein production, leukocytosis and thrombocytosis [58]. Previous studies have indicated that IL-1β may play a role in the pathogenesis of sepsis [48]. For example, deletion of the IL-1 receptor in mice resulted in increased resistance to endotoxin lethality. Furthermore, increased levels of IL-1β have been documented in spontaneous human sepsis and septic shock [48]. Elevated levels of IL-1β have been correlated with increased mortality. TNF-α, and IL-1β are found to be the primary mediators of septic shock [49,58].

IL-1β binds to two membrane receptors, the IL-1 receptor 1 (IL-1RI) and the IL-1 receptor 2 (IL-1RII). However, only IL-1RI transduces the signal to downstream effectors whereas IL-1RII acts as a decoy [59]. Upon IL-1β binding to IL-1RI, IL-1RI will recruit an accessory protein called IL-1RAP through heterodimerization of their common cytoplasmic TIR domains [57]. Consequently, signalling will occur through the recruitment of MyD88, IRAK1, IRAK4 and TRAF6 adaptor proteins as previously described for TLR MyD88 dependent signalling pathway [57]. Therefore, IL-1β will induce the activation of NF-κB and MAPK pathways and amplify the host inflammatory response.

Interferon Gamma

Interferon-γ (IFN-γ), also known as type II interferon, is a pro-inflammatory mediator produced by activated natural killer (NK) cells and T-helper cells [39,48]. Cytokines such as TNF-α, IL-2 and IL-12 will enhance the transcription of IFN-γ [42,60]. IFN-γ is involved in biological processes which include antiviral and immune responses, cell growth and differentiation and apoptosis [39,42]. The main role of this pro-inflammatory cytokine is to activate mononuclear phagocytes [60]. IFN-γ initiates signalling pathways that culminate in the synthesis of enzymes implicated in the respiratory burst and thus allows for macrophages to phagocytose the invading microbe [60]. In addition, IFN-γ activates neutrophils which lead to an up-regulation of the respiratory burst. Also, vascular endothelial cells are activated by IFN-γ thereby stimulating lymphocyte adhesion and extravasation. Further, IFN-γ increases the expression of Class II Major Histocompatibility Complex (MHC) on the surface of a wide variety of cell types [39,60]. As a result, there is enhanced antigen presentation on T cells which amplifies the recognition phase of the immune response. IFN-γ has also been implicated in the pathogenesis of septic shock. *In vitro* administration of LPS, TNF-α, IL-1β and IFN-γ has been shown to have synergistic effects mimicking the adverse hemodynamic effects observed during septic shock [48].

IFN-γ binds to its cognate cell surface receptor which is composed of two subunits; the IFN-γ ligand-binding IFNGR-1 subunit and the accessory IFNGR-2 subunit [39]. Upon ligand binding, heterodimerization of the receptor subunits will occur and allow for subsequent signal transduction through the JAK-STAT pathway. IFNGR-1 subunit will interact with JAK-1 whereas IFNGR-2 interacts with JAK-2 [39]. These kinases will recruit latent cytoplasmic STAT 1 and allow for homodimerization. Subsequently, the dimer will translocate to the nucleus and bind to the GAS sequence and regulate the expression of target genes [39]. Therefore, IFN-γ amplifies the inflammatory response through STAT activation and expression of IFN-induced proteins.

Other Important Cytokines and Mediators

Transforming growth factor-beta (TGF-β) is an important cytokine which has been recently implicated in the pathogenesis of septic shock. TGF-β is expressed in nearly all cell types and plays a pivotal role in the normal development of embryonic tissues including the heart. In adults, TGF-β has been implicated in tissue repair and fibrosis. During sepsis, TGF-β counteracts the actions of TNF-α and IL-1β in the acute phase of the inflammatory response [49]. The synergistic combination of TNF-α and IL-1β has been shown to partially mediate myocardial depression that occurs during septic shock [61]. A recent study demonstrated a dose dependent reversal of TNF-α and/or IL-1β induced myocardial contraction depression by TGF-β [62].

Interleukin 6 (Il-6) is a cytokine that is described as having both pro- and anti-inflammatory effects [63]. IL-6 is synthesized by activated monocytes, macrophages, vascular endothelial cells, fibroblasts and activated B and T lymphocytes [48-50]. IL-6 levels are influenced by other cytokines, including TNF-α and IL-1 in humans [63]. The biological actions of IL-6 are manifested in hepatocytes, B cells and other cell types. IL-6 induces the production of acute-phase hepatic proteins [48-50] and as a result, there is an augmentation in circulating fibrinogen and other proteins that are involved in the transportation and coagulation system. In addition, IL-6 functions as a growth and differentiation factor for activated B cells [50]. IL-6 does not seem to be directly involved in stimulating tissue injury, however, IL-6 plasma levels are steadily elevated in patients with sepsis [48,50,63]. Also, a correlation has been established between the elevated levels of IL-6 and poor outcome in sepsis patients [63] and this implies that IL-6 may be an important marker of systemic inflammation [50].

Macrophage migration inhibitory factor (MIF) is a pro-inflammatory cytokine which is ubiquitously produced in organs including the heart [64]. Upon exposure to a variety of stimuli including endotoxin, toxic shock syndrome toxin 1 and cytokines (TNF-α and IFN-γ), MIF is produced by monocytes and macrophages [49]. MIF will amplify the host response by increasing the production of other pro-inflammatory mediators. Elevated levels of MIF have been documented in patients in septic shock [49] and previous studies have identified MIF as a myocardial depressant factor implicated in LPS induced cardiac dysfunction [64].

High mobility group box 1 protein (HMGB1) has also been implicated in the pathogenesis of sepsis and more specifically, Gram-negative sepsis. HMGB1 was initially identified as a nuclear DNA binding protein and transcription factor [65] and has been recently characterized as a cytokine that binds to cell surface receptors including TLR2 and TLR4. HMGB1 plays a role in the following processes: fever, epithelial barrier dysfunction, chemotactic cell movements, release of pro-inflammatory cytokines and acute inflammation [65,66]. HMGB1 is expressed by macrophages after exposure to early pro-inflammatory cytokines which include TNF-α and IL-1 [49]. Administration of LPS to mice resulted in high levels of HMGB1 in the serum 8-32 h post exposure. Also, elevated levels of HMGB1 were noted in patients succumbing to septic shock [49,67].

In summary, cytokines are rapidly produced in response to invading pathogens by mononuclear phagocytes and other cell types. These cytokines will bind to their cognate cell surface receptors and stimulate the transcription of genes involved in the host response.

Inducible Nitric Oxide Synthase

Nitric oxide (NO) is a cellular mediator that is produced in the endothelium and other cell types [68]. NO is a potent vasodilator that mediates the acute inflammatory response by increasing the blood flow and therefore, optimizes the delivery of soluble mediators and inflammatory cells to the site of infection [68]. NO is produced by the enzyme called NO synthase (NOS) [68] and the three NOS isoforms are endothelial NOS (eNOS), neuronal NOS (nNOS) and inducible NOS (iNOS). eNOS and nNOS are constitutively expressed whereas iNOS is induced in response to a variety of stimuli which include microbial products and cytokines (TNF-α, IL-1, IFN-γ) [50,68]. Monocytes/macrophages, neutrophils, endothelial cells and smooth muscle cells are all capable of producing NO by iNOS. NOS utilizes the essential amino acid L-arginine as its substrate to yield NO and L-citrulline [69]. In the absence of the substrate, NOS generates a product called superoxide. The production of NO, in combination with the vasodilator substance prostaglandins, will subsequently cause smooth muscle relaxation through a guanosine 3', 5'-cyclic monophosphate (cGMP) mechanism [50,68]. Therefore, NO that is generated in the endothelium plays an important role in regulating the vascular tone, blood pressure and blood flow distribution [69,70]. Other major roles of NO are its ability to inhibit leukocyte aggregation as well as reduce platelet aggregation.

During sepsis, LPS and pro-inflammatory cytokines such as TNF-α, IL-1β, and IFN-γ contribute to the induction of iNOS [50,68], consequently, excessive amounts of NO are produced. The role that iNOS plays in the pathogenesis of septic shock has been the subject of controversy for many years. Evidence supporting this hypothesis is found in animal models of sepsis and septic shock which have shown increased plasma levels of NO byproducts such as nitrate and nitrite [68]. Administration of inhibitors to NO synthesis prevented LPS induced hypotension and showed decreased plasma levels of nitrate and nitrite [68]. Interestingly, administration of iNOS inhibitors several hours post-endotoxin introduction were shown to reverse the effects of septic shock which include hypotension and vascular leaks [68]. Other evidence supporting NO's role in the pathogenesis of sepsis include studies using iNOS knock-out mice. LPS and cecal ligation puncture (CLP) models of septic shock both demonstrated increased survival and clinical improvements in the iNOS knockouts when compared to wild type mice [68]. In contrast, several experimental studies have demonstrated that administration of inhibitors of NO synthesis had no beneficial effects. It has also been shown that in human models of sepsis and septic shock, measurements of plasma levels of nitrite and nitrate in patients revealed elevated levels. Treatments with certain inhibitors of NO synthesis in clinical trials have demonstrated short term hemodynamic effects which include improvements in the mean arterial pressure and systemic vascular resistance. However, a decrease in the cardiac output was also noted [68]. Recent clinical trials had to be terminated due to increased mortality in the treatment group. In summary, NO may contribute to the cardiovascular alterations observed during sepsis and septic shock including vasodilation, hyporeactivity to vasopressors and myocardial dysfunction. NO rapidly reacts with other free radicals, for example superoxide radical (O_2^-), to generate highly reactive oxidant species including peroxynitrite that lead to tissue damage. However, blocking NO synthesis as a therapeutic intervention for sepsis is still a matter of debate.

Cell-adhesion Molecules and Chemokines

The acute inflammatory response to infection requires the recruitment of leukocytes to the site of inflammation in order to eradicate the pathogen. Bacterial products and cytokines will induce the expression of adhesion molecules on the surface of endothelial cells and leukocytes to mediate this interaction [60,71]. Adhesion molecules play vital roles in processes which include cell growth and differentiation. However, during an inflammatory response, adhesion molecules are responsible for immune cell transmigration and response [60,71]. Endothelial cells and leukocytes possess complementary surface adhesion molecules which allows the leukocyte to roll along the endothelium following a gradient of chemokines and the leukocyte can eliminate the pathogen by phagocytosis [50,71]. Initially, leukocytes will migrate to the endothelium, also called tethering/rolling, through selectin mediated interactions. Selectins are a family of surface adhesion molecules comprised of three members. These include E-selectin, P-selectin and L-selectin [50,71]. P-selectin is constitutively expressed and stored in endothelial cells and platelets. Upon stimulation by leukotrienes, oxidants and complement, P-selectin will be expressed on the surface of endothelial cells. E-selectin is expressed by endothelial cells in response to LPS and cytokines such as IL-1. L-selectin is expressed on leukocytes and, upon endothelial cell activation, undergoes a conformational change. Once released, L-selectin will accumulate in the plasma as a soluble protein [72]. E-selectin will mediate the initial interaction with activated leukocytes. The leukocytes will tether and roll along the endothelium and subsequently become firmly adhered to the endothelium through integrin binding. Integrins are expressed on the surface of leukocytes and will interact with ligands such as intercellular adhesion molecules (ICAMs) on the surface of endothelial cells [50]. This firm adherence will allow for subsequent diapedesis and chemotaxis [60].

Once the leukocytes are firmly attached to the endothelium, they will penetrate the endothelium and migrate to the extravascular inflammatory area. This process is also mediated by adhesion molecules which include the platelet-endothelial cell adhesion molecule-1 (PECAM-1) [50,60]. Therefore, the tethering/rolling, adherence and diapedesis are mediated by interaction of adhesion molecules on leukocytes and endothelial cells which ultimately leads to the infiltration of leukocytes to the site of infection [50,60].

Adhesion molecules are an integral part of the host response through the recruitment of leukocytes to the site of infection for subsequent elimination of the pathogen. However, excessive expression and activation of these molecules will lead to tissue injury as seen during sepsis and septic shock [73]. During sepsis, E-selectins, P-selectin, ICAM-1, ICAM-2 and VCAM-1 are overexpressed and cause the recruitment of leukocytes at non-infectious locations [73]. Consequently, there is destruction of the endothelial barrier which results in loss of fluid from the circulation into the surrounding tissue. Therefore, during sepsis, the overexpression of adhesion molecules induces the destruction of healthy tissue.

Chemoattractants such as complement components and chemokines also play a pivotal role in the elimination of the invading pathogen [50]. These soluble molecules are necessary for the migration of leukocytes to the site of infection [74]. In addition, chemokines have been shown to be involved in other biological processes including cellular differentiation and activation, angiogenesis/angiostasis, lymphocyte recirculation and immune surveillance [74]. Four families of chemokines have been identified thus far based on the cysteine residues in the N-terminus [50]. These include: CCL1 through CCL28; CXCL1 through CXCL16; XCL1 through XCL2; and CX_3CL1 [74]. Chemokines are further classified as being either constitutive or inducibly expressed. Inducible chemokines play a role in the recruitment of cells to the site of injury [74]. For example, IL-8 (CXCL8), RANTES (CCL5), MIP-1β (CCL4), MCP-1 (CCL2) and IP-10 (CXCL10) are induced in response to stimuli such as LPS, IL-1 and TNF-α [74].

Chemokines have been shown to play a key role during sepsis through immuno-modulating mechanisms. A correlation between chemokines and tissue pathology, organ dysfunction and failure has been documented [74]. Elevated levels of IL-8 (CXCL8) have been measured in the blood of septic patients [74]. IL-8 plays a role in the recruitment of neutrophils. Increased levels of MCP-1 (CCL2) have correlated with the disease course of sepsis. Experimental data have revealed increased expression of MIP-1α (CCL3) and RANTES (CCL5) upon LPS stimulation [74]. Similarly, increased levels of MIP-1α (CCL3) and RANTES (CCL5) have been documented in human septic patients [74]. Therefore, chemokines are also implicated in the pathogenesis of septic shock.

Complement System

The complement system is also an integral part of the innate immune response. Its main role is to promote inflammation and eliminate the pathogen. The complement system is mostly involved in controlling bacterial infections, preferentially Gram-negative infections [75]. Microbes have been shown to activate the complement system during tissue injury and cellular injury associated with burns or traumas. There are three major complement pathways that ultimately lead to the destruction of the invading organism. These include: the classical pathway activated by surface bound IgM or IgG antibodies; the alternative pathway stimulated by microbial surface molecules; and the lectin pathway activated in response to mannose-binding lectin [50]. Consequently, these pathways, through a series of enzymatic cascades, will generate C3a and C3b, cleavage products from the complement component C3 [50,76]. C3a is a neutrophil chemoattractant whereas C3b will promote phagocytosis by binding to the surface of microbes. C3b is able to initiate the cleavage of the complement component termed C5 by associating with other components to form a proteolytic complex [50]. The cleavage products of C5 are C5a and C5b [50]. Similar to C3a, C5a is a chemoattractant for neutrophils and also alters the vascular permeability at the site of infection [76]. The cleaved C5b proteins will form pores on membrane structures by associating with other factors such as C6, C7, C8 and C9 which ultimately lead to cell lysis.

Therefore, an inflammatory response will activate the complement system and produce inflammatory mediators such as C5a and C3a which can activate neutrophils and thus amplify the host response. During sepsis, the complement system is overactive and there is excessive production of C5a [76,77]. Recent evidence has shown that the amount of C5a in the serum correlates to the severity of the disease [77]. The generation of C5a compromises neutrophil function during sepsis.

Coagulation

Recently, the interaction of the coagulation pathway with the inflammatory cascade has been implicated in the pathogenesis of sepsis and Multiple Organ Dysfunction Syndrome (MODS) [78]. The coagulation pathway is activated in response to tissue injury or infection. There are two pathways in the coagulation cascade consisting of the intrinsic pathway and the extrinsic pathway [50]. The intrinsic pathway is generally activated in response to tissue injury whereas the extrinsic pathway is triggered by the production of tissue factor (TF) [50]. These pathways will converge and lead to the increased production of thrombin. Consequently, thrombin will cleave fibrinogen into fibrin forming a blood clot [50].

During sepsis, the extrinsic pathway is the primary mediator of the coagulation response to the infection [50]. TF is produced by endothelial cells and monocytes in response to cytokines (TNF-α, IL-1 and IL-6) and C-reactive protein [7,50,79]. In turn, this triggers the extrinsic coagulation pathway which ultimately leads to the formation of thrombin. In addition, the coagulation cascade will contribute to a pro-inflammatory response. For example, production of thrombin and tissue factor-VIIa can induce the production of pro-inflammatory cytokines such as TNF-α. Therefore, TF plays a key role in mediating the interaction between inflammation and coagulation [50].

As a protective mechanism to prevent an uncontrolled pro-coagulant state, the body produces several factors to counteract the coagulation cascade. The most common factors are anti-thrombin (AT), the protein C and protein S system and the tissue factor pathway inhibitor (TFPI) [50,79-81]. Protein C and its cofactor protein S play an important role in mediating the coagulation and inflammatory responses [50,78,79]. Protein C can inactivate factors which include Va and VIIIa and this inactivation delays the clotting cascade. Furthermore, protein C possesses anti-inflammatory properties by inhibiting the transcription factors NF-κB and AP-1 [78,81]. During sepsis, protein C, protein S and AT are being depleted by the continuous activation of the coagulation pathway and therefore are present in very low levels [50,78,81]. Increased mortality has been shown to correlate with low levels of protein C in septic patients.

In addition, during sepsis, the fibrinolytic system which is involved in mediating the removal of clots or thrombus from injured tissue once the homeostasis is restored, is inhibited [7,79,81-83]. The two main inhibitors of this system are plasminogen activator inhibitor-1 (PAI-1) and thrombin activateable fibrinolysis inhibitor (TAFI). PAI-1 levels are increased in response to LPS thereby leading to impaired fibrinolysis further promoting coagulation [79,83].

The overwhelming immune response to the infection leads to an excess production of many inflammatory and non-inflammatory mediators as described above. These mediators collectively lead to cellular dysfunction. As a result, diffuse tissue injury in both affected and previously unaffected locations will occur. The combination of the uncontrolled inflammatory and coagulation responses will amplify the host response and result in tissue hypoxia and ischemia. Consequently, there will be deterioration of organ function and associated metabolic abnormalities potentially leading to death.

Apoptosis

Apoptosis is a highly regulated physiological process that serves to eliminate unwanted cells. This process, often referred to as programmed cell death, is critical during development and is necessary in maintaining homeostasis and to clear damaged or abnormal cells [84-88]. Apoptosis can be morphologically distinguished from other types of cell death by the presence of membrane blebbing, chromatin condensation, apoptotic body formation and DNA fragmentation [84,86-88]. Apoptosis is a conserved evolutionary mechanism that utilizes proteases called caspases to initiate and execute cell death [89].

Caspases

Caspases play a critical role in the initiation and execution of apoptosis. Currently there are 14 mammalian caspases that have been identified [87,88,90]. Based on homology sequences, the family of caspases has been categorized into three subfamilies. These include the inflammatory mediator caspases 1, 4, 5, 11, 12, 13 and 14; the apoptosis initiator caspases 2, 8, 9 and 10; and the apoptosis executioner caspases 3, 6 and 7 [87,88,90,91]. Each member of the caspase family is constitutively expressed as proenzymes and contains 3 domains; an N-terminal prodomain, a large subunit and a C-terminal small subunit [87,90]. Initiator caspases contain an extended N-terminal prodomain of >90 amino acids. In contrast, effector caspases have a smaller prodomain of approximately 20-30 amino acids [88]. The long prodomain in the initiator caspases contains sequence motifs that allow them to interact with adaptor molecules. For instance, caspases 8 and 10 contain two death effector domains (DED) whereas caspases 2 and 9 contain a caspase-recruitment domain (CARD) [90,92]. Activation of caspases require dimerization and/or proteolytic cleavage [93,94]. For example, caspase 9 is generally present as monomers and must cluster through adaptors such as APAF-1 which is required for activation [87,89]. Proteolytic cleavage of caspases, which is generally required for their activation, are cleaved immediately following an aspartic acid residue [87]. This residue is located between the N-terminal prodomain and the large subunit as well as in the interdomain linker region separating the large and small subunits [87]. The initiator caspases will generally act upstream of the effector caspases [87,95]. An example of a caspase activated cascade is the activation of the initiator caspase 8 which leads to the activation of the effector caspase 3. A substrate of this effector caspase is DNA fragmentation factor (DFF) 45. In the inactive form, DFF 45 is complexed to the nuclease DFF 40 thereby inhibiting its activity [87,95,96]. Upon cleavage by caspase 3, DFF 45 releases DFF 40 and consequently allows DFF 40 to cleave chromatin.

Inhibitors of Apoptosis

Cells are able to negatively regulate the activity of caspases through a family of proteins called inhibitors of apoptosis (IAPs) [97]. Thus far, eight mammalian IAPs have been discovered including XIAP, c-IAP1, c-IAP2, ML-IAP/Livin, ILP-2, NAIP, Bruce/Apollon, and survivin [97-99]. These proteins inhibit apoptosis through several mechanisms, for example ligands and signal transducers in the TNF-α pathway may be blocked by these molecules [87]. In addition, pro-apoptotic members of the Bcl-2 family as well as cytochrome c may be targeted and blocked by IAPs [87]. More importantly, IAPs may directly interact with caspases 3, 7 and 9 [97,100]. For example, caspase 9 is primarily inhibited by a member of the IAP family called XIAP. Similarly, XIAP is the predominant inhibitor for caspases 3 and 7, however, other members of the IAPs such as c-IAP1 and c-IAP2 are also capable of inhibiting the caspase activity but to a lesser extent [88]. IAP family members are characterized by the presence of one to three copies of a baculoviral IAP repeat (BIR) domain [88,97]. XIAP, c-IAP1 and c-IAP2 all contain three tandem BIR copies along with a RING zinc finger domain at their C-termini [88,97]. Despite the presence of three BIR domains, data has shown that these domains are not equivalent in binding and inhibiting caspases. For example, XIAP's BIR3 domain potentially inhibits caspase 9 whereas the linker region between BIR1 and BIR2 inhibit caspase 3 and 7 [88,97,101,102]. RING domains which possess an E3 ubiquitin ligase activity have been shown to be essential in the targeted degradation by the ubiquitin-proteosome system [90,101]. Ubiquitin is covalently attached at a lysine residue of the target protein. The addition of one ubiquitin (monoubiquitination) results in the subcellular localization of the protein. The addition of oligomeric ubiquitin chains (polyubiquitination) results in protein degradation by the 26S proteosome. Ubiquitination is accomplished through a series of steps starting with ubiquitin conjugation to E1, E2 and finally being transferred to the target protein by E3 (ubiquitin ligase enzyme) [100-104]. Therefore, IAPs are able to self target for degradation [100]. For example, the C-terminal RING domain will initiate polyubiquitination at a lysine residue in the XIAP polypeptide. Therefore, XIAP's autoubiquitination would hinder its function of inhibiting apoptosis and allow cell death to occur more readily. In addition, IAPs are also able to degrade other proteins which include caspase 3 and Smac/DIABLO [100,101]. c-IAP1 and c-IAP2 also possess a CARD domain near their RING zinc finger domain which suggests a possible CARD-CARD interaction between c-IAPs and caspases. In summary, IAPs are able to inhibit apoptosis either by binding, inhibiting or degrading caspases.

Intrinsic Pathway

There are two main pathways for the activation of caspases; the intrinsic pathway and the extrinsic pathway. The intrinsic pathway also known as the mitochondrial pathway is activated by intracellular stimuli such as DNA damage (p53), nutrient starvation and growth factor withdrawal [87,88,93]. The mitochondrial pathway serves as an amplification mechanism for the receptor mediated apoptosis. The key element in this pathway is the release of pro-apoptotic factors from the mitochondria which include cytochrome c and apoptosis inducing factor (AIF). When these factors are released, a cytosolic complex called the apoptosome will form in the presence of dATP which consists of cytochrome c, apoptotic protease activating factor-1 (Apaf-1) and procaspase 9 [87,89,93]. More specifically, cytochrome c will bind to the cytosolic protein Apaf-1 and cause a conformational change which leads to Apaf-1 activation. The Apaf-1/cytochrome c will oligomerize into a heptameric structure similar to a wheel in the presence of dATP. The centre of the heptameric structure contains seven N-terminal CARD domains which interacts with the CARD domains of individual procaspase 9 molecules [89]. The proximity of procaspase 9 molecules will lead to dimerization and activation of caspase 9. In turn, procaspase 3 will be recruited to the complex where caspase 9 will cleave the effector caspase into its active form. Caspase 3 will complete the execution phase of apoptosis by cleaving its substrate DFF 45 and allowing the nuclease to fragment the chromosomal DNA [87,89].

Bcl-2 Family

Important regulators of the mitochondrial pathway are the members of the Bcl-2 family which serve as gatekeepers for the mitochondria [88]. Also, they play an essential role in the crosstalk between extrinsic and intrinsic pathways to amplify the response [88,105]. To date, more than two dozen members have been categorized as belonging to the Bcl-2 family [88]. All members contain at least one of four characteristic Bcl-2 homology domains (BH1-BH4) [87,106,107]. According to their structure and function, the Bcl-2 proteins are divided into three categories: 1) pro-survival including Bcl-2, Bcl-X_L, Bcl-w, Mcl-1, A1 (Bfl-1), NR-13 and Boo (Diva) contain all four BH domains [87,92,108]. These anti-apoptotic proteins are inhibitors of Bax/Bak. This inhibitory action is accomplished by either direct binding to Bax/Bak, prevention of Bax/Bak binding to mitochondrial proteins or by counteracting the effects of Bax/Bak on the mitochondria [92,106]; 2) pro-apoptosis including Bax, Bak and Bok (Mtd) which share sequence homology to Bcl-2 in the BH1, BH2 and BH3 domains [87,92,108]. These proteins exist as monomers but upon stimulation, will homo-oligomerize and form pores on the mitochondrial membrane thereby allowing the release of apoptotic factors such as cytochrome c [88,92,106]; 3) BH3-only family including Bid, Bim, Bad, Bik, Blk, Hrk (DP5), Bnip3, Bim_L, and Noxa contain the BH3 domain and exhibit pro-apoptotic activity [107]. BH3-only proteins will act mainly by binding and neutralizing the pro-survival proteins Bcl-2/Bcl-X_L. The inhibition of Bcl-2/Bcl-X_L will allow the activation of Bax/Bak and trigger mitochondrial permeabilization [87,106-108]. The members of the BH3-only family are activated by a variety of stresses. For example, Bid, Bad and Bim are activated by death receptors/Caspase 8, glucose deprivation and cytokine withdrawal, respectively [92]. According to the rheostat model, the levels of Bcl-2/Bcl-XL versus Bax/Bak are at the core of the cell's decision to survive or die [92].

Smac/DIABLO

Smac/DIABLO (second mitochondria-derived activator of caspase/direct IAP-binding protein with low pI) is a protein that is normally localized to the mitochondria. After exposure to lethal apoptotic stimuli, Smac/DIABLO is released into the cytosol from the mitochondria with cytochrome c and mediates apoptosis [88,99]. Cytochrome c activates Apaf-1 whereas Smac/DIABLO promotes caspase activation by associating with the apoptosome and by negatively regulating the IAPs. Smac/DIABLO binds to IAPs thereby preventing the IAP-caspase interaction [88]. For example, Smac/DIABLO binds to the BIR3 domain of XIAP [90]. Since Smac/DIABLO has a dimeric scaffold, it can bind more tightly to XIAP as opposed to the monomeric caspase 9 protein [88]. Consequently, Smac/DIABLO-XIAP complex undergoes ubiquitination and proteosome degradation resulting in significant XIAP depletion and allows for caspase activation [88]. In addition, evidence has been provided that implicates Smac in TNF-α-mediated caspase 8 activation [109].

Caspase 2

Caspase 2 is a unique cysteine protease since it possesses features from both initiator and effector caspases. More specifically, caspase 2 contains a long prodomain which is typical of initiator caspases. In addition, caspase 2 possesses a VDVAD substrate specificity which is common in effector caspases [110]. Caspase 2 is activated in response to a variety of stimuli including UV radiation, treatment with etoposide, trophic factor withdrawal, cytokine deprivation, TNF-α, Fas ligation and more recently, *Salmonella typhimurium* pathogen [111]. Upon activation, caspase 2 will initially be proteolytically cleaved into two fragments consisting of a 32-33 kDa fragment representing the prodomain and large subunit and a 14 kDa fragment containing the small subunit [111]. A second proteolytic step will generate 18 and 12 kDa fragments representing the large and small subunits, respectively. In general, the 32-33 kDa fragment is used as a marker for caspase 2 activation [111].

Recent evidence has shown that caspase 2 is involved in cell death but its exact role is not well understood [112]. Initial studies have indicated that caspase 2 plays a role in TNF-α-induced apoptosis [111]. Caspase 2 contains a CARD domain at the N-terminus which is able to interact with the death adaptor proteins called RAIDD/CRADD [110,111]. RAIDD/CRADD associates with RIP and TRADD and therefore may be responsible for the recruitment of caspase 2 to the TNFR1 [111]. In contrast, Paroni *et al.* have identified caspase 2 as being an effector caspase during TNF-α induced apoptosis [113]. Paroni and colleagues demonstrated that caspase 2 processing is dependent on caspase 3. Furthermore, Paroni and others showed that caspase 2 mediated apoptosis required caspase 9. These results indicate that TNF-α induced apoptosis will activate caspase 8 and 3. Consequently, caspase 3 will process caspase 2. Following its activation, caspase 2 regulates mitochondrial integrity possibly by inducing the translocation of Bid to the mitochondria and allows for the release of cytochrome c and Smac/DIABLO which regulates caspase 9 activation [113]. These results indicate that caspase 2 plays a regulatory function in apoptotic pathways. This regulatory role was also suggested by Droin *et al.* This group reported that caspase 2 can be alternatively spliced into 2 isoforms including a long isoform procaspase-2L and a short isoform caspase-2S. Data generated from Droin and colleagues' study indicated that caspase-2L induces cell death whereas caspase-2S antagonizes cell death induced by DNA damage and serum deprivation [114]. In contrast, Zheng *et al.* have reported caspase 2 involvement in compensatory apoptotic pathways in the absence of other caspases such as caspase 3 and/or caspase 9 during Fas-induced apoptosis [115].

In addition, previous studies have shown that caspase 2 localizes to subcellular compartments including the golgi complex, the mitochondria, the nucleus and the cytoplasm [110,112,116]. Aside from it being constitutively expressed in the nucleus, caspase 2 also contains two nuclear localization sequences in the prodomain. The exact role of caspase 2 is unclear but recent evidence has linked active nuclear caspase 2 to the release of cytochrome c from the mitochondria [112].

Procaspase 2 can also interact with another protein termed ARC through its CARD domain [110,117]. The role of this interaction in modulating apoptosis has been controversial. One study indicated that ARC and procaspase 2 led to inhibition of apoptosis whereas a subsequent study demonstrated that this interaction promoted cell death [110]. In addition, a novel protein called pro-apoptotic caspase adaptor protein (PACAP) has been found to interact with the CARD domain of procaspase 2 and procaspase 9 [110,117]. PACAP possibly triggers apoptosis in a caspase dependent manner. Also, PACAP potentially plays a role similar to the apoptosome where recruitment of procaspase 2 in a complex will lead to its activation [110]. Alternatively, procaspase 2 forms a large complex called the PIDDosome which contains the death domain-containing protein PIDD and RAIDD in response to genotoxic stress [110,117].

Also, caspase 2 interacts with the signal transducer JNK in response to stress activation [110]. In this case, JNK seems to contribute to caspase-2-dependent apoptosis signalling upstream of the mitochondria. Caspase 2 was also found to play a role in STAT1-mediated apoptosis [110].

Furthermore, caspase 2 is suggested to play a role in pro-inflammatory pathways. Recent studies showed that caspase 2 can interact with TRAF2 and RIP1 [117]. RIP1 will associate with NEMO/IKK-γ thereby activating the IKK complex. In turn, IKK will phosphorylate IκB triggering its ubiquitination and degradation for subsequent translocation of NF-κB to the nucleus. Also, RIP1 can phosphorylate p38 MAPK leading to enhanced transcription activation [117]. Therefore, activated caspase 2 can induce the release of mitochondrial components and contribute to cell death. In addition, caspase 2 is involved in the activation of transcription factors.

JNK

Recent evidence has shown JNK to be involved in mediating apoptosis [118]. The JNK group of MAP kinases is comprised of three members including JNK1, JNK2 and JNK3 [119,120]. JNK1 and JNK2 are ubiquitously expressed whereas JNK3 is expressed only in certain tissues which include the heart, brain and testis [118]. Initially, JNK was discovered as a regulator of AP-1 transcription activity. JNK phosphorylates c-Jun in response to extracellular stimuli including cytokines, cell stressors such as UV radiation and viruses [118,120]. In addition, JNK is able to phosphorylate other members of the AP-1 family such as Jun B, Jun D and ATF2 [118-120]. Consequently, the activation of these transcription factors will modulate the expression of genes involved in processes which include cell proliferation, differentiation and apoptosis. The role of JNK in regulating apoptosis is highly controversial because it has been found to have both pro-survival and pro-apoptotic activities and these activities are dependent on the cell type and the stimulus [119]. Recent data has indicated that TNF-α-mediated apoptosis is dependent on JNK [109]. Binding of TNF-α to its cell surface receptor leads to the activation of multiple signalling pathways leading to the activation of NF-κB and JNK. NF-κB is pro-survival and JNK is pro-apoptotic and the balance between NF-κB and JNK will determine the survival outcome of the cell [121]. Activation of NF-κB suppresses apoptosis through the expression of anti-apoptotic genes including cIAPs, XIAP, TRAF-1 and TRAF-2 [28]. In addition, NF-κB has also been shown to induce the expression of c-FLIP which specifically inhibits caspase 8 [121]. Furthermore, Kamata *et al.* have recently demonstrated that NF-κB is responsible for regulating the duration of JNK activation. NF-κB is able to induce the expression of antioxidant proteins such as superoxide dismutase (SOD) 2 which prevents oxidative inhibition of JNK phosphatases thus inhibiting its activation [121,122]. Therefore, inhibition of NF-κB during TNF-α signalling will result in sustained expression of JNK and promote TNF-α-mediated apoptosis. The exact mechanism by which JNK promotes cell death is not fully understood. Initially, it was proposed that TNF-α mediated caspase 8 cleavage required a JNK dependent pathway involving the BH3 protein Bid and Smac/DIABLO [109,121]. More specifically, activated JNK will cleave Bid into a novel form called jBid. jBid can translocate to the mitochondria thereby inducing the release of Smac/DIABLO. Subsequently, Smac/DIABLO will disrupt the c-IAP1-TRAF2 interaction in the TNFR1 complex I [109]. Displacement of c-IAP1 from the complex will relieve the inhibition imposed by c-IAP1-TRAF2 and allow for the activation of caspase 8 [109].

In 2006, Papadakis *et al.* reported that JNK could induce the release of cytochrome c from the mitochondria through the regulation of Bax in response to cadmium stress [123]. These data indicated that JNK will promote the accumulation of Bax at the mitochondria and subsequently, the mitochondria will release cytochrome c into the cytosol. In addition, JNK was identified as being essential for increasing Bax expression [123]. Therefore, Papadakis and colleagues provided a link between JNK and the mitochondrial pathway. Activated JNK will cleave Bid into jBid which induces the release of Smac/DIABLO from the mitochondria and ultimately caspase 8 activation. In type II cells, caspase 8 will amplify the apoptotic response by cleaving Bid into tBid for the release of proapoptotic factors from the mitochondria and induce the intrinsic pathway. Simultaneously, active JNK can increase Bax expression and promote Bax accumulation at the mitochondria and induce mitochondrial apoptosis [123].

In contrast, Chang *et al.* recently demonstrated a novel role for JNK in TNF-α-induced cell death. This group showed that JNK is able to induce c-FLIP degradation [121]. As discussed previously, NF-κB regulates the transcription of the gene *c-Flip*. c-FLIP levels control the progression from TNFR1 complex I to TNFR1 complex II which induces apoptosis. The *c-Flip* gene can express two isoforms including c-FLIP$_L$ and c-FLIP$_S$ [121]. A major difference between these two isoforms is the presence of a caspase-8-like domain (CASP) in the C-terminal region of c-FLIP$_L$ which is lacking in c-FLIP$_S$. Therefore, c-FLIP, more specifically, c-FLIP$_L$ will inhibit caspase 8. It has been shown that in response to TNF-α, JNK will phosphorylate a member of the HECT family of E3 ligases called Itch [121]. Subsequently, Itch will selectively interact with c-FLIP$_L$ through the C-terminal CASP domain and induces its degradation through the ubiquitin-proteosome system. The low levels of c-FLIP$_L$ will promote the formation of complex II and promote TNF-α-mediated apoptosis through activation of caspase 8 [121].

Extrinsic Pathway

The extrinsic pathway is activated by binding of an external stimulus such as Fas ligand (FasL) and TNF-α to its cognate cell surface receptor [92]. These death ligands are homotrimeric and thus cause oligomerization of the receptors upon ligand binding. The receptors contain a death domain (DD) in their intracellular region which will form homotypic interactions with adaptor proteins such as Fas-associated death domain (FADD) and TNFR-associated death domain (TRADD) [87,92]. Subsequently, these receptor-associated adaptor molecules will aggregate and recruit initiator caspases, such as caspase 8 or 10 through their common DED. For example, FasL binding to its receptor Fas will cause a conformational change on the cytoplasmic region of the receptor. The DD of Fas will interact with the DD of FADD. FADD also contains a DED and thus is thought to interact with the prodomain of procaspase 8 which contains two DEDs. Consequently, procaspase 8 will oligomerize and therefore complete the formation of the multicomponent death-inducing signal complex (DISC) [92,93]. Caspase 8 is possibly activated in the DISC by an induced proximity model. This model is based on the theory that a high concentration of procaspase 8 will lead to autoproteolysis and subsequent activation of the caspase cascade [87,92,93]. TNF-α signalling involves the formation of two complexes based on a temporal manner as previously described. Complex I comprised of TNFR1, TRADD, TRAF2, RIP1, and cIAP1 will form upon TNF-α binding to its receptor. TRADD-based complex dissociation from the receptor leads to the recruitment of the FADD and procaspase 8 formation of Complex II [55]. The balance between complex I and complex II is potentially regulated by a protein called c-FLIP [55]. c-FLIP is expressed by NF-κB through the activation of complex I. c-FLIP is a caspase 8 inhibitor since it is structurally similar to caspase 8 but lacks the proteinase site [55,87,93]. Therefore, c-FLIP can be recruited to the complex but is not able to transduce the death inducing signal. Thus, complex II is able to actively participate in cell death when NF-κB activation is inadequate. Activation of caspase 8 leads to the cleavage of downstream effector caspases such as caspase 3 which ultimately leads to DFF 40 activation.

The extrinsic pathway will require the activation of the intrinsic pathway for an amplification mechanism in certain cell types. In Type I cells, the production of caspase 8 is sufficient to activate caspase 3 and the cell is able to directly undergo apoptosis. In Type II cells, the amount of FADD being recruited is very low resulting in a limited amount of active caspase 8 [124]. Therefore, the extrinsic pathway will require the involvement of the mitochondria to execute apoptosis. The crosstalk between these two pathways is dependent in part on the cleavage of the Bcl-2 family member called Bid. Cleavage of the BH3 domain of Bid will release a proapoptotic fragment called truncated Bid (tBid) [105]. Subsequently, Bax/Bak will homooligomerize and allow the release of proapoptotic factors such as cytochrome c ultimately leading to the formation of the apoptosome and subsequent cell death.

Apoptosis and Sepsis

Apoptosis is a process that has been shown to be misregulated during sepsis and septic shock [125-130]. Factors including reactive oxygen species (ROS), endotoxin, Fas ligand and cytokines (TNF-α, IL-1β, IFN-γ, IL-8, IL-16 and IL-10) are important mediators of the apoptotic cascade [94,126,131]. For example, treatment of cardiac myocytes with cytokines led to apoptosis of these cells [132]. More specifically, TNF-α was added at increasing concentrations ranging from 1 to 4 nmol to adult rat cardiac myocytes. These results showed a correlation between increasing TNF-α concentration and increased apoptosis as measured by comet, tunnel and genomic DNA laddering experiments [132]. In addition, Ing *et al.* also observed that cytokines induce apoptosis. A combination of TNF-α (25 ng/mL), IL-1β (5 ng/mL) and IFN-γ (100 U/mL) were used to treat neonatal cardiac myocytes which resulted in apoptosis as determined by propidium iodide staining, genomic DNA laddering and PARP cleavage [133]. In addition, the synergisitic combination of cytokines induced apoptosis in a NOS dependent fashion since this effect was reversible in the presence of the iNOS inhibitor N-methyl arginine (NMA) [133].

Evidence of apoptosis is also observed in cells of the immune system, intestine, liver and lung in experimental live-infection studies of sepsis [134-137]. Genomic DNA laddering was apparent in the liver, lung, kidney and caecum of an endotoxin shock mouse model of sepsis where LPS was injected at a concentration of 5 mg/kg [138]. A confirmatory tunnel assay was employed to reveal LPS induced apoptosis in the liver. Hotchkiss *et al.* utilized a cecal ligation and puncture model of sepsis to report increased lymphocyte apoptosis in the thymus, spleen, ileum, colon, lung, kidney and muscle. Furthermore, Hotchkiss and colleagues showed that apoptosis in lymphoid organs is related to poor outcome in their animal model of sepsis [94,139]. Systemic administration of caspase inhibitors as well as overexpression of the Bcl-2 gene led to an increase in survival rate and decreased apoptosis in the thymus and spleen as determined by DNA laddering and tunnel assay [139]. Neviere *et al.* reported a 3-fold increase in rat myocyte apoptosis and activation of caspase 3, 8 and 9 in an endotoxic shock animal model [140]. Carlson and colleagues recently reported that LPS induces a TNF-α dependent caspase cascade to mediate cardiac apoptosis. Carlson *et al.* performed an experiment which involved the injection of 4 mg/kg of LPS to the intraperitoneum of mice and detected cardiac apoptosis contributing to the cardiac dysfunction [141]. Similarly, Lancel *et al.* reported that endotoxin induced myocardial dysfunction which utilized the death receptor pathway using a rat model of sepsis. Overexpression of Bcl-2 prior to LPS administration prevented myocardial dysfunction in rats [142]. Burn injury, similarly to septic shock, is associated with elevated levels of circulating pro-inflammatory cytokines and reversible myocardial depression [143,144]. Lightfoot *et al.* reported a marked increase in apoptotic cells in the left ventricule of the adult rat heart 24 h and 48 h after burn trauma. The increasing level of myocyte apoptosis was correlated to a decrease in cardiac function [143].

Administration of pan-caspase inhibitors has improved survival by 40-45% in experimental animal models [145]. Oberholzer *et al.* reported that local injections of a specific caspase 9 inhibitor, Z-LEHD-fmk, into thymi of septic mice led to increased survival [146]. Further, introduction of broad spectrum caspase inhibitors z-VAD.fmk and z-DEVD.cmk 2 h post endotoxin challenge led to improved endotoxin-induced myocardial dysfunction, reduced caspase activation and reduced myocyte apoptosis [147]. In addition, intratracheal administration of short interfering RNA (siRNA) directed against Fas (Fas-siRNA) in a septic mouse model decreased pulmonary apoptosis and neutrophil infiltration [148]. *In vivo* delivery of siRNA targeted against Fas or caspase 8 in a CLP mouse model of sepsis led to an increased survival of 50% with a maintained suppressive effect for up to 10 days [145,149].

Here, we present original data demonstrating the involvement of both intrinsic and extrinsic pathways in inducing apoptosis utilizing a well defined tissue culture model of sepsis that mimics the human disease process. Serum was derived from patients who were hospitalized minimally with severe sepsis. The septic sera were obtained at different time points during their hospitalization throughout the course of the disease. These septic sera contained elevated levels of TNF-α, IL-1β and IFN-γ. Human fibroblast cells (2fTGH) were treated for 12 h with either 10% septic sera or normal sera from healthy individuals. After the treatment, the 2fTGH cells were harvested and lysed to obtain protein extracts.

As discussed previously, JNK is a critical mediator in determining the cell's fate between survival and death in TNF-α-mediated apoptosis. Since septic sera have elevated levels of circulating cytokines including TNF-α, we investigated whether JNK was activated in our tissue culture model of sepsis. The cell extracts were fractionated on SDS-PAGE, electrotransferred to PVDF and probed for the phosphorylated active form of JNK and normalized to total JNK expression. Protein immunoblot analysis indicates that septic sera activates JNK when compared to normal serum (Figure 1A-B representative immunoblot). These results indicate that JNK is activated by septic sera and suggests a potential role in promoting cell death.

Our previous study, utilizing the same tissue culture model of sepsis, demonstrated the activation of the extrinsic pathway by septic sera [150 and manuscript submitted]. More specifically, we showed activation of the initiator caspase 8 and the effector caspase 3. Also, septic sera cleaved DFF 45, the inhibitor of DFF 40 suggesting DFF 40 activation. Trypan blue exclusion assay and DNA laddering experiments confirmed the presence of apoptosis in 2fTGH cells treated with septic sera. To determine the possible involvement of other initiator caspases, we examined the processing of procaspase 2 by western immunoblot. Protein immunoblot analysis of 2fTGH cells treated with human septic and normal sera revealed that septic sera is able to generate the 14 kDa fragment representing active caspase 2 (Figure 2 A-D).

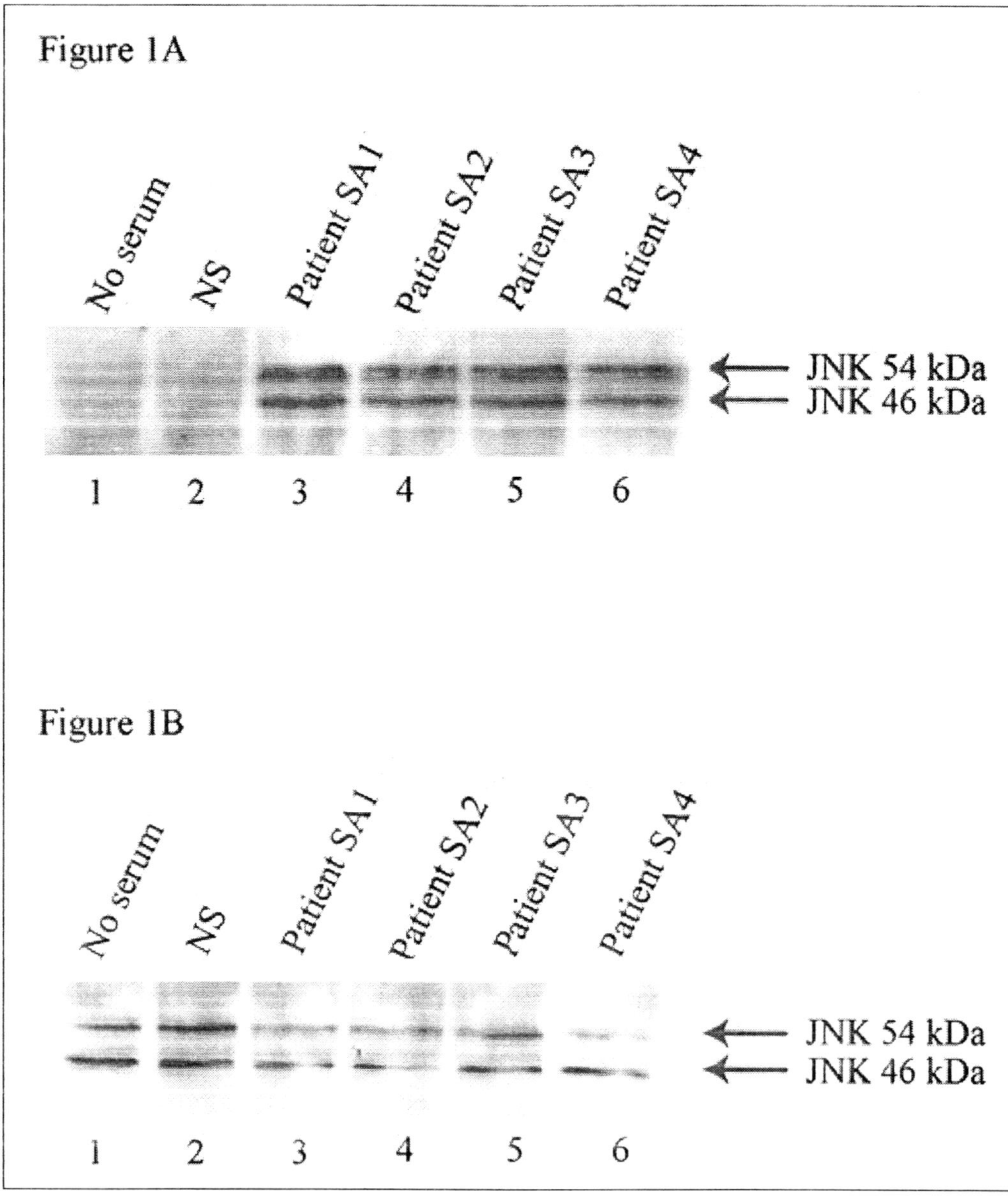

Figure 1. Human septic sera treatment of human fibroblast cells induces the activation of JNK as determined by immunoblot assay. 2fTGH cells were treated for 12 h with either no serum (lane 1), 10% normal serum (lane 2), or 10% septic sera (lanes 3-6). 1A) Western blot was performed using Phospho JNK and 1B) JNK antibodies. Phospho JNK monoclonal antibody detects the endogenous levels of p46 and p54 dually phosphorylated at threonine 183 and tyrosine 185. The JNK monoclonal antibody detects JNK1 at 46 kDa. NS, normal serum; S, septic serum; A, patient designation; 1-4, blood draws.

Therefore, these results reveal that caspase 2 is activated by septic sera and suggests a potential role of caspase 2 in stimulating the pro-inflammatory response as well as amplifying the apoptotic response through the release of mitochondrial components.

To address the involvement of the intrinsic pathway in inducing apoptosis in 2fTGH cells treated with human septic sera, we performed protein immunoblot analysis of caspase 9. Our data indicate that septic sera treatment of fibroblasts initiates signalling that leads to the cleavage of procaspase 9 into a large active 35 kDa fragment which is indicative of activation (Figure 3 representative immunoblot). Activation of caspase 9 occurs after the release of components from the mitochondria such as cytochrome c and Smac/DIABLO into the cytosol. Caspase 2 has also been shown to be able to mediate the release of these components. Similarly, caspase 8 is able to mediate the crosstalk between the extrinsic and intrinsic pathways through the cleavage of Bid into tBid. However, taken together, our results suggest the involvement of both intrinsic and extrinsic pathways in our tissue culture model of sepsis.

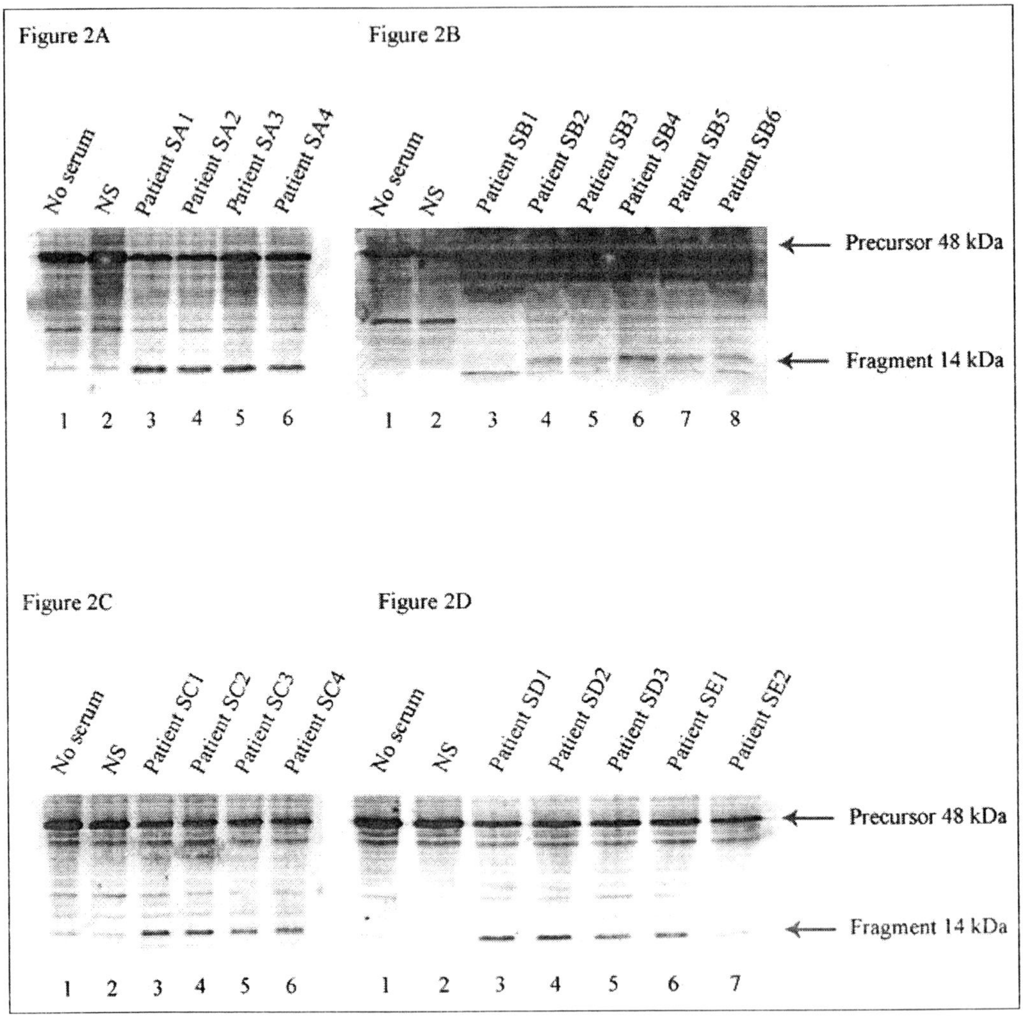

Figure 2. Human septic sera treatment of human fibroblast cells processes procaspase 2 as determined by immunoblot assay. 2fTGH cells were treated for 12 h with either no serum (lane 1), 10% normal serum (lane 2), or 10% septic sera (lanes 3-6). Western blots were performed using Caspase 2 monoclonal antibody which detects endogenous levels of procaspase 2 at 48 kDa as well as the 14 and 12 kDa fragments. NS, normal serum; S, septic serum; A, patient designation; 1-4, blood draws.

Human studies of septic shock have suggested increased apoptosis primarily in cells of the immune system although the phenomenon has also been noted in intestinal epithelial cells [130,145,151-153]. Despite the fact that overt increased myocyte apoptosis has not been reported in the heart in spontaneous disease or in live infection models of sepsis and septic shock, recent publications suggest that preapoptotic signalling may occur contributing to myocyte cellular stress and cytopathic changes in the heart leading to depressed myocardial contractility [130,142,150]. Therefore, widespread preapoptotic signalling may be implicated in the pathogenesis of septic shock by causing cellular dysfunction which progresses to organ failure including myocardial depression.

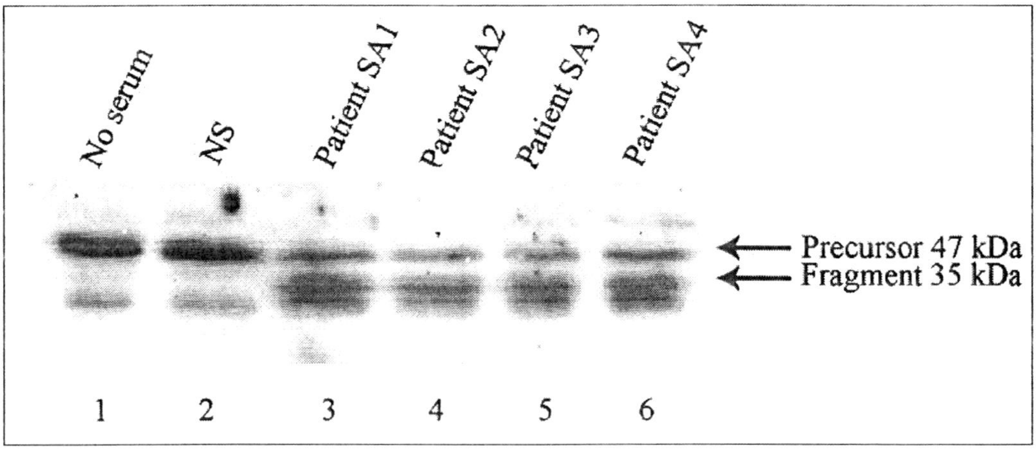

Figure 3. Human septic sera treatment of human fibroblast cells processes procaspase 9 as determined by immunoblot assay. 2fTGH cells were treated for 12 h with either no serum (lane 1), 10% normal serum (lane 2), or 10% septic sera (lanes 3-6). Western blots were performed using Caspase 9 polyclonal antibody which detects the endogenous levels of procaspase 9 at 47 kDa as well as the large active fragments of caspase 9 at 37, 35 and 17 kDa. NS, normal serum; S, septic serum; A-E, patient designation; 1-6, blood draws.

CONCLUSION

Sepsis is characterized as an overwhelming pro-inflammatory systemic response to an infection. Invading pathogens will trigger a host defense response involving the excessive production of inflammatory mediators including cytokines (TNF-α, IL-1β, IFN-γ) from monocytes/macrophages and other cells. Consequently, these cytokines will amplify the host response by stimulating the release of downstream inflammatory mediators such as PAF and NO from leukocytes, macrophages and endothelial cells. Collectively, these mediators produce the pathophysiological abnormalities observed during sepsis and septic shock [1,6,7,50]. These include disseminated intravascular coagulation, infiltration and activation of neutrophils into multiple organs, increased permeability in the endothelial barrier, cardiovascular dysfunction and progressive and sequential dysfunction of various organ systems (MODS) [6,79,154-157]. One of the clinical manifestations of cardiovascular dysfunction observed during septic shock is myocardial depression [2,158,159]. Recently, myocardial depression in human septic shock has been associated with circulating myocardial depressant substances. Cytokines including TNF-α, and IL-1β have been shown to induce a depression in myocardial contractility *in vitro* [160,161]. However, the exact mechanism underlying myocardial depression is yet to be elucidated.

Sepsis and septic shock are the leading causes of morbidity and mortality in the developed world. The pathogenesis of sepsis is very complex and involves overlapping and redundant intracellular signalling pathways leading to the misregulation of processes including cell adhesion molecule regulation, iNOS induction and apoptosis. These processes are broadly regulated by STAT, IRF and NF-κB transcription factors. Modulation of specific members of these families may provide novel targets for altering the disease course. We have shown that caspase dependent apoptosis occurs during sepsis. Inhibition of specific caspases may also alter the disease process. Therefore, sepsis therapies should be designed that target multiple processes that are misregulated during sepsis. The FDA has approved a treatment for patients with severe sepsis, recombinant human activated protein C (rhAPC or Drotrecogin α) [162]. This drug has been found to exhibit anticoagulant and anti-inflammatory properties as well as the ability to downregulate apoptosis [162,163]. rhAPC has been shown to have beneficial effects in relation to patients afflicted with severe sepsis [162-165]. Therefore, a better understanding of the molecular mechanisms involved in the pathogenesis of sepsis and septic shock will contribute to the design of novel therapeutic strategies.

REFERENCES

[1] Annane, D., Bellissant, E., and Cavaillon, J. M., (2005). Septic shock. *Lancet, 365*, 63-78.

[2] Krishnagopalan, S., Kumar, A., and Parrillo, J. E., (2002). Myocardial dysfunction in the patient with sepsis. *Curr Opin Crit Care, 8*, 376-88.

[3] Glauser, M. P., (2000). Pathophysiologic basis of sepsis: considerations for future strategies of intervention. *Crit Care Med, 28*, S4-8.

[4] Riedemann, N. C., Guo, R. F., and Ward, P. A., (2003). The enigma of sepsis. *J Clin Invest, 112*, 460-7.

[5] Hotchkiss, R. S. and Karl, I. E., (2003). The pathophysiology and treatment of sepsis. *N Engl J Med, 348*, 138-50.

[6] Lopez-Bojorquez, L. N., Dehesa, A. Z., and Reyes-Teran, G., (2004). Molecular mechanisms involved in the pathogenesis of septic shock. *Arch Med Res, 35*, 465-79.

[7] Weigand, M. A., Horner, C., Bardenheuer, H. J., and Bouchon, A., (2004). The systemic inflammatory response syndrome. *Best Pract Res Clin Anaesthesiol, 18*, 455-75.

[8] Levy, M. M., Fink, M. P., Marshall, J. C., Abraham, E., Angus, D., Cook, D., Cohen, J., Opal, S. M., Vincent, J. L., and Ramsay, G., (2003). 2001 SCCM/ESICM/ACCP/ATS/SIS International Sepsis Definitions Conference. *Crit Care Med, 31*, 1250-6.

[9] Manocha, S., Feinstein, D., and Kumar, A., (2002). Novel therapies for sepsis: antiendotoxin therapies. *Expert Opin Investig Drugs, 11*, 1795-812.

[10] Murphy, K., Haudek, S. B., Thompson, M., and Giroir, B. P., (1998). Molecular biology of septic shock. *New Horiz, 6*, 181-93.

[11] Guha, M. and Mackman, N., (2001). LPS induction of gene expression in human monocytes. *Cell Signal, 13*, 85-94.

[12] Kaisho, T. and Akira, S., (2000). Critical roles of Toll-like receptors in host defense. *Crit Rev Immunol, 20*, 393-405.

[13] Takeda, K., Kaisho, T., and Akira, S., (2003). Toll-like receptors. *Annu Rev Immunol, 21*, 335-76.

[14] O'Neill, L. A., (2006). How Toll-like receptors signal: what we know and what we don't know. *Curr Opin Immunol, 18*, 3-9.

[15] Fujihara, M., Muroi, M., Tanamoto, K., Suzuki, T., Azuma, H., and Ikeda, H., (2003). Molecular mechanisms of macrophage activation and deactivation by lipopolysaccharide: roles of the receptor complex. *Pharmacol Ther, 100*, 171-94.

[16] Dunne, A. and O'Neill, L. A., (2003). The interleukin-1 receptor/Toll-like receptor superfamily: signal transduction during inflammation and host defense. *Sci STKE, 2003*, re3.

[17] Williams, D. L., Ha, T., Li, C., Kalbfleisch, J. H., Schweitzer, J., Vogt, W., and Browder, I. W., (2003). Modulation of tissue Toll-like receptor 2 and 4 during the early phases of polymicrobial sepsis correlates with mortality. *Crit Care Med, 31*, 1808-18.

[18] Schwandner, R., Dziarski, R., Wesche, H., Rothe, M., and Kirschning, C. J., (1999). Peptidoglycan- and lipoteichoic acid-induced cell activation is mediated by toll-like receptor 2. *J Biol Chem, 274*, 17406-9.

[19] Zhang, G. and Ghosh, S., (2001). Toll-like receptor-mediated NF-kappaB activation: a phylogenetically conserved paradigm in innate immunity. *J Clin Invest, 107*, 13-9.

[20] Moynagh, P. N., (2005). TLR signalling and activation of IRFs: revisiting old friends from the NF-kappaB pathway. *Trends Immunol, 26*, 469-76.

[21] Kirschning, C. J., Wesche, H., Merrill Ayres, T., and Rothe, M., (1998). Human toll-like receptor 2 confers responsiveness to bacterial lipopolysaccharide. *J Exp Med, 188*, 2091-7.

[22] Laflamme, N. and Rivest, S., (2001). Toll-like receptor 4: the missing link of the cerebral innate immune response triggered by circulating gram-negative bacterial cell wall components. *Faseb J, 15*, 155-163.

[23] Lien, E. and Ingalls, R. R., (2002). Toll-like receptors. *Crit Care Med, 30*, S1-S11.

[24] Wagner, H., (2001). Toll meets bacterial CpG-DNA. *Immunity, 14*, 499-502.

[25] Hemmi, H., Takeuchi, O., Kawai, T., Kaisho, T., Sato, S., Sanjo, H., Matsumoto, M., Hoshino, K., Wagner, H., Takeda, K., and Akira, S., (2000). A Toll-like receptor recognizes bacterial DNA. *Nature, 408*, 740-5.

[26] Sen, G. C. and Sarkar, S. N., (2005). Transcriptional signaling by double-stranded RNA: role of TLR3. *Cytokine Growth Factor Rev, 16*, 1-14.

[27] Kariko, K., Buckstein, M., Ni, H., and Weissman, D., (2005). Suppression of RNA recognition by Toll-like receptors: the impact of nucleoside modification and the evolutionary origin of RNA. *Immunity, 23*, 165-75.

[28] Liu, S. F. and Malik, A. B., (2006). NF-kB activation as a pathological mechanism of septic shock and inflammation. *Am J Pathol Lung Cell Mol Physiol, 290*, L622-L645.

[29] Burstein, E. and Duckett, C. S., (2003). Dying for NF-kappaB? Control of cell death by transcriptional regulation of the apoptotic machinery. *Curr Opin Cell Biol*, *15*, 732-7.

[30] Krappmann, D. and Scheidereit, C., (2005). A pervasive role of ubiquitin conjugation in activation and termination of IkappaB kinase pathways. *EMBO Rep*, *6*, 321-6.

[31] Kovalenko, A. and Wallach, D., (2006). If the Prophet Does Not Come to the Mountain: Dynamics of Signaling Complexes in NF-kB Activation. *Molecular Cell*, *22*, 433-436.

[32] Wisdom, R., (1999). AP-1: One Switch for Many Signals. *Exp Cell Res*, *253*, 180-185.

[33] Smoak, K. A. and Cidlowski, J. A., (2004). Mechanisms of glucocorticoid receptor signaling during inflammation. *Mechanisms of Ageing and Development*, *125*, 697-706.

[34] Shaulian, E. and Karin, M., (2001). AP-1 in cell proliferation and survival. *Oncogene*, *20*, 2390-2400.

[35] Ihle, J. N., (1996). STATs: signal transducers and activators of transcription. *Cell*, *84*, 331-4.

[36] Ihle, J. N., (2001). The Stat family in cytokine signaling. *Curr Opin Cell Biol*, *13*, 211-7.

[37] Brierley, M. M. and Fish, E. N., (2005). Stats: multifaceted regulators of transcription. *J Interferon Cytokine Res*, *25*, 733-44.

[38] Takeda, K. and Akira, S., (2000). STAT family of transcription factors in cytokine-mediated biological responses. *Cytokine Growth Factor Rev*, *11*, 199-207.

[39] Samuel, C. E., (2001). Antiviral actions of interferons. *Clin Microbiol Rev*, *14*, 778-809.

[40] Horvath, C. M., (2000). STAT proteins and transcriptional responses to extracellular signals. *Trends Biochem Sci*, *25*, 496-502.

[41] Levy, D. E. and Garcia-Sastre, A., (2001). The virus battles: IFN induction of the antiviral state and mechanisms of viral evasion. *Cytokine Growth Factor Rev*, *12*, 143-56.

[42] Kanno, Y., Levi, B. Z., Tamura, T., and Ozato, K., (2005). Immune cell-specific amplification of interferon signaling by the IRF-4/8-PU.1 complex. *J Interferon Cytokine Res*, *25*, 770-9.

[43] Nguyen, H., Hiscott, J., and Pitha, P. M., (1997). The growing family of interferon regulatory factors. *Cytokine Growth Factor Rev*, *8*, 293-312.

[44] Civas, A., Island, M. L., Genin, P., Morin, P., and Navarro, S., (2002). Regulation of virus-induced interferon-A genes. *Biochimie*, *84*, 643-54.

[45] Taniguchi, T. and Takaoka, A., (2002). The interferon-alpha/beta system in antiviral responses: a multimodal machinery of gene regulation by the IRF family of transcription factors. *Curr Opin Immunol*, *14*, 111-6.

[46] Mamane, Y., Heylbroeck, C., Genin, P., Algarte, M., Servant, M. J., LePage, C., DeLuca, C., Kwon, H., Lin, R., and Hiscott, J., (1999). Interferon regulatory factors: the next generation. *Gene*, *237*, 1-14.

[47] Yanai, H., Mizutani, T., Inuzuka, T., Honda, K., Takaoka, A., and Taniguchi, T., (2005). IRF family transcription factors in type I interferon induction. *International Congress Series*, *1285*, 104-113.

[48] Oberholzer, A., Oberholzer, C., and Moldawer, L. L., (2000). Cytokine signaling--regulation of the immune response in normal and critically ill states. *Crit Care Med, 28*, N3-12.

[49] Zanotti, S. and Kumar, A., (2002). Cytokine modulation in sepsis and septic shock. *Expert Opin Investig Drugs, 11*, 1061-75.

[50] Sherwood, E. R. and Toliver-Kinsky, T., (2004). Mechanisms of the inflammatory response. *Best Pract Res Clin Anaesthesiol, 18*, 385-405.

[51] Baud, V. and Karin, M., (2001). Signal transduction by tumor necrosis factor and its relatives. *Trends Cell Biol, 11*, 372-7.

[52] Baker, S. J. and Reddy, E. P., (1996). Transducers of life and death: TNF receptor superfamily and associated proteins. *Oncogene, 12*, 1-9.

[53] Tewari, M. and Dixit, V. M., (1996). Recent advances in tumor necrosis factor and CD40 signaling. *Curr Opin Genet Dev, 6*, 39-44.

[54] Zhang, B., Huang, Y. H., Chen, Y., Yang, Y., Hao, Z. L., and Xie, S. L., (1998). Plasma tumor necrosis factor-alpha, its soluble receptors and interleukin-1beta levels in critically burned patients. *Burns, 24*, 599-603.

[55] Muppidi, J. R., Tschopp, J., and Siegel, R. M., (2004). Life and death decisions: secondary complexes and lipid rafts in TNF receptor family signal transduction. *Immunity, 21*, 461-5.

[56] Poyet, J. L., Srinivasula, S. M., Lin, J. H., Fernandes-Alnemri, T., Yamaoka, S., Tsichlis, P. N., and Alnemri, E. S., (2000). Activation of the Ikappa B kinases by RIP via IKKgamma /NEMO-mediated oligomerization. *J Biol Chem, 275*, 37966-77.

[57] Daun, J. M. and Fenton, M. J., (2000). Interleukin-1/Toll receptor family members: receptor structure and signal transduction pathways. *J Interferon Cytokine Res, 20*, 843-55.

[58] Dinarello, C. A., (2004). Therapeutic strategies to reduce IL-1 activity in treating local and systemic inflammation. *Curr Opin Pharmacol, 4*, 378-85.

[59] Dinarello, C. A., (1997). Interleukin-1. *Cytokine Growth Factor Rev, 8*, 253-65.

[60] Abbas, A. K., Lichtman, A. H., and Pober, J. S. (1997). *Cellular and Molecular Immunology*: (3rd edition). Toronto, ON: W.B. Saunders Company.

[61] Kumar, A., Thota, V., Dee, L., Olson, J., Uretz, E., and Parrillo, J. E., (1996). Tumor Necrosis Factor alpha and Interleukin 1beta Are Responsible for In Vitro Myocardial Cell Depression Induced by Human Septic Shock Serum. *J. Exp. Med., 183*, 949-58.

[62] Kumar, A., Paladugu, B., Kumar, A., Mensing, J., and Parrillo, J. E., (2007). Transforming Growth Factor-beta Blocks *in vitro* Cardiac Myocyte Depression Induced by Tumor Necrosis Factor alpha, Interleukin-1beta and Human Septic Shock Serum. *Crit Care Med, 35*, 358-64.

[63] Riedemann, N. C., Neff, T. A., Guo, R. F., Bernacki, K. D., Laudes, I. J., Sarma, J. V., Lambris, J. D., and Ward, P. A., (2003). Protective effects of IL-6 blockade in sepsis are linked to reduced C5a receptor expression. *J Immunol, 170*, 503-7.

[64] Garner, L. B., Willis, M. S., Carlson, D. L., DiMaio, J. M., White, M. D., White, D. J., Adams, G. A., IV, Horton, J. W., and Giroir, B. P., (2003). Macrophage migration inhibitory factor is a cardiac-derived myocardial depressant factor. *Am J Physiol Heart Circ Physiol, 285*, H2500-H2509.

[65] Wang, H., Bloom, O., Zhang, M., Vishnubhakat, J. M., Ombrellino, M., Che, J., Frazier, A., Yang, H., Ivanova, S., Borovikova, L., Manogue, K. R., Faist, E., Abraham, E., Andersson, J., Andersson, U., Molina, P. E., Abumrad, N. N., Sama, A., and Tracey, K. J., (1999). HMG-1 as a Late Mediator of Endotoxin Lethality in Mice. *Science*, *285*, 248-251.

[66] Yang, H. and Tracey, K. J., (2005). High mobility group box 1 (HMGB1). *Crit Care Med*, *33*, S472-S474.

[67] Wang, H., H., Y., C.J., C., Sama, A. E., and Tracey, K. J., (2001). HMGB1 as a late mediator of lethal systemic inflammation. *American Journal of Respiratory & Critical Care Medicine*, *164*, 1768-1773.

[68] Kirkeboen, K. A. and Strand, O. A., (1999). The role of nitric oxide in sepsis--an overview. *Acta Anaesthesiol Scand*, *43*, 275-88.

[69] Feihl, F., Waeber, B., and Liaudet, L., (2001). Is nitric oxide overproduction the target of choice for the management of septic shock? *Pharmacol Ther*, *91*, 179-213.

[70] Iskit, A. B. and Guc, O., (2003). Effects of endothelin and nitric oxide on organ injury, mesenteric ischemia, and survival in experimental models of septic shock. *Acta Pharmacol Sin*, *24*, 953-7.

[71] Joyce, D. E., Nelson, D. R., and Grinnell, B. W., (2004). Leukocyte and endothelial cell interactions in sepsis: relevance of the protein C pathway. *Crit Care Med*, *32*, S280-6.

[72] Chandra, A., Enkhbaatar, P., Nakano, Y., Traber, L. D., and Traber, D. L., (2006). Sepsis: emerging role of nitric oxide and selectins. *clinics*, *61*, 71-76.

[73] Nooteboom, A., van der Linden, C. J., and Hendriks, T., (2005). Whole blood-mediated endothelial permeability and adhesion molecule expression: a model study into the effects of bacteria and antibiotics. *J Antimicrob Chemother*, *55*, 150-6.

[74] Coelho, A. L., Hogaboam, C. M., and Kunkel, S. L., (2005). Chemokines provide the sustained inflammatory bridge between innate and acquired immunity. *Cytokine Growth Factor Rev*, *16*, 553-60.

[75] Albrecht, E. A. and Ward, P. A., (2004). Complement-induced impairment of the innate immune system during sepsis. *Curr Allergy Asthma Rep*, *4*, 359-64.

[76] Riedemann, N. C., Guo, R. F., Bernacki, K. D., Reuben, J. S., Laudes, I. J., Neff, T. A., Gao, H., Speyer, C., Sarma, V. J., Zetoune, F. S., and Ward, P. A., (2003). Regulation by C5a of neutrophil activation during sepsis. *Immunity*, *19*, 193-202.

[77] Guo, R.-F. and Ward, P. A., (2006). C5a, a Therapeutic Target in Sepsis. *Recent Patents on Anti-Infective Drug Discovery*, *1*, 57-65.

[78] Diehl, J. L. and Borgel, D., (2005). Sepsis and coagulation. *Curr Opin Crit Care*, *11*, 454-60.

[79] Levi, M., de Jonge, E., and van der Poll, T., (2006). Plasma and plasma components in the management of disseminated intravascular coagulation. *Best Pract Res Clin Haematol*, *19*, 127-142.

[80] LaRosa, S. P., Opal, S. M., Utterback, B., Yan, S. C. B., Helterbrand, J., Simpson, A. J. H., Chaowagul, W., White, N. J., and Fisher, C. J., Jr., (2006). Decreased protein C, protein S, and antithrombin levels are predictive of poor outcome in Gram-negative sepsis caused by Burkholderia pseudomallei. *International Journal of Infectious Diseases*, *10*, 25-31.

[81] Esmon, C. T., (2006). The Protein C Pathway. *Chest, 124*, 26S-32S.

[82] Vallet, B., (2003). Bench-to-bedside review: endothelial cell dysfunction in severe sepsis: a role in organ dysfunction? *Crit Care, 7*, 130-8.

[83] Esmon, C. T., (2005). The interactions between inflammation and coagulation. *Br J Haematol, 131*, 417-30.

[84] Cohen, G. M., (1997). Caspases: the executioners of apoptosis. *Biochem J, 326 (Pt 1)*, 1-16.

[85] Los, M., Wesselborg, S., and Schulze-Osthoff, K., (1999). The Role of Caspases in Development, Immunity, and Apoptotic Signal Transduction: Lessons from Knockout Mice. *Immunity, 10*, 629-639.

[86] Saraste, A. and Pulkki, K., (2000). Morphologic and biochemical hallmarks of apoptosis. *Cardiovasc Res, 45*, 528-37.

[87] Zimmermann, K. C., Bonzon, C., and Green, D. R., (2001). The machinery of programmed cell death. *Pharmacol Ther, 92*, 57-70.

[88] Yan, N. and Shi, Y., (2005). Mechanisms of apoptosis through structural biology. *Annu Rev Cell Dev Biol, 21*, 35-56.

[89] Schafer, Z. T. and Kornbluth, S., (2006). The apoptosome: physiological, developmental, and pathological modes of regulation. *Dev Cell, 10*, 549-61.

[90] Fan, T. J., Han, L. H., Cong, R. S., and Liang, J., (2005). Caspase family proteases and apoptosis. *Acta Biochim Biophys Sin (Shanghai), 37*, 719-27.

[91] Siegel, R. M., (2006). Caspases at the crossroads of immune-cell life and death. *Nat Rev Immunol, 6*, 308-17.

[92] Danial, N. N. and Korsmeyer, S. J., (2004). Cell death: critical control points. *Cell, 116*, 205-19.

[93] Boatright, K. M. and Salvesen, G. S., (2003). Mechanisms of caspase activation. *Curr Opin Cell Biol, 15*, 725-31.

[94] Oberholzer, C., Oberholzer, A., Clare-Salzler, M., and Moldawer, L. L., (2001). Apoptosis in sepsis: a new target for therapeutic exploration. *The FASEB Journal, 15*, 879-892.

[95] Liu, X., Zou, H., Slaughter, C., and Wang, X., (1997). DFF, a Heterodimeric Protein That Functions Downstream of Caspase-3 to Trigger DNA Fragmentation during Apoptosis. *Cell, 89*, 175-184.

[96] Woo, E.-J., Kim, Y., Kim, M., Han, W., Shin, S., Robinson, H., Park, S., and Oh, B., (2004). Structural Mechanism for Inactivation and Activation of CAD/DFF40 in the Apoptotic Pathway. *Molecular Cell, 14*, 531-539.

[97] Schimmer, A. D., (2004). Inhibitor of apoptosis proteins: translating basic knowledge into clinical practice. *Cancer Res, 64*, 7183-90.

[98] Wright, C. W. and Duckett, C. S., (2005). Reawakening the cellular death program in neoplasia through the therapeutic blockade of IAP function. *J Clin Invest, 115*, 2673-8.

[99] Salvesen, G. S. and Duckett, C. S., (2002). IAP proteins: blocking the road to death's door. *Nat Rev Mol Cell Biol, 3*, 401-10.

[100] Liston, P., Fong, W. G., and Korneluk, R. G., (2003). The inhibitors of apoptosis: there is more to life than Bcl2. *Oncogene, 22*, 8568-80.

[101] Salvesen, G. S., Duckett, C.S., (2002). IAP Proteins: Blocking the road to death's door. *Molecular Cell Biology*, *3*, 401-412.

[102] Bergmann, A., Yang, A. Y., and Srivastava, M., (2003). Regulators of IAP function: coming to grips with the grim reaper. *Curr Opin Cell Biol*, *15*, 717-24.

[103] Pickart, C. M., (2001). Mechanisms underlying ubiquitination. *Annu Rev Biochem*, *70*, 503-33.

[104] Pickart, C. M., (2004). Back to the future with ubiquitin. *Cell*, *116*, 181-90.

[105] Yin, X. M., (2006). Bid, a BH3-only multi-functional molecule, is at the cross road of life and death. *Gene*, *369*, 7-19.

[106] Cory, S. and Adams, J. M., (2002). The Bcl2 family: regulators of the cellular life-or-death switch. *Nat Rev Cancer*, *2*, 647-56.

[107] Chen, L., Willis, S. N., Wei, A., Smith, B. J., Fletcher, J. I., Hinds, M. G., Colman, P. M., Day, C. L., Adams, J. M., and Huang, D. C., (2005). Differential targeting of prosurvival Bcl-2 proteins by their BH3-only ligands allows complementary apoptotic function. *Mol Cell*, *17*, 393-403.

[108] Er, E., Oliver, L., Cartron, P. F., Juin, P., Manon, S., and Vallette, F. M., (2006). Mitochondria as the target of the pro-apoptotic protein Bax. *Biochim Biophys Acta*, *1757*, 1301-11.

[109] Deng, Y., Ren, X., Yang, L., Lin, Y., and Wu, X., (2003). A JNK-Dependent Pathway Is Required for TNF-alpha-Induced Apoptosis. *Cell*, *115*, 61-70.

[110] Zhivotovsky, B. and Orrenius, S., (2005). Caspase-2 function in response to DNA damage. *Biochem Biophys Res Commun*, *331*, 859-867.

[111] Shearwin-Whyatt, L. M., Harvey, N., and Kumar, S., (2000). Subcellular localization and CARD-dependent oligomerization of the death adaptor RAIDD. *Cell Death & Differentiation*, *7*, 155-165.

[112] Paroni, G., Henderson, C., Schneider, C., and Brancolini, C., (2002). Caspase-2 Can Trigger Cytochrome c Release and Apoptosis from the Nucleus. *J. Biol. Chem*, *277*, 15147-15161.

[113] Paroni, G., Henderson, C., Schneider, C., and Brancolini, C., (2001). Caspase-2-induced Apoptosis Is Dependent on Caspase-9, but Its Processing during UV-or Tumor Necrosis Factor-dependent Cell Death Requires Caspase-3. *J. Biol. Chem*, *276*, 21907-21915.

[114] Droin, N., Rebe, C., Bichat, F., Hammann, A., Bertrand, R., and Solary, E., (2001). Modulation of apoptosis by procaspase-2 short isoform: selective inhibition of chromatin condensation, apoptotic body formation and phosphatidylserine externalization. *Oncogene*, *20*, 260-269.

[115] Zheng, T. S., Hunot, S., Kuida, K., Momoi, T., Srinivasan, A., Nicholson, D. W., Lazebnik, Y., and Flavell, R. A., (2000). Deficiency in caspase-9 or caspase-3 induces compensatory caspase activation. *Nature Medicine*, *6*, 1241-1247.

[116] Mancini, M., Machamer, C. E., Roy, S., Nicholson, D. W., Thornberry, N. A., Casciola-Rosen, L. A., and Rosen, A., (2000). Caspase-2 Is Localized at the Golgi Complex and Cleaves Golgin-160 during Apoptosis. *Journal of Cell Biology*, *149*, 603-612.

[117] Lamkanfi, M., D'hondt, K., Vande Walle, L., van Gurp, M., Denecker, G., Demeulemeester, J., Kalai, M., Declercq, W., Saelens, X., and Vandenabeele, P., (2005). A Novel Caspase-2 Complex Containing TRAF2 and RIP1. *J. Biol. Chem*, *280*, 6923-6932.

[118] Davis, R. J., (2000). Signal transduction by the JNK group of MAP kinases. *Cell*, *103*, 239-52.

[119] Varfolomeev, E. E. and Ashkenazi, A., (2004). Tumor Necrosis Factor An Apoptosis JuNKie? *Cell*, *116*, 491-497.

[120] Lin, A. and Dibling, B., (2002). The true face of JNK activation in apoptosis. *Aging Cell*, *1*, 112-116.

[121] Chang, L., Kamata, H., Solinas, G., Luo, J., Maeda, S., Venuprasad, K., Liu, Y., and Karin, M., (2006). The E3 Ubiquitin Ligase Itch Couples JNK Activation to TNF-alpha-induced Cell death by Inducing c-FLIP$_L$ Turnover. *Cell*, *124*, 601-613.

[122] Kamata H., Honda, S., Maeda, S., Chang, L., Hirata, H., and Karin, M., (2005). Reactive oxygen species promote TNFalpha-induced death and sustained JNK activation by inhibiting MAP kinase phosphatases. *Cell*, *120*, 649-661.

[123] Papadakis, E. S., Finegan KG, Wang X, Robinson AC, Guo C, Kayahara M, and C., T., (2006). The regulation of Bax by c-Jun N-terminal protein kinase (JNK) is a prerequisite to the mitochondrial-induced apoptotic pathway. *FEBS Lett*, *580*, 1320-1326.

[124] Scaffidi, C., Fulda, S., Srinivasan, A., Friesen, C., Li, F., Tomaselli, K. J., Debatin, K. M., Krammer, P. H., and Peter, M. E., (1998). Two CD95 (APO-1/Fas) signaling pathways. *Embo J*, *17*, 1675-87.

[125] Giamarellos-Bourboulis, E., Routsi, C., Plachouras, D., Markaki, V., Raftogiannis, M., Zervakis, D., Koussoulas, V., Orfanos, S., Kotanidou, A., Armaganidis, A., Roussos, C., and Giamarellou, H., (2006). Early apoptosis of blood monocytes in the septic host: is it a mechanism of protection in the event of septic shock? *Crit Care*, *10*, R76.

[126] Mahidhara, R. and Billiar, T. R., (2000). Apoptosis in sepsis. *Crit Care Med*, *28*, N105-13.

[127] Ayala, A., Lomas, J., Grutkoski, P., and Chung, C., (2003). Pathological aspects of apoptosis in severe sepsis and shock? *The International Journal of Biochemistry & Cell Biology*, *35*, 7-15.

[128] Joshi, V., Kalvakolanu, D., and Cross, A., (2003). Simultaneous activation of apoptosis and inflammation in pathogenesis of septic shock: a hypothesis. *FEBS Lett*, *555*, 180-184.

[129] Wong, H. R., (1998). Potential protective role of the heat shock response in sepsis. *New Horiz*, *6*, 194-200.

[130] Crouser, E. D., (2005). Redefining the roles of apoptosis pathways during sepsis. *Crit Care Med*, *33*, 670-2.

[131] Power, C., N. Fanning, and H.P. Redmond, (2002). Cellular apoptosis and organ injury in sepsis: a review. *Shock*, *18*, 197-211.

[132] Krown, K. A., Page, M. T., Nguyen, C., Zechner, D., Gutierrez, V., Comstock, K. L., Glembotski, C. C., Quintana, P. J., and Sabbadini, R. A., (1996). Tumor necrosis factor

alpha-induced apoptosis in cardiac myocytes. Involvement of the sphingolipid signaling cascade in cardiac cell death. *J. Clin Invest*, *98*, 2854-2865.

[133] Ing, D. J., Zang, J., Dzau, V. J., Webster, K. A., and Bishopric, N. H., (1999). Modulation of cytokine-induced cardiac myocyte apoptosis by nitric oxide, Bak, and Bcl-x. *Circ Res*, *84*, 21-33.

[134] Schulz, R., E. Nava, and S. Moncada, (1992). Induction and potential biological relevance of a Ca(2+) independent nitric oxide synthase in the myocardium. *Br J Pharmacol*, *105*, 575-580.

[135] Liaudet, L., Soriano, F. G., and Szabo, C., (2000). Biology of nitric oxide signaling. *Crit Care Med*, *28*, N37-52.

[136] Chung, C. S., Xu, Y. X., Chaudry, I. H., and Ayala, A., (1998). Sepsis induces increased apoptosis in lamina propria mononuclear cells which is associated with altered cytokine gene expression. *J Surg Res*, *77*, 63-70.

[137] Hiramatsu, M., Hotchkiss, R. S., Karl, I. E., and Buchman, T. G., (1997). Cecal ligation and puncture (CLP) induces apoptosis in thymus, spleen, lung, and gut by an endotoxin and TNF-independent pathway. *Shock*, *7*, 247-253.

[138] Bohlinger, I., Leist, M., Gantner, F., Angermuller, S., Tiegs, G., and Wendel, A., (1996). DNA fragmentation in mouse organs during endotoxic shock. *Am J Pathol*, *149*, 1381-1393.

[139] Hotchkiss, R. S., Swansan, P. E., Knudson, C. M., Chang, K. C., Cobb, J. P., Osborne, D. F., Zollner, K. M., Buchman, T. G., Korsmeyer, S. J., and Karl, I. E., (1999). Overexpression of Bcl-2 in transgenic mice decreases apoptosis and improves survival in sepsis. *J Immunol*, *162*, 4148-4156.

[140] Neviere, R., Fauvel, H., Chopin, C., Formstecher, P., and Marchetti, P., (2001). Caspase inhibition prevents cardiac dysfunction and heart apoptosis in a rat model of sepsis. *Am J Respir Crit Care Med*, *163*, 218-225.

[141] Carlson, D. L., Willis, M. S., White, D. J., Horton, J. W., and Giroir, B. P., (2005). Tumor necrosis factor-alpha-induced caspase activation mediates endotoxin-related cardiac dysfunction. *Crit Care Med*, *33*, 1021-8.

[142] Lancel, S., Petillot, P., Favory, R., Stebach, N., Lahorte, C., Danze, P. M., Vallet, B., Marchetti, P., and Neviere, R., (2005). Expression of apoptosis regulatory factors during myocardial dysfunction in endotoxemic rats. *Crit Care Med*, *33*, 492-6.

[143] Lightfoot, E., Jr., Horton, J. W., Maass, D. L., White, D. J., McFarland, R. D., and Lipsky, P. E., (1999). Major burn trauma in rats promotes cardiac and gastrointestinal apoptosis. *Shock*, *11*, 29-34.

[144] Wang, G. Q., Xia, Z. F., Yu, B. J., Xiao, S. C., and Chen, Y. L., (2001). Cardiac apoptosis in burned rats with delayed fluid resuscitation. *Burns*, *27*, 250-253.

[145] Wesche, D. E., Lomas-Neira, J. L., Perl, M., Chung, C. S., and Ayala, A., (2005). Leukocyte apoptosis and its significance in sepsis and shock. *J Leukoc Biol*, *78*, 325-37.

[146] Oberholzer, C., Tschoeke, S. K., Moldawer, L. L., and Oberholzer, A., (2006). Local thymic caspase-9 inhibition improves survival during polymicrobial sepsis in mice. *J Mol Med*, *84*, 389-395.

[147] Adrie, C. and Pinsky, M. R., (2000). The inflammatory balance in human sepsis. *Intensive Care Med, 26*, 364-375.

[148] Perl, M., Chung, C. S., Lomas-Neira, J., Rachel, T. M., Biffl, W. L., Cioffi, W. G., and Ayala, A., (2005). Silencing of Fas, but not caspase-8, in lung epithelial cells ameliorates pulmonary apoptosis, inflammation, and neutrophil influx after hemorrhagic shock and sepsis. *Am J Pathol, 167*, 1545-59.

[149] Wesche-Soldato, D. E., Chung, C. S., Lomas-Neira, J., Doughty, L. A., Gregory, S. H., and Ayala, A., (2005). In vivo delivery of caspase-8 or Fas siRNA improves the survival of septic mice. *Blood, 106*, 2295-2301.

[150] Kumar, A., Kumar, A., Michael, P., Brabant, D., Parissenti, A. M., Ramana, C. V., Xu, X., and Parrillo, J. E., (2005). Human serum from patients with septic shock activates transcription factors STAT1, IRF1, and NF-kappaB and induces apoptosis in human cardiac myocytes. *J Biol Chem, 280*, 42619-42626.

[151] Hotchkiss, R. S., Swanson, P., Freeman, B., Tinsley, K., Cobb, J., Matuschak, G., Buchman, T., and Karl, I., (1999). Apoptotic cell death in patients with sepsis, shock, and multiple organ dysfunction. *Critical care medicine, 27*, 1230-1251.

[152] Coutinho, H. B., Robalinho, T. I., Coutinho, V. B., Amorim, A. M., Furtado, A. F., Ferraz, A., Ferraz, E., Walker, F., King, G., Sewell, H. F., and Wakelin, D., (1997). Intra-abdominal sepsis: an immunocytochemical study of the small intestine mucosa. *J Clin Pathol, 50*, 294-298.

[153] Hashimoto, S., Kobayashi, A., Kooguchi, K., Kitamura, Y., Onodera, H., and Nakajima, H., (2000). Upregulation of two death pathways of perforin/granzyme and FasL/Fas in septic acute respiratory distress syndrome. *Am J Respir Crit Care Med, 161*, 237-243.

[154] Dellinger, R. P., (2003). Cardiovascular management of septic shock. *Crit Care Med, 31*, 946-55.

[155] Spronk, P. E., Zandstra, D. F., and Ince, C., (2004). Bench-to-bedside review: Sepsis is a disease of the microcirculation. *Crit Care, 8*, 462-468.

[156] Fowler, D. E. and Ping, W., (2002). The cardiovascular response in sepsis: Proposed mechanisms of the beneficial effect of adrenomedullin and its binding protein (Review). *International Journal of Molecular Medicine, 9*, 443-449.

[157] Awad, S. S., (2003). State-of-the-art therapy for severe sepsis and multisystem organ dysfunction. *Am J Surg, 186*, 23S-30S; discussion 31S-34S.

[158] Turner, A., Tsamitros, M., and Bellomo, R., (1999). Myocardial cell injury in septic shock. *Crit Care Med, 27*, 1775-1780.

[159] Kumar, A., Haery, C., and Parrillo, J. E., (2001). Myocardial dysfunction in septic shock: Part I. Clinical manifestation of cardiovascular dysfunction. *J Cardiothorac Vasc Anesth, 15*, 364-76.

[160] Kumar, A., Krieger, A., Symeoneides, S., and Parrillo, J. E., (2001). Myocardial dysfunction in septic shock: Part II. Role of cytokines and nitric oxide. *J Cardiothorac Vasc Anesth, 15*, 485-511.

[161] Court, O., Kumar, A., and Parrillo, J. E., (2002). Clinical review: Myocardial depression in sepsis and septic shock. *Crit Care, 6*, 500-8.

[162] Haley, M., Cui, X., Minneci, P. C., Deans, K. J., Natanson, C., and Eichacker, P. Q., (2004). Activated protein C in sepsis: emerging insights regarding its mechanism of action and clinical effectiveness. *Curr Opin Infect Dis*, *17*, 205-11.

[163] Poulton, B., (2006). Advances in the management of sepsis: the randomised controlled trials behind the Surviving Sepsis Campaign recommendations. *International Journal of Antimicrobial Agents*, *27*, 97-101.

[164] Abraham, E., Laterre, P. F., Garg, R., Levy, H., Talwar, D., Trzaskoma, B. L., Francois, B., Guy, J. S., Bruckmann, M., Rea-Neto, A., Rossaint, R., Perrotin, D., Sablotzki, A., Arkins, N., Utterback, B. G., and Macias, W. L., (2005). Drotrecogin alfa (activated) for adults with severe sepsis and a low risk of death. *N Engl J Med*, *353*, 1332-41.

[165] Parrillo, J. E., (2005). Severe sepsis and therapy with activated protein C. *N Engl J Med*, *353*, 1398-400.

In: Cell Apoptotic Signaling Pathways
Editor: Charles O. Pickens, pp. 41-82
ISBN: 978-1-60021-507-0

Chapter II

The Implication(s) of Mitochondrial Permeability in the Execution Phase of Apoptosis

Gregory Bellot, François M. Vallette and Lisa Oliver*
UMR 601 INSERM, Université de Nantes, 9 Quai Moncousu F-44035 Nantes.
Cedex 01, France.

Abstract

Apoptosis is a cell death program essential for development and tissue homeostasis in metazoan organisms. There are two major intracellular apoptotic signaling pathways defined as the extrinsic pathway or the death receptor pathway and the intrinsic pathway or the mitochondrial pathway. In this chapter we will focus on the key role played by mitochondria during apoptosis. The mitochondrial apoptotic pathway is linked to a change in the permeability of the outer mitochondrial membrane (OMM), which eventually leads to the release of intermembrane space apoptogenic factors such as cytochrome c, Smac/DIABLO, AIF, endoG, HtraA2/OMI and DDP/TIMM8a. Redistribution of these factors to the cytoplasm activates the apoptotic program, this however, is not the result of a simultaneous massive release of all these proteins but as some reports have pointed out, to a scheduled timing of their release. The permeabilization of the OMM is under the control of the Bcl-2 family of proteins and we will discuss the interactions of this family of proteins and the OMM through putative receptors (lipids or proteins). Since important alterations in the mitochondrial morphology and network are observed in the early stages of apoptotic cell death resulting in fragmentation of mitochondrial reticulum and remodeling of mitochondrial cristae, the mechanisms involving the fission/fusion machinery of mitochondria will also be discussed. Moreover some members of the Bcl-2 family of proteins such as Bax, Bak, Bid and Bcl-2 appear to have the capacity to modulate morphological remodeling of

* Correspondence concerning this article should be addressed to Lisa Oliver, Fax: 33-240084082; e-mail: Lisa.Oliver@univ-nantes.fr.

mitochondria suggesting a relationship between the apoptotic molecular events and the dynamics of mitochondrial morphology.

INTRODUCTION

Apoptotic cell death is a conserved mechanism observed in all multi-cellular organisms, implicated in all stages of development and during adult life. Since the first description of apoptosis based essentially on simple morphological changes, its role has been widely investigated in whole tissue and also during embryogenesis (Kerr *et al.,* 1972) (Vaux & Korsmeyer, 1999).

The process of apoptosis is a genetically encoded cell death program, which plays an essential role in embryogenesis, the preservation of tissue homeostasis and elimination of damaged or superfluous cells (Vaux & Korsmeyer, 1999). The dysfunction of the apoptotic pathways is implicated in various pathologies such as autoimmune diseases and immunodeficiency disorders, neurodegenerative diseases (such as Parkinson disease, Huntington disease, Alzheimer's disease, amyotrophic lateral sclerosis) and cancers.

There are two major apoptotic signaling pathways, the initiation and the execution of which is under the control of a family of proteins called protein caspases. The caspases are a family of cysteine proteases that cleave their substrates after an aspartate residue, hence their name (Nicholson, 1999). Twelve human caspases have been identified to date (caspases 1-10, caspase 12 and caspase 14) (Oliver & Vallette, 2005). These caspases can be grouped according to their function. The first group associated with the maturation of cytokines includes caspase 1, 4, 5 and 10 (Los *et al.,* 2003). The second group, which is directly involved in apoptosis can be subdivided into initiator and effector caspases (Oliver & Vallette, 2005). The initiator caspases (such as caspase 8 and 9) are identified by a long prodomain with homotypic protein-protein interaction motifs such as the death effector domain (DED) or the caspase recruitment domain (CARD), which are essential for the transduction of the various signals. The death receptor pathway (or extrinsic pathway) is activated through a ligand binding to receptors such as the tumor necrotic factor (TNF) receptor family and leading to the activation of caspase 8 (Ashkenazi, 2002). The major and probably the unique role of the initiator caspases are to proteolytically activate the effector caspases (caspase 2, 3, 6 and 7) (Salvesen & Abrams, 2004). The effector caspases cleave most key substrates provoking the cellular changes that lead to the apoptotic morphology observed in cells as described by Kerr *et al.* (1972). The mitochondrial pathway (or intrinsic pathway) is initiated by the activation of caspase 9 through the formation of a protein complex called the apoptosome; it is activated following a wide range of stimuli such as deprivation of growth factors, UV, toxins, hypoxia... (Nicholson, 1999). All these stimuli culminate in the activation of a family of proteins called the Bcl-2 family of proteins, which are responsible for the regulation of the apoptotic mitochondrial pathway through the activation of caspases (Antonsson, 2004) (Saelens *et al.*, 2004). Note that the mitochondrial pathway acts as an amplification system for the extrinsic pathway (Li *et al.*, 1998).

1. Proteins released from Mitochondria during Apoptosis

One crucial event that occurs during the mitochondrial apoptotic pathway is the loss of the integrity and the permeabilization of the OMM. Numerous studies have shown that the permeabilization of the OMM is under the control of pro-apoptotic members of Bcl-2 family such as Bax and/or Bak (Kuwana & Newmeyer, 2003) (Kuwana *et al.*, 2002). Three major models have been proposed to explain this increase in the permeability of the OMM. The first model implicates a loss in the mitochondrial membrane potential ($\Delta\Psi m$), resulting in matrix swelling, which eventually leads to the rupture of the OMM after the formation of mega-channels called the so-called permeability transition pores (PTP) allegedly composed of outer and inner membrane proteins (Marzo *et al.*, 1998) (Lemasters *et al.*, 1999). The second model postulates that the association of the pro-apoptotic protein Bax with the voltage-dependent anion channel (VDAC) leads to an increase in VDAC conductance, which is sufficient to induce a large permeability of the OMM (Shimizu *et al.*, 1999). The third model is based on the capacity of pro-apoptotic proteins Bax and/or Bak to form pores large enough to precipitate the release/leakage of the apoptogenic proteins (Saito *et al.*, 2000). However, none of these models fully accounts for all of the permeabilization properties of the OMM during apoptosis. We have recently observed, using Nuclear Magnetic Resonance (NMR) techniques, that the permeabilization of the OMM observed *in vivo* is much more complex than that produced experimentally *in vitro* by the induction of the PTP or the addition of Bax (Pouliquen *et al.*, 2006).

Several proteins released during apoptosis have been identified so far as summarized in Figure 1.

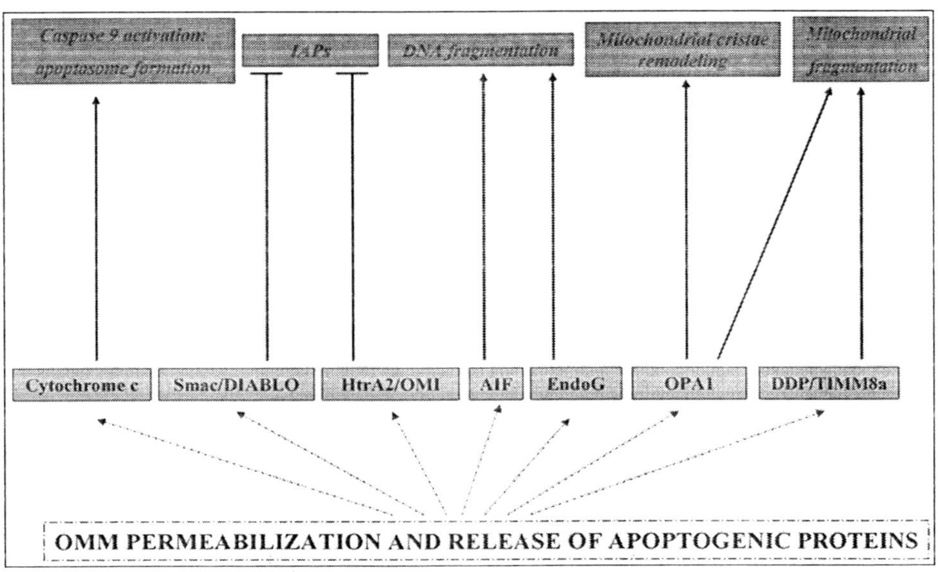

Figure 1. Functions of the apoptogenic factors released from the mitochondria. Functions mediated/triggered by apoptogenic factors released from the IMS downstream of the OMM permeabilization after induction of the mitochondrial apoptotic pathway.

1.1. Cytochrome c

Cytochrome c is a nuclear encoded protein, which is present in the mitochondrial intermembrane space (IMS) and acts as an electron carrier, transferring electrons from complex III to complex IV of the respiratory chain during oxidative phosphorylation. Cytochrome c is translated as an unfolded apoprotein called apocytochrome c containing an internal mitochondrial addressing signal (Diekert *et al.*, 2001). Within the IMS the enzyme cytochrome c heme lyase covalently attaches a reduced heme group to apocytochrome c converting it into the functional globular holocytochrome c. It is holocytochrome c, which is released during apoptosis and acts as an apoptogenic factor within the apoptosome complex.

Cytochrome c was the first protein identified to be released from the mitochondria during apoptosis (Liu *et al.*, 1996). This relocalization occurs early during the permeabilization of the OMM and is inhibited by Bcl-2 (Brustugun *et al.*, 1998) (Kluck *et al.*, 1997a) (Kluck *et al.*, 1997b). There are two pools of cytochrome c within the intermembrane space bound to the IMM: one fraction is loosely attached to the IMM through electrostatic interactions with the anionic phospholipids, primarily the cardiolipin. The second fraction appears tightly bound to the IMM through hydrophobic interactions. The disruption of the association of cytochrome c with the IMM is the first step of the two-step process described by Ott *et al.* (2002) and is followed by the translocation of soluble cytochrome c from the IMS to the cytoplasm. These data suggest that the two pools of cytochrome c could be mobilized by distinct stimuli: the loosely bound pool could be released after electrostatic changes such as variations in pH or changes in temperature while the strongly bound pool will be sensitive to alterations in membrane structure such as lipid peroxydation for example (Nomura *et al.*, 2000).

Once in the cytoplasm, cytochrome c binds to the scaffolding protein apoptotic protease-activating factor-1 (Apaf-1), causing a dATP-dependent conformational change that results in the oligomerization of Apaf-1 into a heptameric complex (Jiang & Wang, 2000). The reorganization of Apaf-1 facilitates the binding of procaspase 9, which is mediated by homotypic interactions between the CARD and results in the formation of a complex called the apoptosome (Hengartner, 2000) (Strasser *et al.*, 2000). This results in the auto-proteolytic activation of procaspase 9 that in turn cleaves and activates effector procaspase 3, 6 and 7 (Li *et al.*, 1997). Thus, cytochrome c participates directly in the caspase activation cascade.

1.2. Smac/DIABLO

Smac/DIABLO (Second mitochondrial activator of caspases/Direct IAP Binding protein of Low PI) is the second protein released from the mitochondria identified as a promoter of caspase activity. However, contrary to cytochrome c it does not function as a direct activator of caspases but acts by antagonizing inhibitors of apoptosis proteins (IAPs) (Du *et al.*, 2000) (Verhagen *et al.*, 2000).

The conserved IAP family of proteins can inhibit the enzymatic activity of certain caspases. (Deveraux *et al.*, 1998) (Deveraux & Reed, 1999) (Holcik & Korneluk, 2001) (Salvesen & Duckett, 2002). At least eight members of the mammalian IAP family have been identified, including X-chromosome-linked IAP (X-IAP), neuronal inhibitor of apoptosis protein (NIAP), cIAP-1, cIAP-2, IAP-like protein (ILP-2), BIR-repeat-containing ubiquitin-conjugating enzyme (Bruce), survivin and ML-IAP/livin. These proteins have been shown to target initiator caspase 9 and effector caspase 3 and 7. The IAPs do not inhibit other caspases, such as caspase 8 or 6. Each IAP protein contains one to three copies of the 80-residue zinc binding baculoviral IAP repeat (BIR). The different BIR domains and segments in the same IAP protein appear to exhibit distinct functions. For example the third BIR domain of X-IAP potently inhibits the activity of active caspase 9 whereas the linker region between BIR-1 and BIR-2 selectively inhibits active caspase 3 and 7 (Shi, 2002) (Shiozaki *et al.*, 2003). In contrast, ML-IAP/livin, which is usually over-expressed in melanomas and lymphomas, contains a single BIR domain shown to inhibit both caspase 3 and 7 (Kasof & Gomes, 2001), while another single BIR protein survivin, does not inhibit any caspase activity *in vitro* but appears to play a role in mitosis (Altieri, 2001). In addition, X-IAP, cIAP-1 and cIAP-2 contain a C-terminal RING-finger domain, which displays an ubiquitin ligase activity that is involved in the degradation of proteins by the proteasome (Jesenberger & Jentsch, 2002) (Saelens *et al.*, 2004).

Smac/DIABLO is a nuclear encoded protein translated into a 29-kDa precursor with an N-terminal mitochondrial targeting sequence. The 23-kDa mature Smac/DIABLO contains a four amino acid sequence in its N-terminus (Ala-Val-Pro-Ile), which forms an IAP binding motif (IBM). The mature Smac/DIABLO forms dimers that are released after permeabilization of the OMM during apoptosis. Once in the cytoplasm these dimers compete with caspases for binding with IAPs through an interaction between the IBM and the BIR domains (Huang *et al.*, 2003), thus enhancing caspase activation. However, as a result of this interaction IAPs, via the RING domain, are able to target Smac/DIABLO for proteasomal degradation (Hu & Yang, 2003). Of note there is no known biological role for Smac/DIABLO in mitochondria contrary to cytochrome c.

1.3. Apoptosis-Inducing Factor (AIF)

The characteristic phenotype of apoptotic cell death includes chromatin condensation and DNA degradation as a result of internucleosomal cleavage. Effector caspases are responsible for this phenotype by activation of proteins such as CAD (caspase activated DNase), which produces DNA fragments, acinus, which stimulates chromatin degradation and HELICARD, a helicase, which accelerates DNA degradation (Saelens *et al.*, 2004).

The capacity of mitochondria to provoke nuclear apoptosis in a caspase-independent manner was first observed using a cell free system in which mitochondria and nuclei were co-incubated (Zamzami *et al.*, 1996). This study proposed that the induced permeability transition of mitochondria could lead to the nuclear alterations observed during apoptosis. The AIF was identified as the protein responsible for this process and was the first mitochondrial factor released from the IMS with the capacity to generate large scale fragmentation of the DNA (50-kb) as well as peripheral chromatin condensation (Susin *et al.*, 1999). However, it does not trigger oligonucleosomal digestion, which seems essentially related to CAD activity.

AIF is related to flavoproteins with oxydoreductase activity (not essential for its pro-apoptotic function) but having no nuclease activity. AIF is synthesized as a 67-kDa precursor with a N-terminal mitochondrial addressing sequence. The mature AIF has a molecular weight of 57-kDa and possesses a C-terminal nuclear translocation sequence. Its structure presents a strong positive electrostatic charge on the surface supporting its capacity to interact with DNA (Ye *et al.*, 2002) after its relocalization from the mitochondria to the nucleus. The apoptotic effects of AIF can be inhibited by an interaction with the cytoprotective factor Hsp70, which sequesters AIF in the cytoplasm. The role of AIF in healthy mitochondria is unknown but it is possible that paradoxically it can prevent death induced by an oxidative stress due to its oxydoreductase activity (Klein *et al.*, 2002). However, the mechanism by which this protein induces the nuclear apoptotic phenotype is not clear but it acts in parallel with the caspase pathway. Indeed, in the nematode *C. elegans* WAH-1, the orthologue of AIF, can stimulate the activity of CPS-6, the counterpart of mammalian endonuclease G, another factor released from mitochondria during apoptotic cell death (Saelens *et al.*, 2004).

1.4. Endonuclease G (endoG)

Endonuclease G (endoG) is another nuclear encoded protein with a N-terminal mitochondrial-addressing sequence. The 30-kDa mature endoG is released from the IMS following mitochondrial permeabilization induced by pro-apoptotic factors such as a truncated Bid (tBid), then translocates to the nucleus where it degrades DNA even in the absence of caspase activation and CAD activity (Li *et al.*, 2001). It is not clear whether endoG induces nucleosomal degradation as described by the group of Wang (Li *et al.*, 2001) or just promotes a higher order of DNA degradation as observed by van Loo *et al.* (2001). In addition, data have been presented that suggest a cooperation of endoG with exonucleases and DNase I (Widlak *et al.*, 2001). The latter observation could explain the complete DNA fragmentation after translocation of endoG into the nucleus observed by Li *et al.* (2001) as well as the data presented by van Loo *et al.* (2001) that endoG alone, is insufficient for a complete DNA digestion. Nevertheless, in *C. elegans* the homolog of endoG, CPS-6 has been observed to play an important role in DNA degradation during apoptosis (Parrish *et al.*, 2001).

1.5. HtrA2/OMI

HtrA2/OMI is a nuclear encoded protein synthesized as a 49-kDa precursor imported into the mitochondria and resides in IMS. It belongs to a conserved family of serine proteases and shares sequence and structural homology with the bacterial endoprotease HtrA/DegP isolated from *E. coli*. HtrA/DegP is localized in the periplasmic space and plays a role in heat and oxidative tolerance. It functions as a chaperone at normal temperatures and develops a serine protease activity under elevated temperatures, resulting in the elimination of misfolded or damaged proteins (Saelens *et al.*, 2004). HtrA2/OMI was shown to be released from the IMS during apoptosis and has the capacity to interact and antagonize all IAPs except survivin (Suzuki *et al.*, 2001). Once the N-terminal extremity is processed, a N-terminal IBM (Ala-Val-Pro-Ser) similar to that detected in Smac/DIABLO is exposed in the mature HtrA2/OMI (37-kDa) protein. HtrA2/OMI acts by antagonizing the IAPs thereby enhancing caspase activity in cytosolic extracts similar to that observed with Smac/DIABLO (van Loo *et al.*, 2002a). Moreover, the serine protease activity of HtrA2/OMI seems capable of inducing cell death in a caspase-independent manner (Hegde *et al.*, 2002).

Besides the IBM and the serine protease domains, HtrA2/OMI also contains a C-terminal Post-synaptic Disc-large Zone (PDZ: comprises 90 amino acids, 2 α-helices and 4 β-sheets) domain, which blocks the access to the catalytic site resulting in an inhibition of the protease activity. The resolution of the structure by crystallography revealed that active HtrA2/OMI is a homo-trimeric assembly mediated by the serine protease domain (Li *et al.*, 2002). The resulting pyramid-shaped structure consists of three IBM domains at the top and three PDZ domains at the bottom. The trimer HtrA2/OMI exerts a low protease activity, which is regulated by the PDZ domains. In mutants lacking the PDZ domain, an increased protease activity can be observed. The group of Shi observed that the homo-trimeric form is necessary for the pro-apoptotic activity and that the monomer does not exert any pro-apoptotic capacity (Li *et al.*, 2002). Surprisingly, HtrA2/OMI variants lacking the IBM domains but retaining the homo-trimeric conformation present similar levels of apoptotic ability and caspase activation as the wild type protein, suggesting that HtrA2/OMI may also promote caspase activation independently of its IAPs binding capacity.

A recent investigation to analyze an apoptotic signal transduction mediated by the human Pag (hPag) using a yeast two-hybrid system identified HtrA2/OMI as a protein partner (Hong *et al.*, 2006). hPag, also known as peroxiredoxin I (PrxI), is a cytosolic member of the TSA/AhpC family, a new form of peroxidases with a cysteine residue at the primary catalytic site, thus referred to as a "thiol peroxidase". PrxI is over-expressed following oxidative stress. The C-terminal domain of hPag interacts with the PDZ domains of HtrA2/OMI increasing its protease activity. Hence, HtrA2/OMI may also be implicated in oxidative tolerance as its prokaryote counterpart. We can thus, conclude that HtrA2/OMI, like Smac/DIABLO, plays a dual role in apoptosis following its release from the IMS: firstly, as an IAP antagonist enhancing caspase activation and secondly, as a serine protease capable of inducing an, as yet, uncharacterized "caspase-independent" cell death.

1.6. DDP/TIMM8a

DDP (Deafness/Dystonia Protein)/TIMM8a was identified as the protein mutated in the human deafness/dystonia syndrome also called Mohr-Tranebjaerg syndrome. This syndrome was later defined as a mitochondrial disease and DDP/TIMM8a was shown to be the homolog of the yeast small Tim protein, Tim8p (Koehler *et al.*, 1999). Small Tim proteins form together with the Tim22 complex the so-called TIM22 (Translocase of Inner Membrane) import apparatus (Curran *et al.*, 2002) (Koehler, 2004b). Like other small Tims, DDP/TIMM8a contains twin CX_3C motif (cysteine-X-X-X-cysteine) and functions as a chaperone in the IMS and is involved in protein importation, assembly and insertion into the IMM (Koehler, 2004a). DDP/TIMM8a exists in the IMS as a 70-kDa complex composed of another small Tim called TIMM13b present in a 3:3 ratio (Roesch *et al.*, 2002). It was demonstrated that a cysteine 66 to tryptophan (C66W) point mutation in the CX_3C motif results in a defect in assembly of the DDP/TIMM8a-TIMM13b complex generating the human deafness/dystonia syndrome (Roesch *et al.*, 2002).

DDP/TIMM8a was identified as a protein released upon the opening of the permeability transition pore complex (Patterson *et al.*, 2000). However, its role in apoptosis was only recently determined. The group of Blackstone showed that the released of DDP/TIMM8a precedes and promotes mitochondrial fission during programmed cell death (Arnoult *et al.*, 2005b). The translocation of DDP/TIMM8a to the cytosol was dependent on the OMM permeabilization by pro-apoptotic proteins Bax and/or Bak and it seems to be co-released with cytochrome c. Once in the cytosol, DDP/TIMM8a binds to Dynamin-related protein-1 (Drp-1), a dynamin-like GTPase involved in the mitochondrial fission process (Frank *et al.*, 2001). The association between DDP/TIMM8a and Drp-1 induces the recruitment to the mitochondria thereby promoting mitochondrial fragmentation (Arnoult *et al.*, 2005b). As a consequence and according to its primary role, DDP/TIMM8a may act as a chaperone for Drp-1 leading to its relocalization to mitochondria. Interestingly, it has been observed that TIMM13b, the IMS partner of DDP/TIMM8a, is also released after OMM permeabilization induced by tBid (van Loo *et al.*, 2002b) although Arnoult *et al.* (2005b) did not observe any interaction between TIMM13b and Drp-1. Consequently, the release of DDP/TIMM8a demonstrates that the mitochondria are not only involved in caspase activation and nuclear fragmentation but may also promote alterations in their own morphology during mitochondrial apoptotic pathway. The latter mechanism could also be important under non-apoptotic conditions during mitochondrial biogenesis.

1.7. Other Proteins released from the IMS during Apoptosis

Numerous other factors are released from the IMS during apoptosis (Patterson *et al.*, 2000). The most recent study performed using a matrix-assisted laser desorption ionization post-source decay (MALDI-PSD) analysis of the supernatant from truncated Bid (tBid)-treated mitochondria (van Loo *et al.*, 2002b), among the proteins identified were candidates such as Acyl-CoA-binding protein (ACBP), WD-40 repeat tumor-specific antigen, Type I RNA helicase Upf1, polypyrimidine tract-binding protein (PTB) and adenylate kinase 2 (AK2).

ACBP is the endogenous ligand of the mitochondrial benzodiazepine receptor and has been described as a factor capable of reducing the calcium concentration required for calpain activation (Melloni *et al.*, 2000). Interestingly, the calpains are a family of calcium-dependent cysteine proteases, which can proteolytically activate pro-apoptotic proteins such as Bid (Chen *et al.*, 2001) and Bax (Cartron *et al.*, 2004b).

The WD-40 repeat tumor-specific antigen is a protein of unknown function and its mitochondrial localization has not been documented. However, WD-40 repeats are present in the C-terminal of Apaf-1 and are important in the regulation of caspase 9 activation. Upf1 is necessary in the control of the total accumulation of a large number of mRNAs in addition to its role in RNA surveillance whereas PTB is a repressive regulator of alternative splicing and seems to be an essential factor involved in the translation of a functional caspase 2 (van Loo *et al.*, 2002b). However, it is important to note that the mitochondrial localization of Upf1 and PTB is only based on computational models. With regards to AK2, it is a protein, which resides in the IMS and catalyses the reversible phosphorylation of nucleotide triphosphates into nucleotide monophosphates (Saelens *et al.*, 2004). Consequently, AK2 could play a role in maintaining the ATP level essential for the early events of mitochondrial apoptotic pathway.

Another protein reported to be released during apoptosis is Dominant Optic Atrophy protein (OPA1) (Arnoult *et al.*, 2005a). OPA1 is a dynamin-like GTPase present in the IMS, which maintains mitochondrial cristae structure and regulates the mitochondrial fusion mechanism (Karbowski & Youle, 2003) (Bossy-Wetzel *et al.*, 2003). It has been observed that the OMM permeabilization by Bax and/or Bak is followed by inhibition of mitochondrial fusion (Karbowski *et al.*, 2004). During apoptosis, mitochondrial fragmentation is facilitated and the structures of the cristae change after the release of OPA1. Moreover, alterations in the cristae organization facilitate cytochrome c liberation (Arnoult *et al.*, 2005b) supporting the two-step process of cytochrome c release as described by Ott *et al.* (2002).

Other proteins have been identified as factors released from IMS following tBid-induced permeabilization of OMM (van Loo *et al.*, 2002b) or activation of permeability transition pore (Patterson *et al.*, 2000). Nevertheless their roles, if any, in apoptosis remain unknown. It is also important to comment that the study of Patterson and colleagues refers to a large number of proteins released from mitochondria after PTP opening, among these proteins are listed numerous enzymes implicated in mitochondrial metabolism (Patterson *et al.,* 2000). These results, for the major part, are probably the consequence of a total damage and rupture of the mitochondrial structure caused by the activation of the PTP instead of a specific release of apoptogenic factors occurring during apoptotic death.

2. Kinetics of the Release of Proteins from the Mitochondrial IMS during Apoptosis

An increase in the permeability of the OMM does not lead to a simultaneous massive release of all the apoptogenic factors but rather to a regulated kinetic process of liberation.

Cytochrome c release occurs very early in the apoptotic process and a cytochrome c-Green Fluorescent Protein (GFP) construct showed that this release preceded phophatidylserine exposure and the loss of plasma membrane integrity in Hela cells (Goldstein *et al.*, 2000) (Gao *et al.*, 2001). The first phase of cytochrome c release occurred within about 5 minutes regardless of the stimulus applied and the duration of this release was not altered by temperatures ranging from 24°C to 37°C. Since the activity of most enzymes changes significantly over 10°C, enzymes cannot be the rate limiting element during cytochrome c release, and as such, support a model of a translocation of cytochrome c through pores formed in the OMM. A loss of $\Delta\Psi m$ accompanies this rapid depletion of the cytochrome c pool. However, the induction of apoptosis with a concomitant inhibition of the proteolytic activation of caspases using a broad range caspase inhibitor (z-Val-Ala-Asp-fluoromethylketone: z-VAD-fmk) does not alter the release of cytochrome c whereas no loss in $\Delta\Psi m$ was detected. These data suggest that the permeability transition is not required for the release of cytochrome c. This study by Goldstein and coworkers present a model of a rapid, total and kinetically invariant cytochrome c release occurring independently of the mitochondrial permeability transition (Goldstein *et al.,* 2000).

Another study by the group of Prehn using TNFα (extrinsic apoptotic pathway), staurosporine (a broad range kinase inhibitor) and valinomycin (an ionophore, which acts on the mitochondria) as inducers of the OMM permeabilization show different patterns of cytochrome c release depending on the type of stimuli (Luetjens *et al.*, 2001). However, it supports the data of Goldstein *et al.* (2000) as the release of cytochrome c induced by TNFα and staurosporine occurs within a similar time range whereas valinomycin, an inducer of PTP opening, induces only a weak release over a much longer period. Moreover, the presence of zVAD-fmk did not block cytochrome c release induced by TNFα or staurosporine but inhibited the effects of valinomycin. These results suggest that the complete translocation of cytochrome c to the cytoplasm occurs in the first steps of the mitochondrial apoptotic pathway and precedes loss of $\Delta\Psi$m and caspase activation.

Topology analysis of the inner membrane reveals that the IMM presents tubular cristae that form compartments conducive to oxidative phosphorylation. Hence, these cristae contain an important fraction of the cytochrome c, which is separated from the IMS by cristae junctions (Mannella *et al.*, 2001). The tightly bound pool of cytochrome c is represented by the fraction contained in the oxidative phosphorylation complexes and thus, sequestered within the cristae. This "cristae pool" of cytochrome c will have to be mobilized in order to permit the complete liberation of cytochrome c implying a remodeling of the internal structure of the mitochondria, and in particular, a remodeling of the mitochondrial cristae. Such a remodeling has been observed in isolated mitochondria pretreated with tBid and wherein a structural reorganization of the IMM caused the fusion of individual cristae culminating in the opening of the cristae space into the IMS (Scorrano *et al.*, 2002). This remodeling process could be the result of a release of OPA1, a protein involved not only in mitochondrial fusion but also in the maintenance of the cristae structure (Arnoult *et al.*, 2005b). The authors of this study demonstrate that a loss of OPA1 seems to facilitate the release of cytochrome c. Indeed, OPA1 appears to be co-released with cytochrome c. The onset of the release of these two proteins seems to be the same although the complete release of OPA1 occurs over a much longer period as compared to cytochrome c. In fact, cytochrome c is completely released within 5 to 10 minutes while duration of OPA1 redistribution occurs over a period of more than 20 minutes. Moreover, the same study shows that mitochondrial fragmentation begins about 10 minutes after cytochrome c release and is complete within about 20 minutes. Therefore, it appears that intrinsic apoptotic stimuli can induce rapid intra-mitochondrial structural changes within 2-5 minutes in order to facilitate the release of all the cytochrome c. The cristae remodeling induced by tBid is also sensitive to cyclosporine A although no major swelling occurs and only a minor depolarization can be observed suggesting a transient opening of the PTP (Scorrano *et al.*, 2002).

The timing of the release of other mitochondrial apoptogenic factors, in particular, Smac/DIABLO has also been investigated. Initially, the liberation of Smac/DIABLO appears to occur concurrently with cytochrome c and precedes the loss in $\Delta\Psi$m (Madesh *et al.*, 2002). Hence, the accumulation of Smac/DIABLO in the cytoplasm is an early rapid event in the mitochondrial apoptotic pathway. However, contrary to the study by Goldstein and colleagues (2002), the group of Hajnóczky observed that cytochrome c translocation to the cytoplasm was sensitive to temperature changes (Madesh *et al.*, 2002). At 35°C tBid induces a complete release of cytochrome c within about 2 minutes whereas at 25°C the duration of this translocation occurs within about 5 minutes. However, this noticeable discrepancy between these two reports is less evident if we focus on the time scales of their observations. For Goldstein and coworkers this release occurs within about 5 +/- 2.5 minutes independently of temperature while the group of Hajnóczky note that the duration for the complete release of cytochrome c is delayed from about 2 minutes at 35°C to about 5 minutes at 25°C. As a result of these observations it is obvious that these two interpretations display the same consequence. Moreover the "astonishing observation" of the temperature insensitive liberation of cytochrome c can be explained by the use of a too low time scale resolution.

Besides, non-sensitivity to temperature variation does not support the model of cristae remodeling. It is obviously improbable that such a mechanism of structural reorganization cannot be altered at low temperatures. Thus it is possible to reconcile the two models and that the delayed cytochrome c release observed by the group of Hajnóczky could be explained by an extension of time required for the cristae remodeling process. Using more precise methods, later investigations on the kinetics of Smac/DIABLO accumulation in the cytosol have detected a more accurate kinetic of Smac/DIABLO release. Data on Smac/DIABLO-YFP (yellow fluorescent protein) release in HeLa cells shows that the efflux of Smac/DIABLO occurs within about 20 minutes (Springs *et al.*, 2002), which is four times longer than the time required for cytochrome c release described by Goldstein *et al.* (2000). In addition compared to other apoptotic markers it seems that the release of Smac/DIABLO takes place before DNA fragmentation but after cell shrinkage. This supports the significant position of Smac/DIABLO in the caspase-induced DNA fragmentation. Another report using real-time single cell analysis compares the kinetics of release of Smac/DIABLO-YFP and cytochrome c-GFP in staurosporine-treated MCF-7 cells and demonstrates that the duration of the cytosolic accumulation of Smac/DIABLO is four times longer than that of cytochrome c. Nevertheless the onset of Smac/DIABLO release is not significantly different from that of cytochrome c (Rehm *et al.*, 2003). This suggests that cytochrome c and Smac/DIABLO can be liberated through a similar mechanism but that their kinetics differs, probably due to a size-dependent process. Since cytochrome c is a small protein (13-kDa) and Smac/DIABLO is a 23-kDa protein located in IMS as a dimer, thus it is four times larger in size than the electron carrier. Moreover, the remarkable demonstration of the group of Prehn that depolarization of mitochondria concurred simultaneously with the release of two apoptogenic factors however, not in the same fashion (Luetjens *et al.*, 2001). While cytochrome c and Smac/DIABLO release follow rapid kinetics, mitochondrial depolarization is progressive and extends over a longer period of time (Rehm *et al.*, 2003). Therefore, significant depolarization is seen only after a complete depletion of mitochondrial Smac/DIABLO and cytochrome c. Furthermore, as with cytochrome c, Smac/DIABLO release occurs in a caspase-independent manner although a previous study shows that an inhibition of caspase activity could block Smac/DIABLO release (Adrain *et al.*, 2001) but not cytochrome c liberation. This could be explained by inhibiting proteasome using a specific inhibitor such as lactacystin, which allows a recovery of the cytoplasmic content of Smac/DIABLO in staurosporine + zVAD-fmk treated cells. Prehn and colleagues suggest that caspase activity stabilizes cytosolic Smac/DIABLO (Luetjens *et al.*, 2001). One explanation can be that zVAD-fmk can enhance the targeting of Smac/DIABLO to the proteasome for degradation by IAPs (Hu & Yang, 2003) as mentioned earlier in this chapter. Interestingly this study also demonstrates for the first time that activation of effector caspases can occur within about 10 minutes after the onset of Smac/DIABLO redistribution. Moreover caspase 3-deficient MCF-7 cells require a complete redistribution of Smac/DIABLO before the detection of caspase activation. A recent report using double immunocytochemistry focusing on the relative timing of redistribution of both cytochrome c and Smac/DIABLO induced by staurosporine in two cellular systems: human osteosarcoma 143B cells and primary cultures of murine embryonic cortical neurons (Lim *et al.*, 2006), show that although the temporal resolution of the release cannot be determined using this type of analysis, the results do show different patterns of release according the cellular type. Neurons present a parallel release of both proteins while 143B cells demonstrate a significant delay in the onset of Smac/DIABLO redistribution. On the other hand different subpopulations of cells can be discerned following exposure to staurosporine. In neurons three subpopulations of cells were discerned: those presenting only (1) a cytochrome c release, (2) a Smac/DIABLO release or (3) both cytochrome c and Smac/DIABLO release. The third group was the most commonly observed. In 143B cells there were only cells showing a single cytochrome c release or both cytochrome c and Smac/DIABLO release. In addition zVAD-fmk did not affect the pattern of release of these proteins in the two types of cells. These findings confirm that diverse cell types can present different timing of release when exposed to the same apoptotic inducer under the same conditions and that the same cells can present dissimilar timing of release when exposed to several apoptotic stimuli.

Concerning other proteins translocated from IMS to cytosol during apoptosis there are less results pointing to their relative timing of release. However, it seems to be clear that apoptogenic factors from IMS can be subdivided in two classes: those liberated independently of caspase activity and those requiring a well established caspase activity for release from IMS. It has been demonstrated that the release of cytochrome c, Smac/DIABLO and HtrA2/OMI does not involve any caspase activity whereas liberation of AIF and endoG depends upon the activation of caspases (Arnoult *et al.*, 2003). Treatment of isolated mitochondria from HeLa cells, in a cell-free system, with tBid or oligomeric Bax results in the redistribution of cytochrome c, Smac/DIABLO and HtrA2/OMI to the supernatant. Moreover a different kinetic of release between these three factors can be distinguished. As expected cytochrome c appears in supernatant after 5 minutes of treatment with oligomeric Bax as for Smac/DIABLO, which becomes visible after 5 minutes of treatment and only clearly materializes after 15 minutes of treatment. With regard to HtrA2/OMI, it appears only after 15 minutes. These findings are in accordance with previous studies describing differences between cytochrome c and Smac/DIABLO release (Springs *et al.*, 2002) (Rehm *et al.*, 2003). Furthermore, it supports the facts that Bax and/or Bak permeabilization of the OMM induces redistribution of these proteins through a similar mechanism and that the variation in the kinetics of release may only be due to a size-dependent mechanism. This idea is consistent with the size of the different proteins: cytochrome c is present as a monomer of 13-kDa, Smac/DIABLO as a dimer assembly of two subunits of 23-kDa and HtrA2/OMI as a homo-trimeric complex of three 37-kDa subunits.

However, for the other factors such as AIF and endoG it was demonstrated that their redistribution to the cytoplasm could not occur without caspase activity (Hsu *et al.*, 1997). Inhibition of caspases by zVAD-fmk prevents the release of AIF and endoG from the IMS. In the same way this translocation is blocked in cells lacking Apaf-1 resulting in a defect in the activation of caspases while release of cytochrome c, Smac/DIABLO and HtrA2/OMI remains unaffected. Thus, these results support the idea that early in the mitochondrial apoptotic pathway cytochrome c, Smac/DIABLO and HtrA2/OMI accumulate in the cytosol; interestingly these three proteins are involved in caspase activation. When caspases are activated then AIF and endoG are released from the IMS and relocated to the nucleus in order to participate in nuclear apoptosis. It confirms the observation that DNA fragmentation follows Smac/DIABLO liberation (Springs *et al.*, 2002).

Concerning mitochondrial fragmentation it seems to begins around 10 minutes after cytochrome c has been released and is complete within about 20 minutes (Arnoult *et al.*, 2005b). Despite the early loss of OPA1, which seems to be involved in cristae remodeling this protein can also be implicated in mitochondrial fragmentation because redistribution of OPA1 to the cytosol induces a block in mitochondrial fusion thus favoring a fission mechanism. The timing of the translocation of OPA1 to the cytoplasm demonstrates that it is co-released with cytochrome c. The onset of the release seems to be the same although release of OPA1 occurs over a much longer period than for cytochrome c; the duration of OPA1 redistribution occurs within about more than 20 minutes. Another IMS factor promoting mitochondrial fission DDP/TIMM8a, is also co-released with cytochrome c and its kinetics of redistribution occurs within around the same time as for OPA1 but with a shorter period. Although DDP/TIMM8a is smaller in size than cytochrome c the duration of its translocation seems to be delayed, one explanation can be the complex state of DDP/TIMM8a, which is responsible for the delay observed for its liberation.

3. Apoptosis, Mitochondria and Bcl-2 Family of Proteins

In the third section of this chapter, we will discuss the relation between the mitochondria and the Bcl-2 family of proteins. This family of proteins controls and regulates the mitochondrial apoptotic pathway by inhibiting or inducing cell death. The effector members of this family are responsible for the permeabilization of the OMM upstream of the events related in the two first parts of this chapter. We will attempt to describe the association of these proteins with the mitochondria and the mechanism by which these proteins control the permeability of the mitochondrial membranes under control and apoptotic conditions. Firstly, in order to fully understand the mechanism of action of these proteins we have to be cognizant of the three dimensional structure. Finally, we will focus on the association/interaction of these proteins with the OMM with a special emphasis on the pro-apoptotic effectors Bax and Bak, which trigger the initiation of the apoptotic intrinsic pathway by increasing the permeability of the OMM. In summary, all the events and actors implicated in the mitochondrial apoptotic pathway are illustrated in Figure 2.

3.1. The Bcl-2 Family of Proteins

The members of the Bcl-2 family of proteins are considered the central regulators of the mitochondrial apoptotic pathway (Cory & Adams, 2002). These proteins control alterations in mitochondrial permeability after an apoptotic stimuli and subsequently are responsible for all the events occurring downstream of the OMM permeabilization. This family of proteins is characterized by the presence of Bcl-2 homology domains or BH domains as illustrated in Figure 3. In mammalians about twenty members of this family are known to date.

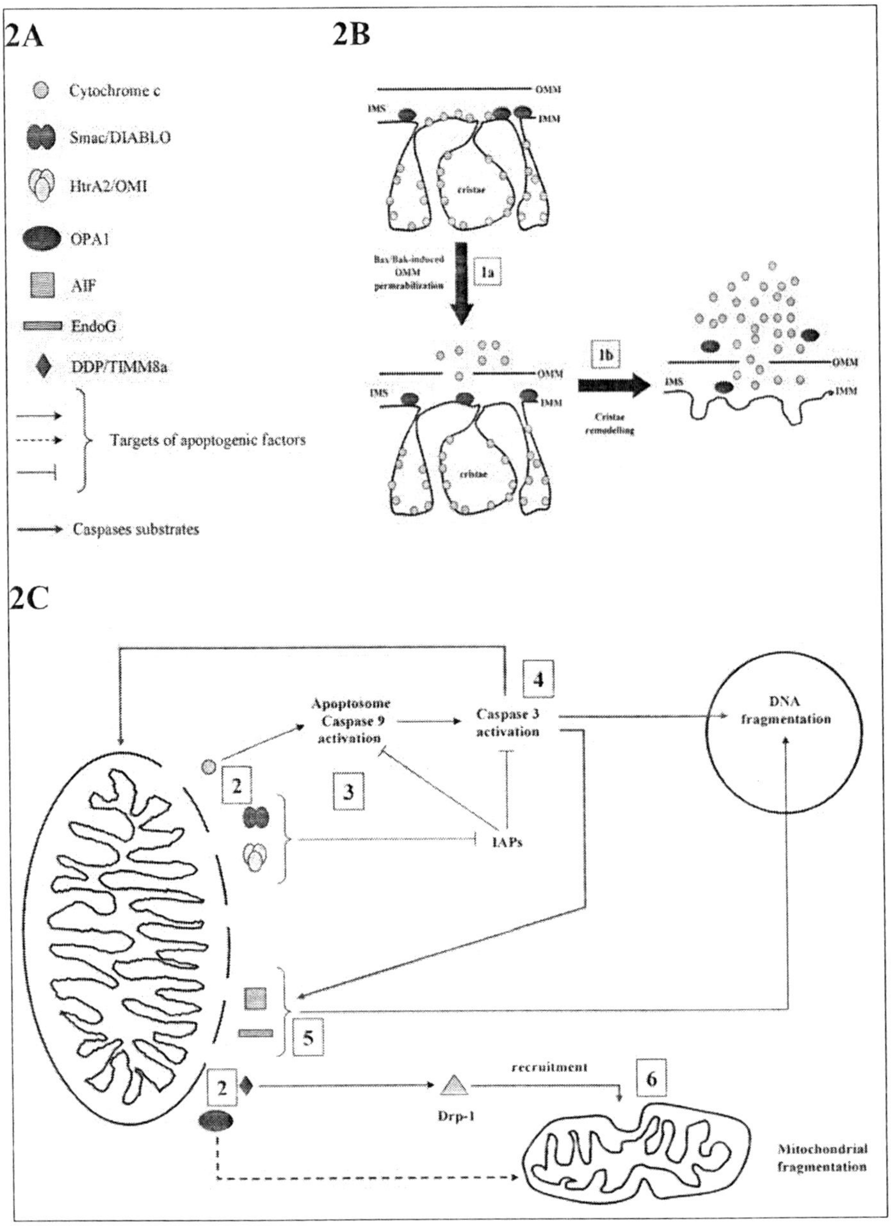

Figure 2. The events of the mitochondrial pathway. A. Symbols used in the figure. B. The two-step process involved in cytochrome c release from mitochondria. (1a) The first step allows the liberation of the loosely bound pool of cytochrome c after permeabilization of the OMM triggered by pro-apoptotic factors Bax and/or Bak. (1b) The second step requires internal structural rearrangements of the IMM in order to mobilize the tightly bound fraction of cytochrome c entrapped in mitochondrial cristae. Cristae remodeling can be induced through the loss of OPA1 resulting in mobilization of both pools of cytochrome c and the complete redistribution to the cytosol. C. (2) The concomitant release of Smac/DIABLO and HtrA2/OMI occurs while cytochrome c is redistributed but over a longer period. Factors such as OPA1 and DDP/TIMM8a are also released at this point. (3) Cytoplasmic cytochrome c participates in the formation of the apoptosome resulting in caspase 9 activation whereas Smac/DIABLO and HtrA2/OMI inhibit the IAPs. (4) The synergistic actions of cytochrome c, Smac/DIABLO and HtrA2/OMI allow for the complete activation of effector caspases such as caspase

3. Active caspase 3 cleaves its substrates resulting in activation of proteins such as CAD and acinus, or inactivation of others such PARP, which trigger chromatin condensation and internucleosomal DNA degradation. At the same time, caspase 3 also targets mitochondria and provokes irreversible damage to the respiratory chain and a sustained loss of $\Delta\Psi m$. (5) Activation of the caspases permits the release of other apoptogenic factors such as AIF and endoG, which then translocate to the nucleus where they induce DNA fragmentation. (6) Cytosolic DDP/TIMM8a induces the recruitment of Drp-1 to the mitochondria where it initiates the fission process. Loss of OPA1 results from a block in the fusion machinery thus promoting mitochondrial fragmentation. This stage leads to a complete fragmentation of the mitochondrial reticulum.

The Bcl-2 family can be subdivided into two subfamilies: the anti-apoptotic proteins and the pro-apoptotic proteins. Among the pro-apoptotic members, the proteins can be separated into two groups: the multi-domain members and the BH-3-only proteins.

The anti-apoptotic members: Bcl-2, Bcl-x_L, Mcl-1, Bcl-w, A1/Bfl-1, Boo/Diva and Nr-13 are characterized by the presence of four BH domains (BH-1, BH-2, BH-3 and BH-4). These proteins act by antagonizing the pro-apoptotic proteins of the Bcl-2 family (Antonsson *et al.*, 1997) (Cheng *et al.*, 2001) and thus function as pro-survival factors. The BH-4 domain, which is present only in this subfamily of Bcl-2 family members seems to be implicated in their anti-apoptotic activity (Huang *et al.*, 1998).

The pro-apoptotic BH-3 only proteins, as suggested by their name, are characterized by the presence of a single BH domain. These proteins can be considered as “stress sensors” and trigger apoptosis by either activating pro-apoptotic multi-domain proteins of the Bcl-2 family or antagonizing the anti-apoptotic members (Juin *et al.*, 2005). Examples of BH-3 only proteins are Bid, Bad, Bim, Bik, Hrk, Bmf, Noxa, PUMA, BNIP3L and BNIP3. Each member of this group of pro-apoptotic factors can be activated following a particular set of stimuli (Huang & Strasser, 2000). Indeed, the BH-3 only proteins cannot induce apoptosis in the absence of the pro-apoptotic multi-domain members, Bax and Bak.

The second group of pro-apoptotic proteins includes the multi-domain members Bax, Bak, Bcl-x_S and Bok. Bax and Bak are considered to be the pro-apoptotic effectors of the Bcl-2 family. These proteins possess three BH domains (BH-1, BH-2 and BH-3). Both proteins are capable of inducing cytochrome c release (Degenhardt *et al.*, 2002) and are the essential actors in the mitochondrial apoptotic pathway. A knockdown of the expression of Bax by the technique of RNA interference or by genetic invalidation (Takeuchi *et al.,* 2005) confers a resistance to apoptosis while a knockdown of Bak shows no discernable effect. However, inactivation of both genes results in a complete impairment of apoptosis and thus, a complete resistance to apoptotic stimuli (Degenhardt *et al.*, 2002) (Cartron *et al.*, 2003a). The apoptotic ability of the pro-apoptotic members of the Bcl-2 family is mediated through their BH-3 domains (Chittenden *et al.*, 1995a).

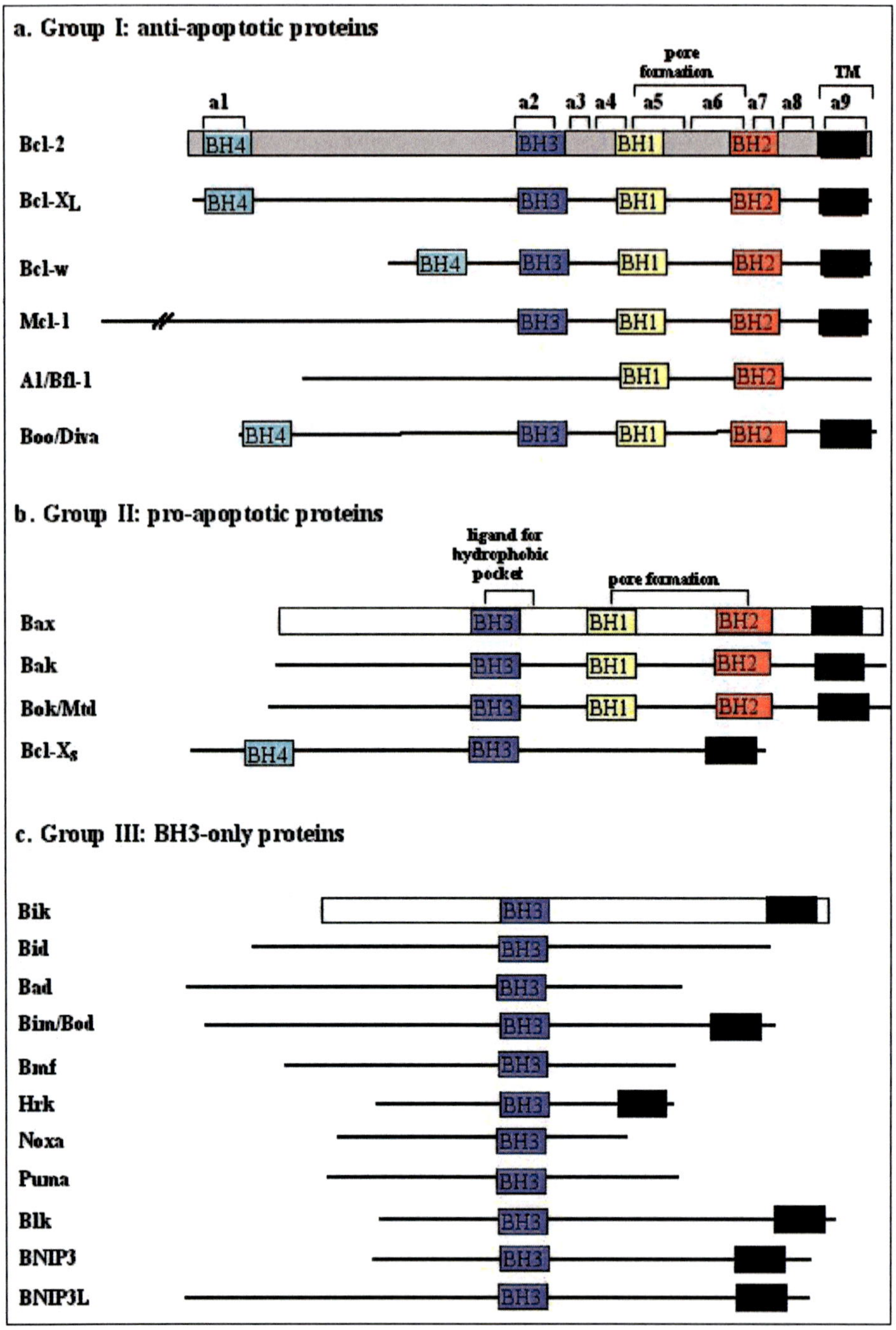

Figure 3. Schematic representation of all the known mammalian Bcl-2 family members. Bcl-2 homology regions 1 - 4 (BH-1 - 4) are indicated. TM indicates a putative transmembrane region that mediates localization to intracellular membranes. α: αhelix. The Hα-5 – Hα-6 overlapping region containing the BH-1 domain corresponds to the pore-forming region based on structural homology with bacterial toxins. The BH-3 domain in the pro-apoptotic members is a ligand for the hydrophobic pocket formed by the BH-1 - BH-3 domains of the anti-apoptotic members.

3.2. The relationship between the Structure and the Function of the Bcl-2 Family of Proteins

Attractive annotations have come after the resolution of the three dimensional structure of member of each category of the Bcl-2 family of proteins. The structural analysis of the proteins such as Bcl-x_L (Muchmore *et al.*, 1996), Bcl-2 (Petros *et al.*, 2001), Bid (Chou *et al.*, 1999) (McDonnell *et al.*, 1999) and Bax (Suzuki *et al.*, 2000) show that these proteins share very similar folds composed of seven to nine α-helices (Hα) although they may have different and/or opposite functions. The overall structure is a globular bunch of amphipathic Hα surrounding the two central hydrophobic Hα-5 and Hα-6.

Interestingly a report by Muchmore *et al.* (1996) on the structure of Bcl-x_L revealed a striking structural homology between the anti-apoptotic proteins and the bacterial α-pore-forming toxins (α-PFTs) such as colicin E1, colicin Ia and diphtheria toxin. These toxins contain a pore-forming domain of ten alpha helices with two central hydrophobic alpha helices that insert into membranes. These toxins possess all of the domains required for binding and pore formation in membranes in a single protein (Zakharov & Cramer, 2002a). The mechanism by which these pore-forming toxins insert into membranes requires a three step action, which includes a binding step, an unfolding step followed by an insertion step (Parker & Feil, 2005) (Tilley & Saibil, 2006). The first step of binding to membranes occurs through a mechanism involving a membrane receptor, which is recognized by a specific region of the toxin. The second step consists of the deployment of the amphipathic helices resulting in the release of the two central hydrophobic helices thereby initiating the insertion of the protein into the membrane (Zakharov & Cramer, 2002b). This model of membrane insertion is commonly called the "umbrella" model. The latter step necessitates a so-called activation of the toxin, a consequence of a change in the conformation of the protein. These modifications in the structure of the toxin can be triggered by variations in the pH as demonstrated for colicin E1, which forms pores following an activation by acid-pH (Musse & Merrill, 2003).

Among the Bcl-2 family members it has also been shown that Bcl-2 (Schendel *et al.*, 1997) (Schlesinger *et al.*, 1997) (Antonsson *et al.*, 1997), Bcl-xL (Minn *et al.*, 1997) and Bax (Schlesinger *et al.*, 1997) (Antonsson *et al.*, 1997) are able to form pores in artificial lipid bilayers. A report resolving the crystal structure of rat Bcl-x_L and using a homology modeling for Bcl-2 and Bax based on this structure have demonstrated that it is the pro-apoptotic Bax, which presents the strongest potential for membrane insertion (Aritomi *et al.*, 1997). This observation that Bax shares a greater homology with bacterial toxins underlines the toxic role of Bax. As observed with bacterial toxins, the pro-apoptotic effectors Bax (Desagher *et al.*, 1999) and Bak (Griffiths *et al.*, 1999) undergo conformational changes after an apoptotic stimuli. In addition, it has been observed for Bak that such a structural reorganization leads to the exposure of its BH-1 domain, which overlaps the N-terminal region containing the core hydrophobic Hα-5 (Griffiths *et al.*, 2001). It has also been shown that a variation in pH can trigger a conformational change resulting in the acquisition of an active form of Bax (Cartron *et al.*, 2004a). Moreover, the multi-domain members of the Bcl-2 family are known to form hetero-oligomers (Sattler *et al.*, 1997) or homo-oligomers (Antonsson *et al.*, 2000) (Antonsson *et al.*, 2001). The hetero-dimerization of Bcl-2 family proteins is a crucial step in the regulation of apoptotic cell death. This hetero-dimerization involves the BH-3 domain of the pro-apoptotic members and a wider surface of the anti-apoptotic members encompassing the BH-1 and BH-2 domains (Petros *et al.*, 2004). The homo-oligomerization of Bax and/or Bak is required for the OMM permeabilization (Antonsson *et al.*, 2001) (Mikhailov *et al.*, 2003). Thus, there exists a strong structural homology and function between the pro-apoptotic proteins Bax and Bak, and some toxins. These toxins target cells preferentially by binding to putative receptors. The question remains, does such receptors exist for the Bcl-2 family of proteins and if so, are these receptors present and active on the OMM?

3.3. Anti-Apoptotic Proteins: Bcl-2 and Bcl-X_L

The Bcl-2 (B cell lymphoma gene 2) protein was discovered through its involvement in a common chromosomal translocation in human follicular B cell lymphomas and was described as a proto-oncogene capable of inhibiting programmed cell death. It was the first member of the so-called Bcl-2 family of proteins to be identified. It is present as an integral protein in three endomembranes: the external membrane of the nuclear envelope, the membrane of endoplasmic reticulum and the mitochondrial outer membrane (Krajewski *et al.*, 1993) (Janiak *et al.*, 1994) (Lithgow *et al.*, 1994). However, it seems that the mitochondrion is the major site for Bcl-2 integration. In the mitochondria, Bcl-2 was initially proposed as an inner membrane protein (Hockenbery *et al.*, 1990) but it was later determined that it was targeted to the OMM through its C-terminal Hα-9 (Nguyen *et al.*, 1993). Data from the laboratory of Shore have demonstrated that Bcl-2 preferentially associates with mitochondria rather than with microsomes and that the association of Bcl-2 with these membranes does not involve the same pathway of importation (Nguyen *et al.*, 1993). Moreover, the OMM targeting of Bcl-2 seems to involve the same import pathway as for the yeast Mas70p protein. The amino acids in the C-terminus of Bcl-2 constitute an anchor signal sequence thus presupposing a model in which Bcl-2 is anchored in the OMM through its C-terminal extremity leaving the bulk of the protein and its N-terminal exposed to the cytoplasm. More recently an OMM receptor of Bcl-2 has been identified in a study investigating the pathway of insertion of Bcl-2 in a yeast model. The authors of this study have shown that Bcl-2 interacts specifically with the import receptor Tom20 (Motz *et al.*, 2002). However, unlike most Tom20-dependent preproteins, the two-lysine residues, which flank the C-terminal hydrophobic anchor of Bcl-2 mediate the interaction with Tom20 via a typical N-terminal presequence binding of Bcl-2 to Tom20. In addition, Bcl-2 interacts exclusively with the Tom20 receptor and no interactions exist with other components of the TOM complex such as Tom22, Tom5 and Tom40 thereby suggesting that there is no translocation of Bcl-2 through the general import pore unlike other proteins such as porin or Tom40. Hence, Bcl-2 is targeted to the mitochondria through the binding of its C-terminal anchor signal sequence to Tom20 and then directly inserts into the OMM through the same hydrophobic C-terminal anchor signal sequence. Recently it has been shown that Bcl-2 undergoes a conformation change during apoptosis precipitating the insertion of the hydrophobic Hα-5 and Hα-6 and this change in conformation can be triggered by the BH-3 domain of BH-3 only proteins such as Bim (Kim *et al.*, 2004). Thus, the anti-apoptotic activity of Bcl-2, which forms hetero-dimers with pro-apoptotic Bcl-2 proteins, results in a change in the conformation of Bcl-2, causing a deeper insertion into the OMM through its C-terminal anchor and the two hydrophobic Hα-5 and Hα-6. As such, it can be suggested that Bcl-2 has the ability to sequester pro-apoptotic proteins such as Bax and Bak in an "inactive" hetero-dimer in the OMM, which will have the consequence of preventing homo-oligomerization of Bax and Bak and the subsequent permeabilization of the OMM.

Unlike Bcl-2, which is exclusively bound to the endomembranes in healthy cells, Bcl-x_L is found both as a soluble form in the cytoplasm and as a mitochondria-associated protein (Hsu *et al.*, 1997). However, under apoptotic conditions Bcl-x_L is relocalized exclusively to the OMM suggesting that it undergoes a structural reorganization during cell death. Thus, despite the presence of a transmembrane domain composed of a Hα-9 similar to that observed in Bcl-2, Bcl-x_L presents a significant cytoplasmic location. Bcl-x_L possesses a C-terminal transmembrane domain but with flanking basic residues at each extremity of this domain, which are required for its mitochondrial targeting. The structure of the Hα-9 surrounded by the basic residues constitutes a complete mitochondrial addressing sequence of the class of tail-anchored mitochondrial proteins. This could explain the exclusive OMM targeting of Bcl-x_L. In comparison, the transmembrane domain of Bcl-2 is flanked by only one residue on either side resulting in a less specific addressing signal, which could explain the wide distribution of Bcl-2 into several different endomembranes (Kaufmann *et al.*, 2003). A study on the insertion of Bcl-x_L into micelles shows that the Hα-1 and Hα-6, and possibly the Hα-5, are partially buried in the hydrophobic core of the micelles (Losonczi *et al.*, 2000). Another study on the insertion of fragments of Bcl-x_L into membranes confirmed that the C-terminal Hα-9 and the Hα-5 could insert into membrane while the Hα-6 could only insert as a component of a Hα-5 and Hα-6 fragment (Garcia-Saez *et al.*, 2006). More recently, data have been presented describing the near complete insertion of Bcl-x_L. As Bcl-x_L can pass from a soluble conformation to an integral membrane conformation, the authors suggested that Bcl-x_L can assume three stable conformations: a soluble form, a C-terminal anchored conformation with the core protein similar to the soluble form and a fully deployed membrane-associated protein (Thuduppathy *et al.*, 2006). These data gave raise to the hypothesis of a four-step model for the translocation of Bcl-x_L to the mitochondria during apoptosis. The cytosolic form can sequester via its C-terminal region, an interaction of this region with its BH-3 binding pocket or through an interaction with another cytoplasmic factor. After an apoptotic stimuli several events could lead to the exposure of the C-terminal anchor signal sequence of Bcl-x_L resulting in its insertion into the OMM through the Hα-9 and probably in the exposure of the BH-3 binding pocket. At this stage, the bulk of the Bcl-x_L protein exposed to the cytosol can bind to other Bcl-2 proteins such as Bax and/or Bak. In the final step, Bcl-x_L is fully deployed and becomes fully included into the OMM with insertion of at least the two hydrophobic Hα-5 and Hα-6. This last event can probably occur in the same way as Bcl-2 after interaction with the BH-3 domain of Bim or through an electrostatic mechanism induced by a lowering of the pH for example. Thus, insertion of Bcl-x_L into the OMM seems to be very similar to that of Bcl-2. However, to date, no known protein of the outer membrane can act as a receptor for the targeting of Bcl-x_L to the mitochondria.

3.4. The Pro-Apoptotic BH-3 only protein, Bid

In healthy cells Bid is present as an inactive cytosolic protein. During apoptosis Bid is cleaved by caspase 8 and its truncated form (tBid) then translocates to the mitochondria (Li *et al.*, 1998). The cleaved active form of Bid can induce the activation of the pro-apoptotic effectors Bax (Cartron *et al.*, 2003a) (Cartron *et al.*, 2004c) and Bak (Wei *et al.*, 2000). This cleavage of Bid by caspase 8 links the extrinsic and intrinsic pathways because activation of caspase 8 usually occurs downstream of the death receptor signaling. However, the inactive form of Bid (the full length 22-kDa: p22) can be activated by other proteases such as calpain (Chen *et al.*, 2001) or granzyme B (Waterhouse *et al.*, 2006) (Cartron *et al.*, 2003a). Interestingly, granzyme B generates two forms of tBid: p15tBid (15-kDa) generated by a direct cleavage by granzyme B and p13tBid (13-kDa) generated indirectly by granzyme B-activated caspase 8. Moreover, it was observed that p15tBid activates preferentially Bak while p13tBid triggers preferentially the activation of Bax (Cartron *et al.*, 2003a).

With a similar overall structure to other Bcl-2 proteins, Bid is composed of eight alpha helices with a core hydrophobic Hα-6 - Hα-7 hairpin but it lacks the C-terminal domain present in Bcl-2 and Bcl-x_L. Following its cleavage the N-terminal helices, Hα-1 and Hα-2, are removed but the overall structure is preserved in tBid. This results in the exposition of the BH-3 domain normally hidden by Hα-1 and probably also the Hα-6 - Hα-7 hairpin (McDonnell *et al.*, 1999) (Chou *et al.*, 1999). The Hα-6 - Hα-7 of Bid is identical to the Hα-5 - Hα-6 hairpin of other Bcl-2 members and can insert into membranes (Garcia-Saez *et al.*, 2004). In addition it has been observed that the Hα-6 has the ability to porate membranes but the pores formed by Bid fragments are not as large as those formed by peptides derived from Bax (Garcia-Saez *et al.*, 2006) supporting the notion that permeabilization of the OMM is not a probable function of Bid. A report analyzing the insertion of tBid into artificial bilayers shows that tBid interacts with membranes via the helix parallel to the surface without α-helices insertion (Gong *et al.*, 2004). However, this lack of helix insertion can be due to the lipidic environment used in this study, since it has been demonstrated that tBid requires cardiolipin, a specific mitochondrial phospholipid, for mitochondrial targeting (Lutter *et al.*, 2000). Cardiolipin is localized to the inner membrane and contact sites (McMillin & Dowhan, 2002) and during apoptosis a degradation of cardiolipin is observed. As we have discussed before, cytochrome c binds to cardiolipin and has to be mobilized before being released from the mitochondria. This mobilization seems to require both peroxidation of cardiolipin abolishing the cytochrome c/IMM interactions and mitochondria cristae remodeling, which can be induced by tBid. Underlining this hypothesis, it has also been shown that Bid displays a lipid transfer activity (Degli Esposti *et al.*, 2001). Interestingly, tBid translocates to mitochondria preferentially at the level of contact sites (Lutter *et al.*, 2001), strongly supporting the idea of a function of tBid in targeting the contact sites, binding to cardiolipin and initiating loss of cardiolipin and cristae remodeling resulting in the mobilization of cytochrome c. Furthermore, it has been reported that Bid cooperates with cardiolipin and Bax to form large membrane openings (Kuwana *et al.*, 2002). This implies that cardiolipins are implicated in mitochondrial targeting of not only Bid but also in Bax-induced permeabilization. However, this appears paradoxical since cardiolipins play a role in tBid targeting to mitochondria while a loss in cardiolipins is described during apoptosis. This problem seems to be solved by a report demonstrating that tBid also binds to cardiolipin metabolites such as monolysocardiolipins (Degli Esposti *et al.*, 2003). This study shows that monolysocardiolipins have the ability to form stable aggregates with Bid and that monolysocardiolipins promote the association of Bid with mitochondria. Indeed, apoptosis-induced cardiolipin degradation results in the diffusion of monolysocardiolipins in both the outer and the inner mitochondrial membranes. These data have led to the hypothesis that the appearance of local monolysocardiolipin-rich areas due to tBid - monolysocardiolipin aggregation could prime the OMM for the subsequent permeabilization by Bax and/or Bak. It has also been shown that tBid/cardiolipin interaction seems to primarily result in perturbations in the mitochondrial respiratory function by inhibiting the rate of phosphorylation and ADP-stimulated respiration (Gonzalvez *et al.*, 2005), which reinforces the idea that tBid may play a role in mitochondrial structure reorganization. Finally a recent report has observed that the binding of tBid to the OMM following TNFα-induced apoptosis resulted in the formation of a 45-kDa cross-linkable complex suggesting a trimerization of tBid or a potential 30-kDa partner for tBid that could serve as a mitochondrial receptor.

This mitochondrial receptor was identified as the integral protein Mtch2 (Grinberg *et al.*, 2005). The Mtch2 (Mitochondrial Carrier Homolog 2) is related to the Mitochondrial Carrier Protein family responsible for metabolite transport across the IMM but the actual function of Mtch2 is unknown. The formation of this complex (Mtch2/tBid) can be abrogated by the over-expression of the anti-apoptotic Bcl-x_L. Gross and coworkers have also observed that Bax co-localizes with the Mtch2/tBid complex (Grinberg *et al.,* 2005). Thus the association of tBid with mitochondria requires similar structural changes denoted in other Bcl-2 proteins. Moreover this targeting requires cardiolipins and/or its metabolites for an optimal and effective function of tBid, which seems to catalyze cristae remodeling, activation of Bax and probably play a role in the loss of cardiolipin. Furthermore, tBid binding to mitochondria requires a mitochondrial protein, which can act as a receptor involved in the anchoring and the accurate insertion of tBid into the OMM. Finally, this interaction of tBid with cardiolipins and Mtch2 seems to precipitate the formation of domains conducive to Bax for proper activation and subsequent pore formation.

3.5. The Pro-Apoptotic Effectors Bax and Bak

Among the pro-apoptotic multi-domain members of the Bcl-2 family, the functional mechanism of cell death induced by Bax has been the most widely investigated. In healthy cells Bax is essentially present as a soluble protein (Figure 4-1) but a minor fraction of Bax can also be found associated with mitochondria (Hsu *et al.*, 1997). After an apoptotic stimuli, Bax undergoes a conformational change to facilitate its translocation to mitochondria where it provokes a permeabilization of the OMM (Figure 4-2, 3) leading to the release of the apoptogenic factors (Degenhardt *et al.*, 2002). This permeabilization of the OMM is associated with the formation of Bax oligomers (Antonsson *et al.*, 2001), which culminates in the formation of channels in the mitochondrial membranes (Schlesinger *et al.*, 1997) and this channel forming activity can be inhibited by Bcl-2 (Antonsson *et al.*, 1997). This capacity to generate channels requires the homo-oligomerization of a membrane-bound Bax (Antonsson *et al.*, 2000). As described by Aritomi *et al.* (1997) using structure modelization based on the crystal structure of Bcl-x_L, of the Bcl-2 family, Bax presents the strongest potential for membrane insertion.

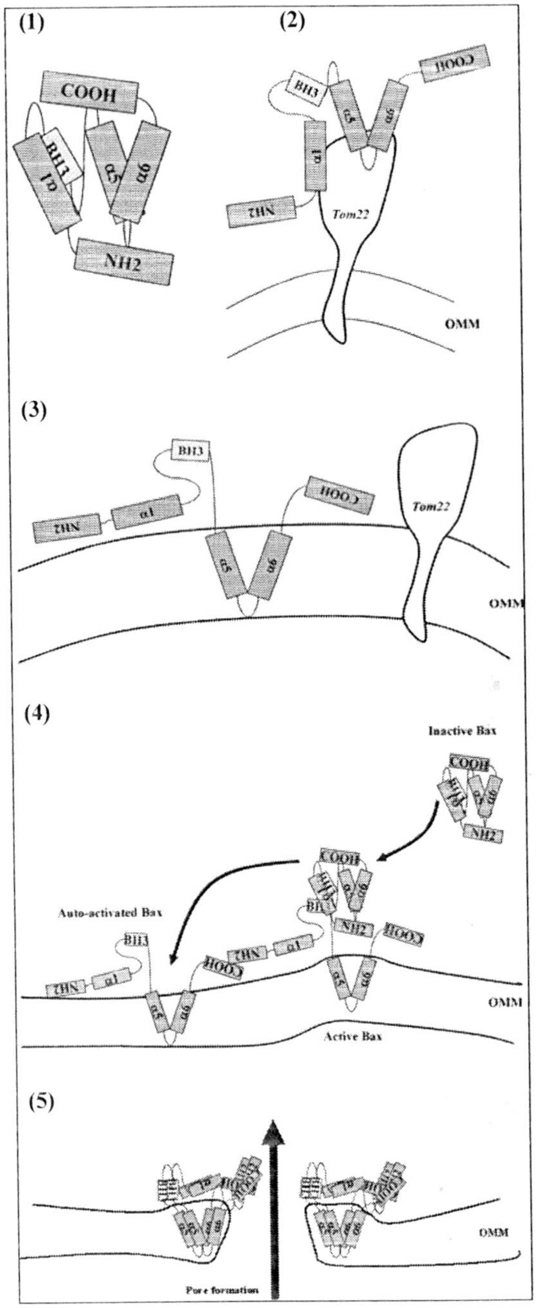

Figure 4. Model of Bax insertion and oligomerization into the OMM. (1) In normal cells Bax is present as a soluble monomer. (2) After induction of apoptosis Bax undergoes a conformational change resulting in the exposure of the N-terminal and C-terminal extremities. Bax is targeted to mitochondria via the Tom22 receptor through the Hα1 and the linker region between the Hα-5 and Hα-6. (3) Bax inserts into OMM through the hydrophobic Hα-5 and Hα-6 with its BH-3 domain exposed to the cytosol. (4) Bax auto-activation mediated by the exposed BH-3. (5) Oligomerization into the OMM and pore formation.

As for other members of the Bcl-2 family, the inactive cytosolic Bax exists as a globular protein composed of nine Hα with the Hα-5 - Hα-6 hairpin and the BH-3 domain buried within the core of the protein while the C-terminal Hα-9 hides the BH-3 domain and exposes only its polar amino acids to the cytoplasm (Suzuki *et al.*, 2000). The NMR structural resolution of Bax reveals that the N-terminal and the C-terminal regions exhibit an elevated number of potential positions suggesting a latent high mobility of these domains. Therefore, the translocation of Bax from the cytoplasm to the OMM during apoptotic cell death implies an extensive conformational change in Bax leading to the exposition of domains responsible for the interaction with mitochondria and also for the integration into the OMM. Under apoptotic conditions, the activation of Bax results in the unfolding of the transmembrane Hα-9 and in a modification in the conformation of the N-terminal domain (Nechushtan *et al.*, 1999). However, unlike the other multi-domains Bcl-2 proteins such as Bcl-2 and Bcl-x_L, the association of Bax with the OMM does not require a single domain for both the targeting and the primary integration into the OMM. Indeed although the C-terminal region was first described as an "anchoring domain" further studies involving the N-terminal region (the first 20 amino acids) of Bax in the regulation of Bax targeting to mitochondria defined this sequence as an Apoptosis Regulating Targeting (ART) sequence (Goping *et al.*, 1998) (Cartron *et al.*, 2002a). Moreover, the identification of a variant of Bax called Bax Ψ, which lacks the first 19 amino acids reveals that Bax Ψ is a constitutive OMM associated form of Bax with a higher toxicity than the full length Bax (Cartron *et al.*, 2002b). It has also been shown that a deletion of the C-terminal extremity of Bax (BaxΔC) does not alter the mitochondrial association and that BaxΔC exhibits a similar potential for cell death as the full length Bax (Tremblais *et al.*, 1999) (Priault *et al.*, 1999) (Oliver *et al.*, 2000). A sequence in the N-terminal region of the Hα-1 of Bax, immediately after the ART domain, contains a mitochondrial addressing signal, which is sufficient to trigger mitochondrial association of a cytosolic reporter protein while the Hα-9, which is present in the C-terminal region is not capable of doing (Cartron *et al.*, 2003b). Thus, it is clear that N-terminal domain of Bax is essentially implicated in the regulation of mitochondrial addressing. Furthermore, we have recently demonstrated that distinct domains regulate the targeting and the insertion of Bax into mitochondria (Cartron *et al.*, 2005). Indeed, while exposure of the ART domain displays the mitochondrial addressing sequence the C-terminal domain controls the exposition of the Hα-5 - Hα-6 responsible for the insertion into the OMM. Note that the Hα-1 has been shown to be involved in the activation of Bax through an interaction with the BH-3 domain of Bid (Cartron *et al.*, 2004c).

Thus as for other Bcl-2 proteins, the Hα-5 - Hα-6 of Bax is responsible for its insertion into the OMM, indeed a Hα-5 alone or a Hα-5 - Hα-6 peptide possesses an intrinsic capacity to insert into membranes (Garcia-Saez *et al.*, 2004). Moreover, it has been observed that the Hα-5 peptide has the same capacity as full length Bax to trigger pore formation in membranes and with similar characteristics such as large conductivity and ionic selectivity depending of the lipid-charge content (Garcia-Saez *et al.*, 2005). This study suggests that the Hα-5 peptide forms a toroidal structure in membranes combining peptides and lipids. Furthermore, comparable results can be obtained using a Hα-6 peptide of Bax. Thus, Hα-5 and Hα-6 permeabilize membranes and this pore-forming activity is sensitive to the lipid curvature emphasizing the importance of the lipid content, which is in accordance with the observations of the association of Bid with the OMM as reported above. These peptides generate pores smaller than the full length Bax and the increase in permeability induced by the Hα-5 and Hα-6 peptide is associated with transbilayer redistribution supporting a toroidal pore model. Hence, the formation of a pore caused by an activated Bax is mediated through the Hα-5 and Hα-6 domain. This model is supported by a previous study on full length Bax, which demonstrates that the Hα-5 and Hα-6 domain of Bax but neither its transmembrane domain nor its BH-3 are required for the subsequent release of cytochrome c from isolated mitochondria (Heimlich *et al.*, 2004). A deletion of this pore-forming domain completely impairs the release of cytochrome c and Smac/DIABLO.

As we have previously stated, the insertion of Bax into mitochondrial membrane is also associated with the formation of homo-oligomers. This has been substantiated by the data that apoptosis induced in FL5.12 cells by a withdrawal of interleukin 3 results in the formation of cross-linkable homo-dimers of Bax in the OMM. Moreover the enforced dimerization of Bax monomers results in their translocation to the OMM and the subsequent apoptosis (Gross *et al.*, 1998). These data suggest that the active form of Bax is essentially an integral mitochondrial membrane dimer, which is the basic structure of the high molecular weight oligomer/complex detected under apoptotic conditions. Recently the observation was made that monomeric Bax was inserted into the OMM prior to the oligomerization process of Bax (Annis *et al.*, 2005). A study by Tan *et al.* (2005) using a remarkable system of liposomes with an entrapped fluorescent CB-dextran containing active membrane-bound Bax, activated by a Hα-2 - Hα-3 fragment. These liposomes are then subjected to a sucrose gradient float-up centrifugation in order to isolate liposomes with an inserted Bax and to ensure the elimination of the activator peptide and inactive soluble or aggregated Bax. At this stage the release of CB-dextran has just started and the addition of monomeric Bax results in an enhancement of CB-dextran release whereas addition of more activator peptide has no effect. The transmembrane domain of Bcl-2 was shown to inhibit this mechanism (Tan *et al.*, 2006). This supports the notion of a tight regulation of Bax activity at the OMM level depending on the ratio Bcl-2/active Bax, which reflects the degree of Bcl-2/Bax heterotypic and Bax/Bax homotypic interactions. When this ratio is high the hetero-dimers are dominant in the OMM and Bcl-2 prevents Bax oligomerization. On the other hand when this ratio is low the amount of Bax inserted into mitochondria overwhelms the anti-apoptotic capacity of Bcl-2 thereby favoring a homo-dimer formation that results in a cascade of auto-activation of Bax (Figure 4-4).

As we have stated previously in this chapter, Bax is capable of forming channels in lipidic bilayers. In comparison to other channels formed by other Bcl-2 proteins the "Bax-channels" seems to be the most potent candidates for triggering the release of apoptogenic factors (Schlesinger *et al.*, 1997). However, the exact mechanism by which Bax induces an increase in mitochondrial permeability remains unclear. Therefore, several hypotheses have been proposed to explain this Bax-induce permeabilization. Amid all these theories we can find the opening of the permeability transition pore (PTP) induced by Bax through a direct interaction with PTP components. This permeability transition induces matrix swelling and rupture of the OMM. The PTP components are proteins from both the inner and the outer mitochondrial membranes forming a mega-complex consisting of the ANT (Adenine Nucleotide Translocator), VDAC (Voltage-Dependent Anionic Channel), hexokinase, cyclophilin D and PBR (Peripheral Benzodiazepin Receptor). It has been observed that Bax interacts with the ANT during apoptosis (Marzo *et al.*, 1998). However, this model proposes a non-specific permeabilization of the OMM whereas since the beginning of this chapter we have seen that the intrinsic pathway implies a really well organized hierarchy of events, which probably cannot support such a non-specific permeabilization of the mitochondria. Moreover, a comparison between the activation of the PTP and a Bax induced permeabilization in isolated mitochondria demonstrates that these two mechanisms are strictly distinct (Pouliquen *et al.*, 2006).

Another model involves a direct interaction between Bax and VDAC (Shimizu *et al.*, 1999). Indeed, this work by the group of Tsujimoto suggests that both Bak and Bcl-x_L interact with VDAC. Using liposomes containing VDAC these authors observed that incubation of FITC-cytochrome c and Bax induced the accumulation of cytochrome c in the liposomes while this mechanism was inhibited by the addition of Bcl-x_L. Moreover electrophysiological studies reveal that the interaction of Bax with VDAC results in an increase in the level conductance of VDAC allowing the passage of cytochrome c while Bcl-x_L induces the closing of VDAC (Shimizu *et al.*, 2000). Hence, this model implies the formation of a specific pore permitting the translocation of cytochrome c independently of PTP opening, however, these results remain controversial and no further studies have observed such a modulation of VDAC by Bax. In addition, as other reports have argued Bax alone is sufficient to trigger membrane permeabilization suggesting the possibility that this model could be an experimental artifact. The last model, and actually the predominant, implies the formation of a pore mediated by the oligomerization of Bax in a similar manner to that formed by bacterial toxins (Figure 4-5). As we have discussed before, oligomerization of Bax is a necessary step for the appearance of channel activity (Antonsson *et al.*, 2000) and this is associated with the translocation of Bax to the OMM and subsequently, cell death (Gross *et al.*, 1998) (Antonsson *et al.*, 2001). Underlining these observations is the Bax-dependent release of cytochrome c, which has been reproduced in pure liposomes. Interestingly these pores show a Bax concentration-dependent progression from a pore size of about 11 μm to a pore dimension of about 22 μm consisting of two and four Bax molecules respectively (Saito *et al.*, 2000). The larger pore observed in this study is sufficient to allow the translocation of cytochrome c. Thus, Bax alone can be sufficient for the membrane permeabilization observed during cell death. Besides, Bax has been described as an inducer of stress curvature in planar bilayers by reducing the linear tension of the membrane thus resulting in formation of pores similar to those detected by the group of Korsmeyer (Saito *et al.*, 2000) (Basanez *et al.*, 1999). These results have formed the basis for the hypothesis that Bax can generate lipidic pores in a comparable manner to the model described later with Hα-5 and Hα-6 of Bax (Garcia-Saez *et al.*, 2005). This hypothesis has been supplemented since by reports, which point out the cooperation between Bid, Bax and cardiolipin triggering the supramolecular openings in the OMM (Kuwana *et al.*, 2002) and the observation that a combined action of Bax and tBid could cause the formation of lipidic pores (Terrones *et al.*, 2004). At this time only Bax-induced permeabilization of artificial membranes constituted the essential observations of the pore forming ability of Bax. However, electrophysiological studies using liposomes formed with pure mitochondrial outer membranes from apoptotic cells revealed the appearance of high conductance channels (Pavlov *et al.*, 2001). This megachannel was called MAC (Mitochondrial Apoptotic Channel) and its higher conductance levels involved an estimated pore diameter of about 4 nm, which is a consistent size for the passage of cytochrome c. Further investigations on the activity of MAC confirmed its capacity to permit the translocation of cytochrome c (Guo *et al.*, 2004). Furthermore, oligomeric Bax was identified as a component of this channel putatively involved in cytochrome c release (Dejean *et al.*, 2005). To further underline these observations, MAC has been observed in an *in vivo* model of apoptotic rat liver (Guihard *et al.*, 2004). Surprisingly, three populations of mitochondria have been isolated from apoptotic liver: a high-density population (HAM: Heavy Apoptotic Mitochondria) similar to control mitochondria (HCM: Heavy Control Mitochondria), a low-density population composed of fully swollen mitochondria (LAM2: Light Apoptotic Mitochondria 2) and an intermediate density population (named LAM1). These populations may represent three chronological stages of the apoptotic mitochondrial pathway since the biochemical characterization show a progressive release of cytochrome c from the HAM to the LAM2 coinciding with an accumulation and an oligomerization of Bax in mitochondria. However, the activity of MAC is only found in the fully swollen mitochondria (LAM2 fraction) but in this fraction a significant amount of cytochrome c has been already released. Therefore, although MAC is a putative candidate for the transport of cytochrome c during apoptosis, in this model it appears that MAC is a late apoptotic event observed only in strongly impaired mitochondria. One explanation could be that MAC corresponds to an activity generated by high degrees of oligomerization of Bax while smaller oligomerization levels of Bax are sufficient to trigger cytochrome c release without producing a MAC related activity.

With regards to the existence of putative mitochondrial receptors for Bax we have discussed above that interactions of Bax with proteins such as VDAC and ANT have been described thus identifying them as potential candidates for this role. Further investigations using a bacterial two-hybrid assay have allowed us to identify a Bax-interacting protein required for the first steps of the association of Bax with mitochondria. Interestingly, the protein identified in this study is the import receptor Tom22, a component of the importation apparatus. Indeed, preventing the Bax/Tom22 interaction through microinjection of anti-Tom22 antibodies or silencing of Tom22 expression inhibits the Bax-induced apoptosis. Mapping the domains of Bax involved in this interaction reveals that two domains of Bax are implicated. The first domain is the Hα-1, thus confirming the role of this domain in the mitochondrial addressing of Bax. Interestingly, the second domain corresponds to the linker region between the two hydrophobic Hα-5 and Hα-6 responsible for the insertion of Bax into the OMM (Bellot *et al.*, 2006). Thus, Bax is translocated to mitochondria through the mitochondrial protein import pathway as a substrate of the preprotein receptor Tom22. Although this stage of Bax translocation is transient, the tethering to Tom22 is critical because it is essential for promoting the subsequent permeabilization of the OMM. Disruption of this step causes a considerable decrease in Bax relocalization, which consequently leads to the inhibition of the mitochondrial pathway of the apoptosis. Besides it may also represent an early key regulator point for the membrane insertion of Bax since the domain linking Hα-5 and Hα-6 interacts directly with Tom22.

In summary the pro-apoptotic Bax is addressed to mitochondria via the protein importation machinery in a similar but yet different mechanism than Bcl-2, which associates with the OMM through the Tom20 receptor pathway. These mechanisms of mitochondrial targeting constitute critical points for the regulation of the functions of the Bcl-2 proteins. These pathways require further investigations in order to target the design of new therapeutics agents with pro- or anti-apoptotic effects.

3.5.1. Bak

The Bcl-2 Antagonist Killer (Bak) was identified by complementary DNA cloning of Bcl-2 homologues (Chittenden *et al.*, 1995b) (Kiefer *et al.*, 1995) and as a protein interacting with E1B 19K, a viral anti-apoptotic Bcl-2 homolog (Farrow *et al.*, 1995). Bak shares similar sequence homology with the other members of the Bcl-2 family as it contains three BH domains (BH-1, BH-2 and BH-3) and a C-terminal hydrophobic transmembrane domain indicating that Bak may exist as an integral membrane protein. Although Bax and Bak are homologs it seems that they have non redundant roles (Cartron *et al.*, 2003a). These observations were assessed through the phenotype of their respective knockout mice. Indeed, Bak knockout mice are phenotypically normal (Lindsten *et al.*, 2000) whereas Bax knockout mice display some phenotypic abnormalities such as hyperplasia of thymocytes and B cells and infertility in males (Knudson *et al.*, 1995). However, unlike its widely studied counterpart Bax, the mechanism of cell death induced by the pro-apoptotic effector Bak remains poorly documented. The cause is essentially that the three dimensional structure of Bak is not known thereby increasing the difficulty to determine exactly how Bak functions in provoking cell death. We will attempt to summarize the current knowledge about this pro-apoptotic multi-domain member of the Bcl-2 family.

The first important information concerns the localization of Bak. As opposed to Bax, Bak is mainly present in the mitochondrial membrane in both healthy and apoptotic cells (Griffiths *et al.*, 1999) (Wei *et al.*, 2000) and it can localize to the membrane of endoplasmic reticulum as well as (Scorrano *et al.*, 2003). Thus, surprisingly Bak resides in the OMM, the same membrane it causes damage to during apoptosis. This is surprising because ectopic enforced expression of Bak has been reported to induce a rapid and extensive apoptosis in fibroblasts and sympathetic neurons (Chittenden *et al.*, 1995b) (Farrow *et al.*, 1995). This suggests that Bak is not completely inactive in non-apoptotic cells as compared to soluble monomeric Bax, which is completely inert under these conditions. As a consequence the association of Bak with the OMM may require some additional factors required to maintain a restrain on its pro-apoptotic activity. In order to monitor the possible post-translational changes involved in the non-apoptotic or apoptotic conformations of Bak, Griffiths and colleagues performed two-dimensional gel electrophoresis analysis and concluded that no discernable modifications in Bak occur during apoptosis induced by staurosporine (Griffiths *et al.*, 1999). However, the data did show that Bak exists fundamentally in multiple isoforms. In fact it was later demonstrated that Bak resides in mitochondrial membranes in normal cells as part of large complexes in a broad range from 43 to 5000-kDa (Mikhailov *et al.*, 2003). These complexes probably regulate the intrinsic killing activity of Bak.

On the other hand examination of the molecular events committing cells to apoptosis reveals that Bak undergoes a conformational change allowing the exposure of its N-terminal extremity before apoptosis induced by agents such as staurosporine and etoposide (Griffiths *et al.*, 1999). The action of tBid also results in the formation of an "open conformer" of Bak with the N-terminal exposed (Ruffolo & Shore, 2003). In addition a second structural reorganization follows the N-terminal exposure before the onset of cell death, which leads to the exposure of a second epitope: the BH-1 domain of Bak (Griffiths *et al.*, 2001). Interestingly, according to the known structures of the Bcl-2 proteins, the BH-1 domain overlaps the N-terminal region of the hydrophobic Hα-5 responsible, with Hα-6, for the insertion of Bcl-2, Bcl-x_L and Bax into the OMM. Therefore, like Bax, Bak appears to require a structural change, which can lead to the exposure of the Hα-5 - Hα-6, a crucial step for mitochondrial insertion of the Bcl-2 proteins at the onset of apoptosis. Griffiths and colleagues also have demonstrated that Bak associates with the mitochondrial-associated Bcl-x_L in untreated cells and dissociates from this Bcl-x_L after the induction of apoptosis. However, the over-expression of Bcl-x_L has no effect on the kinetics of the exposure of the N-terminal epitope of Bak whereas it delays the exposure of the BH-1 domain. It has been previously determined that a peptide derived from the BH-3 domain of Bak interacts with the hydrophobic pocket formed by the BH-1, BH-2 and BH-3 domains of Bcl-x_L (Sattler *et al.*, 1997). Moreover, the helical structure of the C-terminal region of Bak, a peptide of remarkable hydrophobicity, has the ability to interact with membranes. The C-terminal peptide of Bak breaks the barrier properties of lipidic membranes and induces carboxyfluorescein leakage from phospholipids vesicles (Martinez-Senac *et al.*, 2002) thereby possibly indicating that this domain of Bak may act as a membrane anchor.

The following stage of the activation of the pro-apoptotic function of Bak results in the formation of oligomers. The oligomerization of Bak can be induced by truncated Bid (Wei *et al.*, 2000). We have already observed that Bak associates with mitochondrial complexes in healthy cells but when apoptosis occurs Bak rearranges itself in smaller complexes ranging from 67 to 700-kDa (Mikhailov *et al.*, 2003). Cross-linking studies performed by Mikhailov *et al.* (2003) observed that monomeric Bak is embedded in complexes in normal mitochondria while the rearrangement that occurs after an apoptotic stimuli constitutes essentially the homo-oligomerization of Bak. The tBid-induced "open conformer" of Bak with its N-terminal exposed appears prior to the oligomerization. Interestingly this "open conformer" of Bak is able to induce the auto-oligomerization of Bak with the closed conformation (Ruffolo & Shore, 2003). This cascade is also blocked by Bcl-2, which hetero-dimerizes with the open conformer of Bak emphasizing the role of the hetero-dimerization in the negative regulation of the Bax/Bak-induced cell death.

The potential partners of Bak in the OMM of healthy cells have just recently been identified. The first category of these factors sequestering Bak in an inactive state involved an association with anti-apoptotic members of the Bcl-2 family. Under viable conditions, Bak interacts directly and specifically with the pro-survival Mcl-1 and Bcl-x_L, which maintains Bak as an inactive protein (Willis *et al.*, 2005). Bak-dependent apoptosis requires the release of Bak from either Mcl-1 or Bcl-x_L in order to be fully effective. A disruption of these hetero-complexes requires the intervention of BH3-only proteins (Moreau *et al.*, 2003). As demonstrated by Willis *et al.* (2005) Noxa binds to Mcl-1 and induces its degradation via the proteasomal pathway thereby liberating Bak. With regards to the Bak/Bcl-x_L complex, this complex would require another BH-3 only protein for dissociation since Noxa does not bind Bcl-x_L. The truncated form of Bid is important for activation of Bak (Wei *et al.*, 2000) (Ruffolo & Shore, 2003) but this seems activate Bak more than to disrupt inactivating complexes. However, conformational changes causes by its activation by tBid could result in the dissociation of Bak from Bcl-x_L (Griffiths *et al.*, 2001) while enhancing the possible formation of Bak/Bcl-2 heterodimers (Ruffolo & Shore, 2003). According to these data Bak appears to be strongly regulated by the anti-apoptotic members of the Bcl-2 family. In addition another protein not belonging to the Bcl-2 family has been demonstrated to interact with Bak in healthy cells. Indeed, cross-linking experiments have shown the existence of a 60-kDa complex composed of Bak and the isoform 2 of VDAC (Cheng *et al.*, 2003). Analyses of the functionality of this complex through silencing of the VDAC2 gene have revealed that VDAC2$^{-/-}$ cells were more responsive to staurosporine treatment demonstrating an accelerated release of cytochrome c. This interaction involves the BH-1 and BH-3 domains of Bak. Thus VDAC2 emerges as a regulator in Bak-dependent cell death pathway. There is another pathway of regulation of Bak activity, which seems to engage its pro-apoptotic counterpart Bax. Indeed, Bax and Bak both form homo-oligomers during apoptosis and induce cytochrome c release. No hetero-oligomers between Bax and Bak seem to exist but an association between the homo-oligomers can be observed (Mikhailov *et al.*, 2003). An induction of ATP depletion in Bax$^{-/-}$ cells has no effect on the oligomerization of Bak and the release of cytochrome c while in Bak$^{-/-}$ cells, a depletion of ATP, causes the oligomerization of Bax to be delayed, however, apoptosis occurs normally (Mikhailov *et al.*, 2003). Additionally a recent study shows that a Bax deficiency diminishes the complex formed between Bak and VDAC2 (Chandra *et al.*, 2005). This report supports the idea that Bak becomes the major pro-apoptotic effector in the absence of Bax as already described (Cartron *et al.*, 2003a). The absence of Bax results in a loosely regulated binding of Bak by VDAC2. Consequently, there may exist a Bax-related regulation of Bak function and/or association with mitochondria. In summary Bak is strongly regulated and/or sequestered at the mitochondrial level possibly because of its intrinsic capacity to kill. This tight regulation of the action of Bak is probably the reason why its ability to induce apoptosis appear to be less efficient than Bax.

Conclusion

In this chapter we have demonstrated the central role played by mitochondria in the regulation of apoptotic cell death. Indeed the mitochondrion contains all the components necessary to trigger directly or indirectly the impairment of the vital functions of the cell including its own destruction. In addition, the mitochondria can target the nucleus directly or indirectly through the activation of the caspases thereby inducing chromatin condensation and internucleosomal DNA degradation. Furthermore, the morphological changes in the mitochondria have dramatic consequences on the mitochondrial function. The remodeling of cristae probably alters the respiratory function and oxidative phosphorylation since cristae are the siege of these mechanisms. Moreover fragmentation of the mitochondrial network results in defects in energy delivering and Ca^{2+} mobilization in the cell. Consequently the potential killing ability of the mitochondria necessitates a strict control of its death potential. The members of the Bcl-2 family, which regulates the mitochondrial cell death pathway, mediate this control. Indeed, it appears that this system has followed an evolutionary progress from an initial system of bacterial toxins triggering another bacterium to a complex mechanism involving a large family of proteins, which can interact which each other for negative or positive regulation of cell death. Thus cells produce their own endotoxins, the pro-apoptotic effectors Bax and Bak, but they also synthesized the anti-apoptotic Bcl-2 and Bcl-x_L, which can counteract the action of Bax and Bak. In addition cells also synthesize the BH3-only proteins, which act as stress sensors activating the pro-apoptotic proteins or antagonizing the anti-apoptotic Bcl-2 family members. In addition we have also shown that the Bcl-2 family of proteins modulates morphological modifications in mitochondria and that it is an evolutionary mechanism of programmed cell death. However, the complexity of the control of apoptosis by mitochondria is strict since it can implicate a role for the endoplasmic reticulum and Ca^{2+}. Recent evidence have been reported that some Bcl-2 family members such as Bcl-2, Bik, Bax and Bak acts at the level of the endoplasmic reticulum where these proteins control the release of Ca^{2+}. Since mitochondria and endoplasmic reticulum are intimately connected and the release of Ca^{2+} from the endoplasmic reticulum leads to the uptake of Ca^{2+} by the mitochondria it would appear that Ca^{2+} signaling could play a role in apoptosis, notably by modulating caspase activity (Juin *et al.,* 1998) (Pelletier *et al.,* 2005). Finally we can conclude that mitochondria are key regulators of the apoptotic cell death and that the control points of their death function remain highly unclear.

REFERENCES

Adrain, C., Creagh, E. M. & Martin, S.J. (2001). Apoptosis-associated release of Smac/DIABLO from mitochondria requires active caspases and is blocked by Bcl-2. *EMBO J, 20*, pp. 6627-6636.

Altieri, D. C. (2001). Cytokinesis, apoptosis and survivin: three for tango? *Cell Death Differ, 8*, pp. 4-5.

Annis, M. G., Soucie, E. L., Dlugosz, P. J., Cruz-Aguado, J. A., Penn, L. Z., Leber, B. & Andrews, D.W. (2005). Bax forms multispanning monomers that oligomerize to permeabilize membranes during apoptosis. *EMBO J, 24*, pp. 2096-2103.

Antonsson, B. (2004). Mitochondria and the Bcl-2 family proteins in apoptosis signaling pathways. *Mol Cell Biochem, 256-257*, pp. 141-155.

Antonsson, B., Conti, F., Ciavatta, A., Montessuit, S., Lewis, S., Martinou, I., Bernasconi, L., Bernard, A., Mermod, J. J., Mazzei, G., Maundrell, K., Gambale, F., Sadoul, R. & Martinou, J.C. (1997). Inhibition of Bax channel-forming activity by Bcl-2. *Science, 277*, pp. 370-372.

Antonsson, B., Montessuit, S., Lauper, S., Eskes, R. & Martinou, J.C. (2000). Bax oligomerization is required for channel-forming activity in liposomes and to trigger cytochrome c release from mitochondria. *Biochem J, 345 Pt 2*, pp. 271-278.

Antonsson, B., Montessuit, S., Sanchez, B. & Martinou, J.C. (2001). Bax is present as a high molecular weight oligomer/complex in the mitochondrial membrane of apoptotic cells. *J Biol Chem, 276*, pp. 11615-11623.

Aritomi, M., Kunishima, N., Inohara, N., Ishibashi, Y., Ohta, S. & Morikawa, K. (1997). Crystal structure of rat Bcl-xL. Implications for the function of the Bcl-2 protein family. *J Biol Chem, 272*, pp. 27886-27892.

Arnoult, D., Gaume, B., Karbowski, M., Sharpe, J. C., Cecconi, F. & Youle, R.J. (2003). Mitochondrial release of AIF and EndoG requires caspase activation downstream of Bax/Bak-mediated permeabilization. *EMBO J, 22*, pp. 4385-4399.

Arnoult, D., Grodet, A., Lee, Y., Estaquier, J. & Blackstone, C. (2005b). Release of OPA1 during apoptosis participates in the rapid and complete release of cytochrome c and subsequent mitochondrial fragmentation. *J Biol Chem, 280*, pp. 35742-35750.

Arnoult, D., Rismanchi, N., Grodet, A., Roberts, R. G., Seeburg, D. P., Estaquier, J., Sheng, M. & Blackstone, C. (2005a). Bax/Bak-dependent release of DDP/TIMM8a promotes Drp1-mediated mitochondrial fission and mitoptosis during programmed cell death. *Curr Biol, 15*, pp. 2112-2118.

Ashkenazi, A. (2002). Targeting death and decoy receptors of the tumour-necrosis factor superfamily. *Nat Rev Cancer, 2*, pp. 420-430.

Basanez, G., Nechushtan, A., Drozhinin, O., Chanturiya, A., Choe, E., Tutt, S., Wood, K. A., Hsu, Y., Zimmerberg, J. & Youle, R.J. (1999). Bax, but not Bcl-xL, decreases the lifetime of planar phospholipid bilayer membranes at subnanomolar concentrations. *Proc Natl Acad Sci U S A, 96*, pp. 5492-5497.

Bellot, G., Cartron, P., Oliver, L., Juin, P., Armstrong, L., Bornstein, P., Mihara, K., Manon, S. & Vallette, F.M. (2006). Tom22, a core component of the mitochondria outer membrane protein translocation pore, is a mitochondrial receptor for the pro-apoptotic protein Bax. *Cell Death Differ (in print)*.

Bossy-Wetzel, E., Barsoum, M. J., Godzik, A., Schwarzenbacher, R. & Lipton, S.A. (2003). Mitochondrial fission in apoptosis, neurodegeneration and aging. *Curr Opin Cell Biol, 15*, pp. 706-716.

Brustugun, O. T., Fladmark, K. E., Doskeland, S. O., Orrenius, S. & Zhivotovsky, B. (1998). Apoptosis induced by microinjection of cytochrome c is caspase-dependent and is inhibited by Bcl-2. *Cell Death Differ, 5*, pp. 660-668.

Cartron, P. F., Oliver, L., Mayat, E., Meflah, K. & Vallette, F.M. (2004a). Impact of pH on Bax alpha conformation, oligomerisation and mitochondrial integration. *FEBS Lett, 578*, pp. 41-46.

Cartron, P., Arokium, H., Oliver, L., Meflah, K., Manon, S. & Vallette, F.M. (2005). Distinct domains control the addressing and the insertion of Bax into mitochondria. *J Biol Chem, 280*, pp. 10587-10598.

Cartron, P., Gallenne, T., Bougras, G., Gautier, F., Manero, F., Vusio, P., Meflah, K., Vallette, F. M. & Juin, P. (2004c). The first alpha helix of Bax plays a necessary role in its ligand-induced activation by the BH3-only proteins Bid and PUMA. *Mol Cell, 16*, pp. 807-818.

Cartron, P., Juin, P., Oliver, L., Martin, S., Meflah, K. & Vallette, F.M. (2003a). Nonredundant role of Bax and Bak in Bid-mediated apoptosis. *Mol Cell Biol, 23*, pp. 4701-4712.

Cartron, P., Moreau, C., Oliver, L., Mayat, E., Meflah, K. & Vallette, F.M. (2002a). Involvement of the N-terminus of Bax in its intracellular localization and function. *FEBS Lett*, *512*, pp. 95-100.

Cartron, P., Oliver, L., Juin, P., Meflah, K. & Vallette, F.M. (2004b). The p18 truncated form of Bax behaves like a Bcl-2 homology domain 3-only protein. *J Biol Chem*, *279*, pp. 11503-11512.

Cartron, P., Oliver, L., Martin, S., Moreau, C., LeCabellec, M., Jezequel, P., Meflah, K. & Vallette, F.M. (2002b). The expression of a new variant of the pro-apoptotic molecule Bax, Baxpsi, is correlated with an increased survival of glioblastoma multiforme patients. *Hum Mol Genet*, *11*, pp. 675-687.

Cartron, P., Priault, M., Oliver, L., Meflah, K., Manon, S. & Vallette, F.M. (2003b). The N-terminal end of Bax contains a mitochondrial-targeting signal. *J Biol Chem*, *278*, pp. 11633-11641.

Chandra, D., Choy, G., Daniel, P. T. & Tang, D.G. (2005). Bax-dependent regulation of Bak by voltage-dependent anion channel 2. *J Biol Chem*, *280*, pp. 19051-19061.

Chen, M., He, H., Zhan, S., Krajewski, S., Reed, J. C. & Gottlieb, R.A. (2001). Bid is cleaved by calpain to an active fragment in vitro and during myocardial ischemia/reperfusion. *J Biol Chem*, *276*, pp. 30724-30728.

Cheng, E. H. Y., Sheiko, T. V., Fisher, J. K., Craigen, W. J. & Korsmeyer, S.J. (2003). VDAC2 inhibits BAK activation and mitochondrial apoptosis. *Science*, *301*, pp. 513-517.

Cheng, E. H., Wei, M. C., Weiler, S., Flavell, R. A., Mak, T. W., Lindsten, T. & Korsmeyer, S.J. (2001). BCL-2, BCL-X(L) sequester BH3 domain-only molecules preventing BAX- and BAK-mediated mitochondrial apoptosis. *Mol Cell*, *8*, pp. 705-711.

Chittenden, T., Flemington, C., Houghton, A. B., Ebb, R. G., Gallo, G. J., Elangovan, B., Chinnadurai, G. & Lutz, R.J. (1995a). A conserved domain in Bak, distinct from BH1 and BH2, mediates cell death and protein binding functions. *EMBO J*, *14*, pp. 5589-5596.

Chittenden, T., Harrington, E. A., O'Connor, R., Flemington, C., Lutz, R. J., Evan, G. I. & Guild, B.C. (1995b). Induction of apoptosis by the Bcl-2 homologue Bak. *Nature*, *374*, pp. 733-736.

Chou, J. J., Li, H., Salvesen, G. S., Yuan, J. & Wagner, G. (1999). Solution structure of BID, an intracellular amplifier of apoptotic signaling. *Cell*, *96*, pp. 615-624.

Cory, S. & Adams, J.M. (2002). The Bcl2 family: regulators of the cellular life-or-death switch. *Nat Rev Cancer*, *2*, pp. 647-656.

Curran, S. P., Leuenberger, D., Schmidt, E. & Koehler, C.M. (2002). The role of the Tim8p-Tim13p complex in a conserved import pathway for mitochondrial polytopic inner membrane proteins. *J Cell Biol*, *158*, pp. 1017-1027.

Degenhardt, K., Sundararajan, R., Lindsten, T., Thompson, C. & White, E. (2002). Bax and Bak independently promote cytochrome C release from mitochondria. *J Biol Chem*, *277*, pp. 14127-14134.

Degli Esposti, M., Cristea, I. M., Gaskell, S. J., Nakao, Y. & Dive, C. (2003). Proapoptotic Bid binds to monolysocardiolipin, a new molecular connection between mitochondrial membranes and cell death. *Cell Death Differ*, *10*, pp. 1300-1309.

Degli Esposti, M., Erler, J. T., Hickman, J. A. & Dive, C. (2001). Bid, a widely expressed proapoptotic protein of the Bcl-2 family, displays lipid transfer activity. *Mol Cell Biol, 21*, pp. 7268-7276.

Dejean, L. M., Martinez-Caballero, S., Guo, L., Hughes, C., Teijido, O., Ducret, T., Ichas, F., Korsmeyer, S. J., Antonsson, B., Jonas, E. A. & Kinnally, K.W. (2005). Oligomeric Bax is a component of the putative cytochrome c release channel MAC, mitochondrial apoptosis-induced channel. *Mol Biol Cell, 16*, pp. 2424-2432.

Desagher, S., Osen-Sand, A., Nichols, A., Eskes, R., Montessuit, S., Lauper, S., Maundrell, K., Antonsson, B. & Martinou, J.C. (1999). Bid-induced conformational change of Bax is responsible for mitochondrial cytochrome c release during apoptosis. *J Cell Biol, 144*, pp. 891-901.

Deveraux, Q. L. & Reed, J.C. (1999). IAP family proteins-suppressors of apoptosis. *Genes Dev, 13*, pp. 239-252.

Deveraux, Q. L., Roy, N., Stennicke, H. R., Van Arsdale, T., Zhou, Q., Srinivasula, S. M., Alnemri, E. S., Salvesen, G. S. & Reed, J.C. (1998). IAPs block apoptotic events induced by caspase-8 and cytochrome c by direct inhibition of distinct caspases. *EMBO J, 17*, pp. 2215-2223.

Diekert, K., de Kroon, A. I., Ahting, U., Niggemeyer, B., Neupert, W., de Kruijff, B. & Lill, R. (2001). Apocytochrome c requires the TOM complex for translocation across the mitochondrial outer membrane. *EMBO J, 20*, pp. 5626-5635.

Du, C., Fang, M., Li, Y., Li, L. & Wang, X. (2000). Smac, a mitochondrial protein that promotes cytochrome c-dependent caspase activation by eliminating IAP inhibition. *Cell, 102*, pp. 33-42.

Farrow, S. N., White, J. H., Martinou, I., Raven, T., Pun, K. T., Grinham, C. J., Martinou, J. C. & Brown, R. (1995). Cloning of a bcl-2 homologue by interaction with adenovirus E1B 19K. *Nature, 374*, pp. 731-733.

Frank, S., Gaume, B., Bergmann-Leitner, E. S., Leitner, W. W., Robert, E. G., Catez, F., Smith, C. L. & Youle, R.J. (2001). The role of dynamin-related protein 1, a mediator of mitochondrial fission, in apoptosis. *Dev Cell, 1*, pp. 515-525.

Gao, W., Pu, Y., Luo, K. Q. & Chang, D.C. (2001). Temporal relationship between cytochrome c release and mitochondrial swelling during UV-induced apoptosis in living HeLa cells. *J Cell Sci, 114*, pp. 2855-2862.

Garcia-Saez, A. J., Coraiola, M., Dalla Serra, M., Mingarro, I., Menestrina, G. & Salgado, J. (2005). Peptides derived from apoptotic Bax and Bid reproduce the poration activity of the parent full-length proteins. *Biophys J, 88*, pp. 3976-3990.

Garcia-Saez, A. J., Coraiola, M., Serra, M. D., Mingarro, I., Muller, P. & Salgado, J. (2006). Peptides corresponding to helices 5 and 6 of Bax can independently form large lipid pores. *FEBS J, 273*, pp. 971-981.

Garcia-Saez, A. J., Mingarro, I., Perez-Paya, E. & Salgado, J. (2004). Membrane-insertion fragments of Bcl-xL, Bax, and Bid. *Biochemistry, 43*, pp. 10930-10943.

Goldstein, J. C., Waterhouse, N. J., Juin, P., Evan, G. I. & Green, D.R. (2000). The coordinate release of cytochrome c during apoptosis is rapid, complete and kinetically invariant. *Nat Cell Biol, 2*, pp. 156-162.

Gong, X., Choi, J., Franzin, C. M., Zhai, D., Reed, J. C. & Marassi, F.M. (2004). Conformation of membrane-associated proapoptotic tBid. *J Biol Chem, 279*, pp. 28954-28960.

Gonzalvez, F., Pariselli, F., Dupaigne, P., Budihardjo, I., Lutter, M., Antonsson, B., Diolez, P., Manon, S., Martinou, J., Goubern, M., Wang, X., Bernard, S. & Petit, P.X. (2005). tBid interaction with cardiolipin primarily orchestrates mitochondrial dysfunctions and subsequently activates Bax and Bak. *Cell Death Differ, 12*, pp. 614-626.

Goping, I. S., Gross, A., Lavoie, J. N., Nguyen, M., Jemmerson, R., Roth, K., Korsmeyer, S. J. & Shore, G.C. (1998). Regulated targeting of BAX to mitochondria. *J Cell Biol, 143*, pp. 207-215.

Griffiths, G. J., Corfe, B. M., Savory, P., Leech, S., Esposti, M. D., Hickman, J. A. & Dive, C. (2001). Cellular damage signals promote sequential changes at the N-terminus and BH-1 domain of the pro-apoptotic protein Bak. *Oncogene, 20*, pp. 7668-7676.

Griffiths, G. J., Dubrez, L., Morgan, C. P., Jones, N. A., Whitehouse, J., Corfe, B. M., Dive, C. & Hickman, J.A. (1999). Cell damage-induced conformational changes of the pro-apoptotic protein Bak in vivo precede the onset of apoptosis. *J Cell Biol, 144*, pp. 903-914.

Grinberg, M., Schwarz, M., Zaltsman, Y., Eini, T., Niv, H., Pietrokovski, S. & Gross, A. (2005). Mitochondrial carrier homolog 2 is a target of tBID in cells signaled to die by tumor necrosis factor alpha. *Mol Cell Biol, 25*, pp. 4579-4590.

Gross, A., Jockel, J., Wei, M. C. & Korsmeyer, S.J. (1998). Enforced dimerization of BAX results in its translocation, mitochondrial dysfunction and apoptosis. *EMBO J, 17*, pp. 3878-3885.

Guihard, G., Bellot, G., Moreau, C., Pradal, G., Ferry, N., Thomy, R., Fichet, P., Meflah, K. & Vallette, F.M. (2004). The mitochondrial apoptosis-induced channel (MAC) corresponds to a late apoptotic event. *J Biol Chem, 279*, pp. 46542-46550.

Guo, L., Pietkiewicz, D., Pavlov, E. V., Grigoriev, S. M., Kasianowicz, J. J., Dejean, L. M., Korsmeyer, S. J., Antonsson, B. & Kinnally, K.W. (2004). Effects of cytochrome c on the mitochondrial apoptosis-induced channel MAC. *Am J Physiol Cell Physiol, 286*, pp. C1109-C1117.

Hegde, R., Srinivasula, S. M., Zhang, Z., Wassell, R., Mukattash, R., Cilenti, L., DuBois, G., Lazebnik, Y., Zervos, A. S., Fernandes-Alnemri, T. & Alnemri, E.S. (2002). Identification of Omi/HtrA2 as a mitochondrial apoptotic serine protease that disrupts inhibitor of apoptosis protein-caspase interaction. *J Biol Chem, 277*, pp. 432-438.

Heimlich, G., McKinnon, A. D., Bernardo, K., Brdiczka, D., Reed, J. C., Kain, R., Kronke, M. & Jurgensmeier, J.M. (2004). Bax-induced cytochrome c release from mitochondria depends on alpha-helices-5 and -6. *Biochem J, 378*, pp. 247-255.

Hengartner, M. O. (2000). The biochemistry of apoptosis. *Nature, 407*, pp. 770-776.

Hockenbery, D., Nunez, G., Milliman, C., Schreiber, R. D. & Korsmeyer, S.J. (1990). Bcl-2 is an inner mitochondrial membrane protein that blocks programmed cell death. *Nature, 348*, pp. 334-336.

Holcik, M. & Korneluk, R.G. (2001). XIAP, the guardian angel. *Nat Rev Mol Cell Biol, 2*, pp. 550-556.

Hong, S., Cha, M. & Kim, I. (2006). Specific protein interaction of human Pag with Omi/HtrA2 and the activation of the protease activity of Omi/HtrA2. *Free Radic Biol Med, 40*, pp. 275-284.

Hsu, Y. T., Wolter, K. G. & Youle, R.J. (1997). Cytosol-to-membrane redistribution of Bax and Bcl-X(L) during apoptosis. *Proc Natl Acad Sci U S A, 94*, pp. 3668-3672.

Hu, S. & Yang, X. (2003). Cellular inhibitor of apoptosis 1 and 2 are ubiquitin ligases for the apoptosis inducer Smac/DIABLO. *J Biol Chem, 278*, pp. 10055-10060.

Huang, D. C. & Strasser, A. (2000). BH3-Only proteins-essential initiators of apoptotic cell death. *Cell, 103*, pp. 839-842.

Huang, D. C., Adams, J. M. & Cory, S. (1998). The conserved N-terminal BH4 domain of Bcl-2 homologues is essential for inhibition of apoptosis and interaction with CED-4. *EMBO J, 17*, pp. 1029-1039.

Huang, Y., Rich, R. L., Myszka, D. G. & Wu, H. (2003). Requirement of both the second and third BIR domains for the relief of X-linked inhibitor of apoptosis protein (XIAP)-mediated caspase inhibition by Smac. *J Biol Chem, 278*, pp. 49517-49522.

Janiak, F., Leber, B. & Andrews, D.W. (1994). Assembly of Bcl-2 into microsomal and outer mitochondrial membranes. *J Biol Chem, 269*, pp. 9842-9849.

Jesenberger, V. & Jentsch, S. (2002). Deadly encounter: ubiquitin meets apoptosis. *Nat Rev Mol Cell Biol, 3*, pp. 112-121.

Jiang, X. & Wang, X. (2000). Cytochrome c promotes caspase-9 activation by inducing nucleotide binding to Apaf-1. *J Biol Chem, 275*, pp. 31199-31203.

Juin, P., Cartron, P. F. & Vallette, F.M. (2005). Activation of Bax by BH3 domains during apoptosis: the unfolding of a deadly plot. *Cell Cycle, 4*, pp. 637-642.

Juin P., Tremblais K., LeCabellec M. T., Gregoire M., Meflah K. & Vallette F.M. (1998). Potentiation of apoptosis by mitochondria in a cell-free system. *Biochem Biophys Res Commun, 253,* pp. 185-191.

Karbowski, M. & Youle, R.J. (2003). Dynamics of mitochondrial morphology in healthy cells and during apoptosis. *Cell Death Differ, 10*, pp. 870-880.

Karbowski, M., Arnoult, D., Chen, H., Chan, D. C., Smith, C. L. & Youle, R.J. (2004). Quantitation of mitochondrial dynamics by photolabeling of individual organelles shows that mitochondrial fusion is blocked during the Bax activation phase of apoptosis. *J Cell Biol, 164*, pp. 493-499.

Kasof, G. M. & Gomes, B.C. (2001). Livin, a novel inhibitor of apoptosis protein family member. *J Biol Chem, 276*, pp. 3238-3246.

Kaufmann, T., Schlipf, S., Sanz, J., Neubert, K., Stein, R. & Borner, C. (2003). Characterization of the signal that directs Bcl-x(L), but not Bcl-2, to the mitochondrial outer membrane. *J Cell Biol, 160*, pp. 53-64.

Kerr, J. F., Wyllie, A. H. & Currie, A.R. (1972). Apoptosis: a basic biological phenomenon with wide-ranging implications in tissue kinetics. *Br J Cancer, 26*, pp. 239-257.

Kiefer, M. C., Brauer, M. J., Powers, V. C., Wu, J. J., Umansky, S. R., Tomei, L. D. & Barr, P.J. (1995). Modulation of apoptosis by the widely distributed Bcl-2 homologue Bak. *Nature, 374*, pp. 736-739.

Kim, P. K., Annis, M. G., Dlugosz, P. J., Leber, B. & Andrews, D.W. (2004). During apoptosis bcl-2 changes membrane topology at both the endoplasmic reticulum and mitochondria. *Mol Cell, 14*, pp. 523-529.

Klein, J. A., Longo-Guess, C. M., Rossmann, M. P., Seburn, K. L., Hurd, R. E., Frankel, W. N., Bronson, R. T. & Ackerman, S.L. (2002). The harlequin mouse mutation downregulates apoptosis-inducing factor. *Nature, 419*, pp. 367-374.

Kluck, R. M., Bossy-Wetzel, E., Green, D. R. & Newmeyer, D.D. (1997a). The release of cytochrome c from mitochondria: a primary site for Bcl-2 regulation of apoptosis. *Science, 275*, pp. 1132-1136.

Kluck, R. M., Martin, S. J., Hoffman, B. M., Zhou, J. S., Green, D. R. & Newmeyer, D.D. (1997b). Cytochrome c activation of CPP32-like proteolysis plays a critical role in a Xenopus cell-free apoptosis system. *EMBO J, 16*, pp. 4639-4649.

Knudson, C. M., Tung, K. S., Tourtellotte, W. G., Brown, G. A. & Korsmeyer, S.J. (1995). Bax-deficient mice with lymphoid hyperplasia and male germ cell death. *Science, 270*, pp. 96-99.

Koehler, C. M. (2004a). The small Tim proteins and the twin Cx3C motif. *Trends Biochem Sci, 29*, pp. 1-4.

Koehler, C. M. (2004b). New developments in mitochondrial assembly. *Annu Rev Cell Dev Biol, 20*, pp. 309-335.

Koehler, C. M., Leuenberger, D., Merchant, S., Renold, A., Junne, T. & Schatz, G. (1999). Human deafness dystonia syndrome is a mitochondrial disease. *Proc Natl Acad Sci U S A, 96*, pp. 2141-2146.

Krajewski, S., Tanaka, S., Takayama, S., Schibler, M. J., Fenton, W. & Reed, J.C. (1993). Investigation of the subcellular distribution of the bcl-2 oncoprotein: residence in the nuclear envelope, endoplasmic reticulum, and outer mitochondrial membranes. *Cancer Res, 53*, pp. 4701-4714.

Kuwana, T. & Newmeyer, D.D. (2003). Bcl-2-family proteins and the role of mitochondria in apoptosis. *Curr Opin Cell Biol, 15*, pp. 691-699.

Kuwana, T., Mackey, M. R., Perkins, G., Ellisman, M. H., Latterich, M., Schneiter, R., Green, D. R. & Newmeyer, D.D. (2002). Bid, Bax, and lipids cooperate to form supramolecular openings in the outer mitochondrial membrane. *Cell, 111*, pp. 331-342.

Lemasters, J. J., Qian, T., Bradham, C. A., Brenner, D. A., Cascio, W. E., Trost, L. C., Nishimura, Y., Nieminen, A. L. & Herman, B. (1999). Mitochondrial dysfunction in the pathogenesis of necrotic and apoptotic cell death. *J Bioenerg Biomembr, 31*, pp. 305-319.

Li, H., Zhu, H., Xu, C. J. & Yuan, J. (1998). Cleavage of BID by caspase 8 mediates the mitochondrial damage in the Fas pathway of apoptosis. *Cell, 94*, pp. 491-501.

Li, L. Y., Luo, X. & Wang, X. (2001). Endonuclease G is an apoptotic DNase when released from mitochondria. *Nature, 412*, pp. 95-99.

Li, P., Nijhawan, D., Budihardjo, I., Srinivasula, S. M., Ahmad, M., Alnemri, E. S. & Wang, X. (1997). Cytochrome c and dATP-dependent formation of Apaf-1/caspase-9 complex initiates an apoptotic protease cascade. *Cell, 91*, pp. 479-489.

Li, W., Srinivasula, S. M., Chai, J., Li, P., Wu, J., Zhang, Z., Alnemri, E. S. & Shi, Y. (2002). Structural insights into the pro-apoptotic function of mitochondrial serine protease HtrA2/Omi. *Nat Struct Biol*, *9*, pp. 436-441.

Lim, M. L. R., Chen, B., Beart, P. M. & Nagley, P. (2006). Relative timing of redistribution of cytochrome c and Smac/DIABLO from mitochondria during apoptosis assessed by double immunocytochemistry on mammalian cells. *Exp Cell Res*, *312*, pp. 1174-1184.

Lindsten, T., Ross, A. J., King, A., Zong, W. X., Rathmell, J. C., Shiels, H. A., Ulrich, E., Waymire, K. G., Mahar, P., Frauwirth, K., Chen, Y., Wei, M., Eng, V. M., Adelman, D. M., Simon, M. C., Ma, A., Golden, J. A., Evan, G., Korsmeyer, S. J., MacGregor, G. R. & Thompson, C.B. (2000). The combined functions of proapoptotic Bcl-2 family members bak and bax are essential for normal development of multiple tissues. *Mol Cell*, *6*, pp. 1389-1399.

Lithgow, T., van Driel, R., Bertram, J. F. & Strasser, A. (1994). The protein product of the oncogene bcl-2 is a component of the nuclear envelope, the endoplasmic reticulum, and the outer mitochondrial membrane. *Cell Growth Differ*, *5*, pp. 411-417.

Liu, X., Kim, C. N., Yang, J., Jemmerson, R. & Wang, X. (1996). Induction of apoptotic program in cell-free extracts: requirement for dATP and cytochrome c. *Cell*, *86*, pp. 147-157.

Los, M., Burek, C. J., Stroh, C., Benedyk, K., Hug, H. & Mackiewicz, A. (2003). Anticancer drugs of tomorrow: apoptotic pathways as targets for drug design. *Drug Discov Today*, *8*, pp. 67-77.

Losonczi, J. A., Olejniczak, E. T., Betz, S. F., Harlan, J. E., Mack, J. & Fesik, S.W. (2000). NMR studies of the anti-apoptotic protein Bcl-xL in micelles. *Biochemistry*, *39*, pp. 11024-11033.

Luetjens, C. M., Kogel, D., Reimertz, C., Dussmann, H., Renz, A., Schulze-Osthoff, K., Nieminen, A. L., Poppe, M. & Prehn, J.H. (2001). Multiple kinetics of mitochondrial cytochrome c release in drug-induced apoptosis. *Mol Pharmacol*, *60*, pp. 1008-1019.

Lutter, M., Fang, M., Luo, X., Nishijima, M., Xie, X. & Wang, X. (2000). Cardiolipin provides specificity for targeting of tBid to mitochondria. *Nat Cell Biol*, *2*, pp. 754-761.

Lutter, M., Perkins, G. A. & Wang, X. (2001). The pro-apoptotic Bcl-2 family member tBid localizes to mitochondrial contact sites. *BMC Cell Biol*, *2*, p. 22.

Madesh, M., Antonsson, B., Srinivasula, S. M., Alnemri, E. S. & Hajnóczky, G. (2002). Rapid kinetics of tBid-induced cytochrome c and Smac/DIABLO release and mitochondrial depolarization. *J Biol Chem*, *277*, pp. 5651-5659.

Mannella, C. A., Pfeiffer, D. R., Bradshaw, P. C., Moraru, I. I., Slepchenko, B., Loew, L. M., Hsieh, C. E., Buttle, K. & Marko, M. (2001). Topology of the mitochondrial inner membrane: dynamics and bioenergetic implications. *IUBMB Life*, *52*, pp. 93-100.

Martinez-Senac, M. D. M., Corbalan-Garcia, S. & Gomez-Fernandez, J.C. (2002). The structure of the C-terminal domain of the pro-apoptotic protein Bak and its interaction with model membranes. *Biophys J*, *82*, pp. 233-243.

Marzo, I., Brenner, C., Zamzami, N., Jurgensmeier, J. M., Susin, S. A., Vieira, H. L., Prevost, M. C., Xie, Z., Matsuyama, S., Reed, J. C. & Kroemer, G. (1998). Bax and adenine nucleotide translocator cooperate in the mitochondrial control of apoptosis. *Science*, *281*, pp. 2027-2031.

McDonnell, J. M., Fushman, D., Milliman, C. L., Korsmeyer, S. J. & Cowburn, D. (1999). Solution structure of the proapoptotic molecule BID: a structural basis for apoptotic agonists and antagonists. *Cell, 96*, pp. 625-634.

McMillin, J. B. & Dowhan, W. (2002). Cardiolipin and apoptosis. *Biochim Biophys Acta, 1585*, pp. 97-107.

Melloni, E., Averna, M., Salamino, F., Sparatore, B., Minafra, R. & Pontremoli, S. (2000). Acyl-CoA-binding protein is a potent m-calpain activator. *J Biol Chem, 275*, pp. 82-86.

Mikhailov, V., Mikhailova, M., Degenhardt, K., Venkatachalam, M. A., White, E. & Saikumar, P. (2003). Association of Bax and Bak homo-oligomers in mitochondria. Bax requirement for Bak reorganization and cytochrome c release. *J Biol Chem, 278*, pp. 5367-5376.

Minn, A. J., Velez, P., Schendel, S. L., Liang, H., Muchmore, S. W., Fesik, S. W., Fill, M. & Thompson, C.B. (1997). Bcl-x(L) forms an ion channel in synthetic lipid membranes. *Nature, 385*, pp. 353-357.

Moreau, C., Cartron, P., Hunt, A., Meflah, K., Green, D. R., Evan, G., Vallette, F. M. & Juin, P. (2003). Minimal BH3 peptides promote cell death by antagonizing anti-apoptotic proteins. *J Biol Chem, 278*, pp. 19426-19435.

Motz, C., Martin, H., Krimmer, T. & Rassow, J. (2002). Bcl-2 and porin follow different pathways of TOM-dependent insertion into the mitochondrial outer membrane. *J Mol Biol, 323*, pp. 729-738.

Muchmore, S. W., Sattler, M., Liang, H., Meadows, R. P., Harlan, J. E., Yoon, H. S., Nettesheim, D., Chang, B. S., Thompson, C. B., Wong, S. L., Ng, S. L. & Fesik, S.W. (1996). X-ray and NMR structure of human Bcl-xL, an inhibitor of programmed cell death. *Nature, 381*, pp. 335-341.

Musse, A. A. & Merrill, A.R. (2003). The molecular basis for the pH-activation mechanism in the channel-forming bacterial colicin E1. *J Biol Chem, 278*, pp. 24491-24499.

Nechushtan, A., Smith, C. L., Hsu, Y. T. & Youle, R.J. (1999). Conformation of the Bax C-terminus regulates subcellular location and cell death. *EMBO J, 18*, pp. 2330-2341.

Nguyen, M., Millar, D. G., Yong, V. W., Korsmeyer, S. J. & Shore, G.C. (1993). Targeting of Bcl-2 to the mitochondrial outer membrane by a COOH-terminal signal anchor sequence. *J Biol Chem, 268*, pp. 25265-25268.

Nicholson, D. W. (1999). Caspase structure, proteolytic substrates, and function during apoptotic cell death. *Cell Death Differ, 6*, pp. 1028-1042.

Nomura K., Imai H., Koumura T., Kobayashi T. & Nakagawa Y. (2000). Mitochondrial phospholipid hydroperoxide glutathione peroxidase inhibits the release of cytochrome c from mitochondria by suppressing the peroxidation of cardiolipin in hypoglycaemia-induced apoptosis. *Biochem J, 351*, pp. 183-193.

Oliver, L. & Vallette, F.M. (2005). The role of caspases in cell death and differentiation. *Drug Resist Updat, 8*, pp. 163-170.

Oliver, L., Priault, M., Tremblais, K., LeCabellec, M., Meflah, K., Manon, S. & Vallette, F.M. (2000). The substitution of the C-terminus of bax by that of bcl-xL does not affect its subcellular localization but abrogates its pro-apoptotic properties. *FEBS Lett, 487*, pp. 161-165.

Ott, M., Robertson, J. D., Gogvadze, V., Zhivotovsky, B. & Orrenius, S. (2002). Cytochrome c release from mitochondria proceeds by a two-step process. *Proc Natl Acad Sci U S A*, *99*, pp. 1259-1263.

Parker, M. W. & Feil, S.C. (2005). Pore-forming protein toxins: from structure to function. *Prog Biophys Mol Biol*, *88*, pp. 91-142.

Parrish, J., Li, L., Klotz, K., Ledwich, D., Wang, X. & Xue, D. (2001). Mitochondrial endonuclease G is important for apoptosis in C. elegans. *Nature*, *412*, pp. 90-94.

Patterson, S. D., Spahr, C. S., Daugas, E., Susin, S. A., Irinopoulou, T., Koehler, C. & Kroemer, G. (2000). Mass spectrometric identification of proteins released from mitochondria undergoing permeability transition. *Cell Death Differ*, *7*, pp. 137-144.

Pavlov, E. V., Priault, M., Pietkiewicz, D., Cheng, E. H., Antonsson, B., Manon, S., Korsmeyer, S. J., Mannella, C. A. & Kinnally, K.W. (2001). A novel, high conductance channel of mitochondria linked to apoptosis in mammalian cells and Bax expression in yeast. *J Cell Biol*, *155*, pp. 725-731.

Pelletier M., Oliver L., Meflah K. & Vallette F.M. (2005). Caspase-3 can be pseudo-activated by a Ca2+-dependent proteolysis at a non-canonical site. *FEBS Lett, 579,* pp. 2364-2368.

Petros, A. M., Medek, A., Nettesheim, D. G., Kim, D. H., Yoon, H. S., Swift, K., Matayoshi, E. D., Oltersdorf, T. & Fesik, S.W. (2001). Solution structure of the antiapoptotic protein bcl-2. *Proc Natl Acad Sci U S A*, *98*, pp. 3012-3017.

Petros, A. M., Olejniczak, E. T. & Fesik, S.W. (2004). Structural biology of the Bcl-2 family of proteins. *Biochim Biophys Acta*, *1644*, pp. 83-94.

Pouliquen, D., Bellot, G., Guihard, G., Fichet, P., Meflah, K. & Vallette, F.M. (2006). Mitochondrial membrane permeabilization produced by PTP, Bax and apoptosis: a 1H-NMR relaxation study. *Cell Death Differ*, *13*, pp. 301-310.

Priault, M., Camougrand, N., Chaudhuri, B. & Manon, S. (1999). Role of the C-terminal domain of Bax and Bcl-XL in their localization and function in yeast cells. *FEBS Lett*, *443*, pp. 225-228.

Rehm, M., Dussmann, H. & Prehn, J.H.M. (2003). Real-time single cell analysis of Smac/DIABLO release during apoptosis. *J Cell Biol*, *162*, pp. 1031-1043.

Roesch, K., Curran, S. P., Tranebjaerg, L. & Koehler, C.M. (2002). Human deafness dystonia syndrome is caused by a defect in assembly of the DDP1/TIMM8a-TIMM13 complex. *Hum Mol Genet*, *11*, pp. 477-486.

Ruffolo, S. C. & Shore, G.C. (2003). BCL-2 selectively interacts with the BID-induced open conformer of BAK, inhibiting BAK auto-oligomerization. *J Biol Chem*, *278*, pp. 25039-25045.

Saelens, X., Festjens, N., Vande Walle, L., van Gurp, M., van Loo, G. & Vandenabeele, P. (2004). Toxic proteins released from mitochondria in cell death. *Oncogene*, *23*, pp. 2861-2874.

Saito, M., Korsmeyer, S. J. & Schlesinger, P.H. (2000). BAX-dependent transport of cytochrome c reconstituted in pure liposomes. *Nat Cell Biol*, *2*, pp. 553-555.

Salvesen, G. S. & Abrams, J.M. (2004). Caspase activation - stepping on the gas or releasing the brakes? Lessons from humans and flies. *Oncogene*, *23*, pp. 2774-2784.

Salvesen, G. S. & Duckett, C.S. (2002). IAP proteins: blocking the road to death's door. *Nat Rev Mol Cell Biol*, *3*, pp. 401-410.

Sattler, M., Liang, H., Nettesheim, D., Meadows, R. P., Harlan, J. E., Eberstadt, M., Yoon, H. S., Shuker, S. B., Chang, B. S., Minn, A. J., Thompson, C. B. & Fesik, S.W. (1997). Structure of Bcl-xL-Bak peptide complex: recognition between regulators of apoptosis. *Science*, *275*, pp. 983-986.

Schendel, S. L., Xie, Z., Montal, M. O., Matsuyama, S., Montal, M. & Reed, J.C. (1997). Channel formation by antiapoptotic protein Bcl-2. *Proc Natl Acad Sci U S A*, *94*, pp. 5113-5118.

Schlesinger, P. H., Gross, A., Yin, X. M., Yamamoto, K., Saito, M., Waksman, G. & Korsmeyer, S.J. (1997). Comparison of the ion channel characteristics of proapoptotic BAX and antiapoptotic BCL-2. *Proc Natl Acad Sci U S A*, *94*, pp. 11357-11362.

Scorrano, L., Ashiya, M., Buttle, K., Weiler, S., Oakes, S. A., Mannella, C. A. & Korsmeyer, S.J. (2002). A distinct pathway remodels mitochondrial cristae and mobilizes cytochrome c during apoptosis. *Dev Cell*, *2*, pp. 55-67.

Scorrano, L., Oakes, S. A., Opferman, J. T., Cheng, E. H., Sorcinelli, M. D., Pozzan, T. & Korsmeyer, S.J. (2003). BAX and BAK regulation of endoplasmic reticulum Ca2+: a control point for apoptosis. *Science*, *300*, pp. 135-139.

Shi, Y. (2002). Mechanisms of caspase activation and inhibition during apoptosis. *Mol Cell*, *9*, pp. 459-470.

Shimizu, S., Ide, T., Yanagida, T. & Tsujimoto, Y. (2000). Electrophysiological study of a novel large pore formed by Bax and the voltage-dependent anion channel that is permeable to cytochrome c. *J Biol Chem*, *275*, pp. 12321-12325.

Shimizu, S., Narita, M. & Tsujimoto, Y. (1999). Bcl-2 family proteins regulate the release of apoptogenic cytochrome c by the mitochondrial channel VDAC. *Nature*, *399*, pp. 483-487.

Shiozaki, E. N., Chai, J., Rigotti, D. J., Riedl, S. J., Li, P., Srinivasula, S. M., Alnemri, E. S., Fairman, R. & Shi, Y. (2003). Mechanism of XIAP-mediated inhibition of caspase-9. *Mol Cell*, *11*, pp. 519-527.

Springs, S. L., Diavolitsis, V. M., Goodhouse, J. & McLendon, G.L. (2002). The kinetics of translocation of Smac/DIABLO from the mitochondria to the cytosol in HeLa cells. *J Biol Chem*, *277*, pp. 45715-45718.

Strasser, A., O'Connor, L. & Dixit, V.M. (2000). Apoptosis signaling. *Annu Rev Biochem*, *69*, pp. 217-245.

Susin, S. A., Lorenzo, H. K., Zamzami, N., Marzo, I., Snow, B. E., Brothers, G. M., Mangion, J., Jacotot, E., Costantini, P., Loeffler, M., Larochette, N., Goodlett, D. R., Aebersold, R., Siderovski, D. P., Penninger, J. M. & Kroemer, G. (1999). Molecular characterization of mitochondrial apoptosis-inducing factor. *Nature*, *397*, pp. 441-446.

Suzuki, M., Youle, R. J. & Tjandra, N. (2000). Structure of Bax: coregulation of dimer formation and intracellular localization. *Cell*, *103*, pp. 645-654.

Suzuki, Y., Imai, Y., Nakayama, H., Takahashi, K., Takio, K. & Takahashi, R. (2001). A serine protease, HtrA2, is released from the mitochondria and interacts with XIAP, inducing cell death. *Mol Cell*, *8*, pp. 613-621.

Takeuchi O., Fisher J., Suh H., Harada H., Malynn B. A. & Korsmeyer S.J. (2005). Essential role of BAX, BAK in B cell homeostasis and prevention of autoimmune disease. *Proc Natl Acad Sci U S A*, *102*, pp. 11272-11277.

Tan, C., Dlugosz, P. J., Peng, J., Zhang, Z., Lapolla, S. M., Plafker, S. M., Andrews, D. W. & Lin, J. (2006). Auto-activation of the apoptosis protein Bax increases mitochondrial membrane permeability and is inhibited by Bcl-2. *J Biol Chem, 281*, pp. 14764-14775.

Tan, K. O., Fu, N. Y., Sukumaran, S. K., Chan, S., Kang, J. H., Poon, K. L., Chen, B. S. & Yu, V.C. (2005). MAP-1 is a mitochondrial effector of Bax. *Proc Natl Acad Sci U S A, 102*, pp. 14623-14628.

Terrones, O., Antonsson, B., Yamaguchi, H., Wang, H., Liu, J., Lee, R. M., Herrmann, A. & Basanez, G. (2004). Lipidic pore formation by the concerted action of proapoptotic BAX and tBID. *J Biol Chem, 279*, pp. 30081-30091.

Thuduppathy, G. R., Craig, J. W., Kholodenko, V., Schon, A. & Hill, R.B. (2006). Evidence that Membrane Insertion of the Cytosolic Domain of Bcl-x(L) is Governed by an Electrostatic Mechanism. *J Mol Biol, 359*, pp. 1045-1058.

Tilley, S. J. & Saibil, H.R. (2006). The mechanism of pore formation by bacterial toxins. *Curr Opin Struct Biol, 16*, pp. 230-236.

Tremblais, K., Oliver, L., Juin, P., Le Cabellec, T. M., Meflah, K. & Vallette, F.M. (1999). The C-terminus of bax is not a membrane addressing/anchoring signal. *Biochem Biophys Res Commun, 260*, pp. 582-591.

van Loo, G., Demol, H., van Gurp, M., Hoorelbeke, B., Schotte, P., Beyaert, R., Zhivotovsky, B., Gevaert, K., Declercq, W., Vandekerckhove, J. & Vandenabeele, P. (2002b). A matrix-assisted laser desorption ionization post-source decay (MALDI-PSD) analysis of proteins released from isolated liver mitochondria treated with recombinant truncated Bid. *Cell Death Differ, 9*, pp. 301-308.

van Loo, G., Schotte, P., van Gurp, M., Demol, H., Hoorelbeke, B., Gevaert, K., Rodriguez, I., Ruiz-Carrillo, A., Vandekerckhove, J., Declercq, W., Beyaert, R. & Vandenabeele, P. (2001). Endonuclease G: a mitochondrial protein released in apoptosis and involved in caspase-independent DNA degradation. *Cell Death Differ, 8*, pp. 1136-1142.

van Loo, G., van Gurp, M., Depuydt, B., Srinivasula, S. M., Rodriguez, I., Alnemri, E. S., Gevaert, K., Vandekerckhove, J., Declercq, W. & Vandenabeele, P. (2002a). The serine protease Omi/HtrA2 is released from mitochondria during apoptosis. Omi interacts with caspase-inhibitor XIAP and induces enhanced caspase activity. *Cell Death Differ, 9*, pp. 20-26.

Vaux, D. L. & Korsmeyer, S.J. (1999). Cell death in development. *Cell, 96*, pp. 245-254.

Verhagen, A. M., Ekert, P. G., Pakusch, M., Silke, J., Connolly, L. M., Reid, G. E., Moritz, R. L., Simpson, R. J. & Vaux, D.L. (2000). Identification of DIABLO, a mammalian protein that promotes apoptosis by binding to and antagonizing IAP proteins. *Cell, 102*, pp. 43-53.

Waterhouse, N. J., Sedelies, K. A., Sutton, V. R., Pinkoski, M. J., Thia, K. Y., Johnstone, R., Bird, P. I., Green, D. R. & Trapani, J.A. (2006). Functional dissociation of DeltaPsim and cytochrome c release defines the contribution of mitochondria upstream of caspase activation during granzyme B-induced apoptosis. *Cell Death Differ, 13*, pp. 607-618.

Wei, M. C., Lindsten, T., Mootha, V. K., Weiler, S., Gross, A., Ashiya, M., Thompson, C. B. & Korsmeyer, S.J. (2000). tBID, a membrane-targeted death ligand, oligomerizes BAK to release cytochrome c. *Genes Dev, 14*, pp. 2060-2071.

Wei, M. C., Zong, W. X., Cheng, E. H., Lindsten, T., Panoutsakopoulou, V., Ross, A. J., Roth, K. A., MacGregor, G. R., Thompson, C. B. & Korsmeyer, S.J. (2001). Proapoptotic BAX and BAK: a requisite gateway to mitochondrial dysfunction and death. *Science*, *292*, pp. 727-730.

Widlak, P., Li, L. Y., Wang, X. & Garrard, W.T. (2001). Action of recombinant human apoptotic endonuclease G on naked DNA and chromatin substrates: cooperation with exonuclease and DNase I. *J Biol Chem*, *276*, pp. 48404-48409.

Willis, S. N., Chen, L., Dewson, G., Wei, A., Naik, E., Fletcher, J. I., Adams, J. M. & Huang, D.C.S. (2005). Proapoptotic Bak is sequestered by Mcl-1 and Bcl-xL, but not Bcl-2, until displaced by BH3-only proteins. *Genes Dev*, *19*, pp. 1294-1305.

Ye, H., Cande, C., Stephanou, N. C., Jiang, S., Gurbuxani, S., Larochette, N., Daugas, E., Garrido, C., Kroemer, G. & Wu, H. (2002). DNA binding is required for the apoptogenic action of apoptosis inducing factor. *Nat Struct Biol*, *9*, pp. 680-684.

Zakharov, S. D. & Cramer, W.A. (2002a). Colicin crystal structures: pathways and mechanisms for colicin insertion into membranes. *Biochim Biophys Acta*, *1565*, pp. 333-346.

Zakharov, S. D. & Cramer, W.A. (2002b). Insertion intermediates of pore-forming colicins in membrane two-dimensional space. *Biochimie*, *84*, pp. 465-475.

Zamzami, N., Susin, S. A., Marchetti, P., Hirsch, T., Gomez-Monterrey, I., Castedo, M. & Kroemer, G. (1996). Mitochondrial control of nuclear apoptosis. *J Exp Med*, *183*, pp. 1533-1544.

In: Cell Apoptotic Signaling Pathways
Editor: Charles O. Pickens, pp. 83-114
ISBN: 978-1-60021-507-0

Chapter III

ROLE FOR NA^+/H^+ EXCHANGER 1 (NHE1) IN THE CONTROL OF APOPTOTIC PATHWAYS

Dominique Lagadic-Gossmann[1,*]***, Laurence Huc***[2] ***and Xavier Terkli***[2]
[1]Inserm U620, Faculté de Pharmacie, Université Rennes I, 2 av Prof Léon Bernard, 35043 Rennes cedex, France;
[2]Burdon Sanderson Cardiac Science Centre, Department of Physiology, Anatomy and Genetics, Parks Road, Oxford, OX1 3PT, UK.

ABSTRACT

Intracellular pH plays an important role in the maintenance of normal cell function and has therefore to be tightly controlled. This control is largely based upon the activity of membrane transporters, among which the ubiquitous Na^+/H^+ exchanger isoform 1 (NHE1). Besides its role as a pH and volume regulator, more recent works have suggested NHE1 as a multifaceted regulator, notably via its scaffolding and cytoskeleton anchoring properties. Such properties appear to be important notably for activation of survival pathways. Indeed, it has been recently shown that NHE1 forms a signalling complex through binding to ezrin/radixin/moiesin (ERM) proteins, which allows activation of the PI3K/Akt pathway and hence survival. Besides this role, the exchanger can be modulated by diverse extracellular or intracellular stimuli, leading to pH_i changes that have been implicated in cell proliferation and migration, as well as in cell death. Regarding proliferation, it has thus been shown that NHE1 expression is often up-regulated in cancer cells, possibly resulting from reactive oxygen species (ROS) disturbances. Due to this up-regulation, inhibition of NHE1 has been proposed as an alternative for cancer therapy. Indeed, intracellular acidification, such as that resulting from NHE1 inhibition, seems to be a common apoptotic feature helping for caspase activation. Also, a few diverse apoptotic stimuli have been reported to lead to NHE1 inhibition, notably through cleavage of the transporter by caspases. Despite several

* Correspondence concerning this article should be addressed to Dominique Lagadic-Gossmann, Tel: +33(0)223234837; Fax: +33(0)223234794; E-mail: dominique.lagadic@rennes.inserm.fr.

studies indicating activation of NHE1 during proliferation processes, such an effect has also been observed upon apoptotic stimuli such as toxic xenobiotics (eg. polycyclic aromatic hydrocarbons) or cytokine deprivation. However, the precise role for such an early activation in the apoptotic cascade has still to be determined and might involve mitochondrial dysfunction. The aim of this review is to present and discuss new findings implicating NHE1 as a central signalling event activated by stress conditions and modulating cell proliferation and death processes.

INTRODUCTION

The extra- and intracellular concentration of free H^+ ions is one of the most tightly controlled parameters in mammals. Despite their low concentrations in most body compartments, ~ nM levels compared to mM levels for Na^+ and K^+, H^+ ions have striking effects on a wide variety of biological processes such as *eg.* cell proliferation, cell volume regulation, contractile activity of cardiac cells,.... These effects are facilitated by the high reactivity of H^+ ions and the abundance of H^+ binding sites on macromolecules and metabolites.

Variations in ambient pH alter the occupancy of acidic and basic groups on proteins and other molecules. The resulting change in net charge affects interactions with other molecules and the distribution of intra-molecular charge. Thus the acid-base status of the extra- and intracellular milieu influences numerous cellular properties, including the conformational state of proteins, ligand-receptor interactions, enzyme activity, transmembrane distribution of charged molecules including drugs, etc. In this context, intracellular pH (pH_i) has to be tightly controlled in order for the cells to function properly.

At the cellular level, several ion transport systems operate to maintain steady-state pH_i and to restore this parameter following acute acid-base challenges. Some of these transporters are stimulated upon an alkaline load and carry protons into cells whereas the others allow an acid efflux upon an acid load. The steady-state pH_i is given by the balance between acid influxes and acid effluxes (Leem *et al.*, 1999). As a consequence, any change in the activity of either transporter will then result in an alteration of pH_i; such an alteration might then serve as an intracellular signal to regulate physiological processes or, in some circumstances, might be deleterious for the cell, ultimately leading to cell death *eg.* through apoptotic processes. Regarding this latter aspect, the role of pH_i in apoptosis has been previously reviewed (Lagadic-Gossmann *et al.*, 2004; Matsuyama and Reed, 2000).

In mammalian as well as other eukaryotic cells, restoration of steady-state pH_i following acidification (such as that produced by cell metabolism) is achieved most efficiently by the electroneutral exchange of intracellular H^+ for extracellular Na^+, a process that is mediated by a family of integral cation transporters commonly referred to as Na^+/H^+ antiporters or exchangers (NHE). The first known isoform of the protein (NHE1) was cloned in 1989 (Sardet *et al.*, 1989) followed by identification of eight other isoforms, NHE2-NHE9 (Brett *et al.*, 2005). The NHEs exhibit different tissue and cellular localization patterns, and mediate a wide variety of physiological functions [for a review, see (Orlowski and Grinstein, 2004)]. The plasma membrane NHE isoform, NHE1, is the most widely expressed and highly conserved across vertebrate species, and provides the principal pathway for acid extrusion in most mammalian cells (reviewed in Karmazyn *et al.*, 2003). Its main function is to regulate intracellular pH and cell volume, while the other isoforms are involved in more specialized functions, such as acid-base vectorial transport across epithelia. The present review will focus only on NHE-1, which is also the most well-characterized NHE isoform.

An intringuing feature of NHE1 is that its activity is controlled by various endogenous and exogenous molecules, including hormones, growth factors, pharmacological agents, and mechanical stimuli. NHE1 response to intracellular protons is thus modulated via the involvement of a variety of signaling molecules, including calcineurin B-homologous protein (CHP), Ca^{2+}/calmodulin, 14-3-3 protein, and diverse protein kinases (*eg.* Nck-interacting kinase [NIK], $p90^{RSK}$ and Extracellular Regulated Kinase [ERK]) (reviewed in Karmazyn *et al.*, 2003). Although the interrelationships among these signalling effectors are not totally known, their net action is to transmit signals to the more general regulator, the intracellular proton. This is made possible by the modular nature of NHE-1 regulation (Counillon and Pouyssegur, 2000), which integrates these different stimuli by changing the apparent affinity of the transporter for intracellular H^+, according to a kinetic Monod-Whyman-Changeux mechanism recently described (Lacroix *et al.*, 2004). The resulting increases in intracellular pH upon endogenous molecules may in turn affect diverse signalling pathways which remain to be precised, thereby affecting *eg.* cell growth. Aside from pH regulation, Na^+/H^+ exchange activity also assists in regulating sodium fluxes and cell volume after osmotic shrinkage [reviewed in (Pedersen *et al.*, 2006)]; and a few works also indicate that NHE1 might play an important role as an anchoring and scaffolding protein [reviewed in (Baumgartner *et al.*, 2004)].

Besides its physiological roles, the Na^+/H^+ exchanger has also been reported to be an important mediator of the damage that occurs in some pathological situations such as during cardiac ischaemia and reperfusion. In this pathological situation, such a role relies upon the fact that NHE activation increases $[Na^+]_i$ which, in heart muscle, can elevate $[Ca^{2+}]_i$ via effects on sarcolemmal Na^+/Ca^{2+} exchange. The excessive activation of cardiac NHE that accompanies postischemic reperfusion can cause abnormally high $[Ca^{2+}]_i$ resulting in myocardial injury and arrhythmias, these effects being minimized upon inhibition of NHE (Avkiran and Marber, 2002; Karmazyn *et al.*, 2003).

The Na^+/H^+ exchanger is also important in abnormal growth of cells. In relation to this, it is worth noting that this transport mechanism has been implicated in myocardial hypertrophy (Chen *et al.*, 2004) as well as in cancer development (Cardone *et al.*, 2005b); in this context, it has for instance been proposed that inhibition of NHE1 could help treating some cancers (Harguindey *et al.*, 2005). It has also been suggested that the development of cardiac as well as hepatic fibrosis would involve activation of this transporter (Benedetti *et al.*, 2001; Di Sario *et al.*, 2003; Haussinger, 2001; Linz and Busch, 2003; Svegliati-Baroni *et al.*, 1999). Other processes affected by NHE1 activation are related to cell death. However, the roles for NHE1 notably in the apoptotic processes are poorly known.

In this review, we will present the current knowledge regarding the role of NHE1 in proliferation, migration and cell death processes, and will discuss possible mechanisms underlying the ambivalent role of NHE1 in both proliferation and apoptosis.

A. Role of NHE1 in Cell Proliferation and Migration

A.1. Implication of NHE1 in Proliferation

Proliferation of sea urchin eggs, lymphocytes, fibroblasts and malignant cells can be triggered by modest alkalinization (Grinstein *et al.*, 1985; Pouyssegur *et al.*, 1982; Reichert *et al.*, 2002; Reshkin *et al.*, 2000a; Reshkin *et al.*, 2000b). In line with this, diverse proliferative stimuli such as epidermal growth factor (EGF), colony-stimulating factor-1 (CSF-1) or phorbol esters have thus been reported to induce an increase of pH_i through early activation of NHE1, whose inhibition hampers the related proliferation process (Besson *et al.*, 1998; L'Allemain *et al.*, 1984; Lee *et al.*, 2003; Ulrich-Baker *et al.*, 1988; Vairo *et al.*, 1992). Note also that activation of this exchanger has been associated with liver regeneration (Dallenbach *et al.*, 1994), a situation where cell proliferation is massive (Fausto *et al.*, 2006); and inhibiting NHE1 in this situation prevents DNA synthesis and liver regeneration after partial hepatectomy (Suzuki and Tsukamoto, 2004). Besides, cellular alkalinization resulting from NHE1 activation is similarly important in oncogenic transformation and is necessary for both the development and maintenance of some transformed phenotypes (Cardone *et al.*, 2005b; Reshkin *et al.*, 2000b). Furthermore, preliminary experiments conducted with Ras-transformed fibroblasts deleted in *nhe1* have established proof of principle for tumor regression by manipulating tumor pH. Indeed, tumor cells lacking *nhe1* showed a drastic reduction in tumor incidence unless the production of lactic acid was genetically suppressed (Pouyssegur *et al.*, 2001). Altogether these data clearly indicate the implication of NHE1 in proliferative processes (Figure 1).

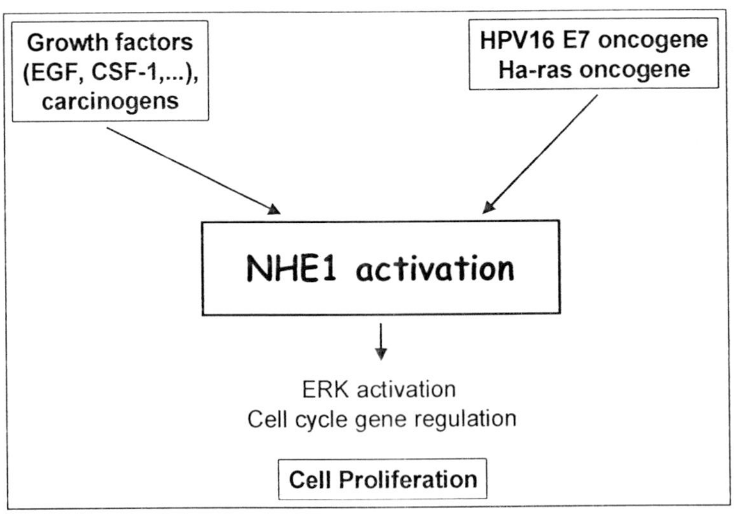

Figure 1. NHE1 activation is involved in proliferation processes induced by growth factors, carcinogens (*eg.* phorbol esters) or oncogenes. EGF: epidermal growth factor; CSF-1: colony-stimulating factor-1.

Regarding the mechanisms underlying such an implication, it is noteworthy that a recent study on cell cycle reports that the transient increase in pH_i induced by NHE1 may provide an important signal for efficient progression from S phase to G2/M (Putney and Barber, 2003). More specifically, these authors have found that ion translocation by NHE1 is necessary for controlling both the activity and the expression of various M phase regulators, such as cyclin B1 and the kinase Cdc2, thereby promoting mitotic entry. Moreover, a recent transcriptomic study performed by the same group further supports the primordial role of NHE1 in cell growth regulation (Putney and Barber, 2004). Indeed, comparing steady-state gene expression between fibroblasts stably expressing a mutant NHE1 lacking ion translocation activity and fibroblasts stably expressing wild-type NHE1, they have found that in the absence of NHE1 activity, significant changes occurred in the expression pattern of genes related to growth factor signalling, growth and oncogenesis, and DNA synthesis and cell cycle control. Interestingly, several genes involved in carbohydrate metabolism, electron transport and oxidative phosphorylation were also affected by the absence of NHE1. This point is worth emphasizing especially in the context of cancer. Indeed, independently of genetic variability, cellular pathological alkalosis, together with an abnormally high glycolytic metabolism, are two of the principal factors characterizing cancer cells and malignant tumors in general (Gatenby and Gawlinski, 2003; Gatenby and Gillies, 2004; Harguindey *et al.*, 1995; Harguindey *et al.*, 2005). More precisely, the pH_i of tumors could become highly acidic as a result of the overproduction of acid-equivalent species by glycolytic metabolism (protons, lactic acid). To survive and proliferate, cells must then extrude these acids, through the activity of several pumps, exchangers and transporters. In this context, it is not surprising to find not only an activation of NHE1 but also an up-regulation of its expression in many tumor cells (Cardone *et al.*, 2005b; Harguindey *et al.*, 2005). Considering this latter aspect together with data from Barber's group (Putney and Barber, 2004), one might then propose that NHE1 might also exert a feedback control on the energy metabolism of tumor cells through an effect on the expression of the metabolism-related enzymes. Such an interrelationship between energy metabolism and NHE1 will deserve further investigation in the future. With respect to the regulation of gene expression by NHE1, up to now, very little is known about the underlying mechanisms. Nevertheless, two recent observations could help defining these mechanisms: (1) NHE1 activation controls activation of the MAPK (Mitogen activated protein kinase) ERK (Mukhin et al., 2004); (2) BCL6 and E4BP4 are two transcription suppressors shown to mediate the inhibitory effect on different genes of NHE1 null mutation (Zhou *et al.*, 2004).

Concerning the up-regulation of NHE1 expression in cancer, the mechanisms underlying this phenomenon remain also to be precised. A recent work by Clement's group indicates a possible role for cell redox status in this up-regulation (Akram *et al.*, 2006). One might also suppose a role for hypoxia-inducible factor 1 (HIF-1) since this factor, which has become recognized as a strong promoter of tumor growth [reviewed in (Pouyssegur *et al.*, 2006)], has recently been shown to up-regulate NHE1 mRNA and protein expression in pulmonary arterial smooth muscle cells, thereby inducing an intracellular alkalinization in response to hypoxia (Shimoda *et al.*, 2006).

The experiments performed by Barber's group clearly demonstrate that ion translocation function of NHE1 is important for controlling gene expression (Putney and Barber, 2004). However, one might also suppose the involvement of the scaffolding function of the transporter in modulating the pattern of genes controlled by NHE1; indeed, as it will be described below, a recent study has shown that NHE1, by functionning as a scaffold, might help the activation of signalling pathways as important as the PI3K/Akt pathway (Wu *et al.*, 2004), which is known to control the expression of diverse genes [reviewed in (Brunet *et al.*, 2001; Cantley, 2002; Chang *et al.*, 2003)]. When considering the ion translocation function of NHE1, besides affecting intracellular H^+ concentration, it also affects extracellular H^+ concentration, especially close to the plasma membrane; such an effect has to be kept in mind, especially in situations such as cancers. Indeed, tumor cells are most often characterized by an acidic interstitial extracellular pH (6.2-6.9 compared with 7.3-7.4 in normal tissues) *versus* an alkaline intracellular pH (7.12-7.65 compared with 6.99-7.2 in normal tissues) (Gillies *et al.*, 2002). So the tumor extracellular environment is more acidic than the intracellular medium creating a reverse pH gradient across the cell membrane that increases as the tumor progresses. The development and maintenance of this gradient is directly due to the ability of the tumor cells to secrete protons and acidify their extracellular environment, which increases with tumor aggressiveness (Montcourrier *et al.*, 1997; Parkins *et al.*, 1997). Besides interfering with chemotherapy by modifying the action of many anticancer drugs *(Gerweck et al.*, 2006; Kozin *et al.*, 2001), the acidic extracellular pH (pH_e) may also affect cell response to endogenous molecules (such as the cytokine TRAIL [TNF-related apoptosis-inducing ligand]; (Meurette *et al.*, 2005)) or gene expression (Bumke *et al.*, 2003; Duggan *et al.*, 2006). Regarding this point, it is interesting to emphasize that proteases of the extracellular matrix (*eg.* MMP [matrix metalloproteinase] 1 and 9) are among the genes regulated by acidification of pH_e (Kato *et al.*, 2005; Shimokawa *et al.*, 2005). This latter study further showed that phospholipase D, MAPKs and the transcription factor NFκB were involved in the regulation of MMP-9 by acidic pH_e. Based upon these works showing that pH_e controls MMP expression and activity, it is not surprising to find NHE1 involved in cell migration processes, especially as this protein has been shown to play an important role in the formation of focal adhesions.

A.2. Implication of NHE1 in Cell Adhesion and Migration

A.2.1. Cell Adhesion

Cell contacts with the extracellular matrix (ECM) are important determinants of cell growth, differentiation, and migration (Clark and Brugge, 1995) These contacts or focal adhesions are mediated by integrin family of cell surface receptor. Focal adhesion contacts are stable cell-substrate interactions evolving from focal complexes, which are characterized as small and transient cell interactions between the cell and ECM (Friedl and Wolf, 2003). Focal contacts include integrins, the focal adhesion kinase (FAK), talin, vinculin paxilin and other proteins attached to the actin filament network (Brakebusch and Fassler, 2003). In the assembly of focal adhesions, integrins link to actin filaments through intracellular cytoskeletal complexes (Craig and Johnson, 1996). The link between the actin cytoskeleton and the plasma membrane is facilitated by members of the band 4.1 superfamily of proteins, which includes band 4.1 and the ezrin/radixin/moesin (ERM) proteins (Takeuchi *et al.*, 1994; Tsukita and Yonemura, 1997). The carboxyl terminus of band 4.1 associates indirectly with actin through its direct binding to spectrin (Ungewickell *et al.*, 1979) and the carboxyl termini of ERM proteins associate directly with actin (Turunen *et al.*, 1994). The amino termini of band 4.1 and ERM proteins bind directly integral membrane proteins thereby linking the actin cytoskeleton to the plasma membrane. For example, the band 4.1 binds to the anion exchanger 1 (AE1) at a motif of clustered positively charged amino acids in the juxtamembrane region of the cytoplasmic domain (Jons and Drenckhahn, 1992). A similar motif is present in CD44 for binding band 4.1 and ezrin (Legg and Isacke, 1998); this motif is also present in NHE1 (Denker *et al.*, 2000)

NHE1 is colocalized with both ezrin and radixin in lamellipodia of fibroblasts (Denker *et al.*, 2000), where it functions as a plasma anchor for actin filaments by its direct binding with ERM proteins. Disrupting the ERM binding site on NHE1 altered the distribution of ezrin and paxilin in response to integrin activation, and impairs formation of focal adhesions (Denker *et al.*, 2000). ERM and cytoskeletal anchoring by NHE1 are thus required for integrin and Rho GTP-dependent assembly of focal adhesions and actin stress fibers (Denker *et al.*, 2000). One possible link between the NHE1/ERM association in lamellipodia and the assembly of proteins at focal adhesion sites may be the formation of new actin stress fibers. Indeed, upon RhoA activation, the related bundling of actin filaments into stress fibers promotes the aggregation of integrins and the assembly of focal adhesions (Burridge and Chrzanowska-Wodnicka, 1996). Moreover, NHE1-ezrin interaction has been shown to be a key determinant in the induction of stress fibers in response to activation of RhoA by lysophosphatidic acid (LPA) (Denker *et al.*, 2000). It is likely that RhoA and the RhoA kinase ROKα or p160ROCK (ROCK) regulate the association of NHE1 and ERM proteins. RhoA-dependent signalling pathways activate ERM proteins by phosphorylating a conserved C-terminal threonine residue (Matsui *et al.*, 1998), which has been suggested to result in their unfolding and association with the plasma membrane (Gary and Bretscher, 1995). Additionally, ROCK directly phosphorylates C-terminal threonines of all three ERM proteins *in vitro* (Matsui *et al.*, 1998). RhoA and ROCK also act upstream NHE1 to increase its activity, and ROCK directly phosphorylates NHE1, which could also induce a conformational change in the NHE1 C terminus that unmasks the juxtamembrane ERM binding site (Denker *et al.*, 2000).

A.2.2. Cell Migration

Cell migration plays a basic role in many physiological and pathophysiological processes among which metastasis. It requires: (1) reorganization of the cytoskeleton (Cramer and Mitchison, 1993); (2) activity of ion channels and transporters (Denker and Barber, 2002); (3) membrane recycling by endocytosis and exocytosis (Bretscher and Aguado-Velasco, 1998); (4) a coordinated formation of focal adhesion contacts to ECM mediated by receptors such as integrins (Beningo et al., 2001) or CD44 (Verfaillie *et al.*, 1994).

Several studies have led to the same conclusion, that is, NHE1 is necessary for directed locomotion. Indeed, its pharmacological inhibition by cariporide or amiloride derivatives impairs migration of endothelial cells (Bussolino *et al.*, 1989), fibroblasts (Denker and Barber, 2002), Madin-Darby canine kidney (MDCK) cells (Schwab *et al.*, 2005), human leukocytes (Ritter *et al.*, 1998) and neutrophils (Rosengren *et al.*, 1994). Actually, NHE1 affects cell migration by several mechanisms: (a) linking cell volume and directed locomotion (Ritter *et al.*, 1998); (b) stabilizing the cytoskeleton to the plasma membrane via regulation of intracellular pH; (c) serving as an anchor to attach the cytoskeleton to the plasma membrane (Denker and Barber, 2002); (d) exporting H^+ across the plasma membrane (Denker and Barber, 2002); (e) interacting with the signalling molecules (Putney and Barber, 2004); (f) controlling gene expression. As these mechanisms have been recently reviewed (Stock and Schwab, 2006), only main traits will be summarized here (Figure 2).

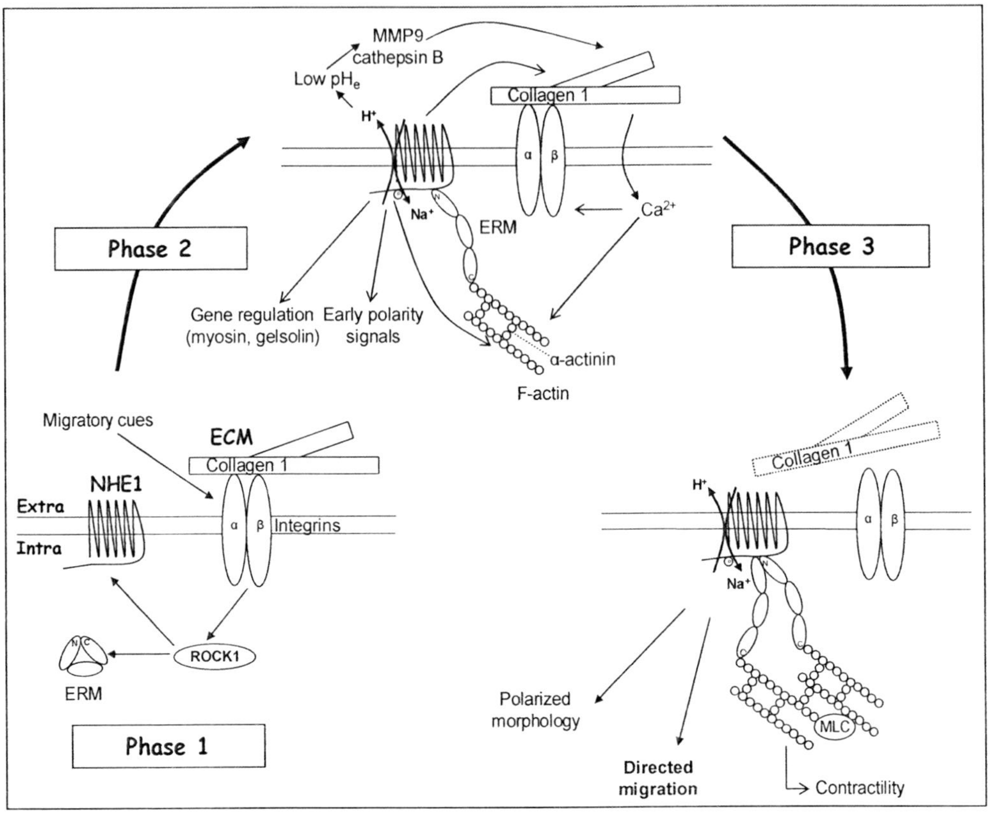

Figure 2. NHE1 activation plays an essential role in migration processes. *Phase 1*, Upon cell exposure to migratory cues, the kinase ROCK1 is activated notably via integrin receptor, leading to phosphorylation of both NHE1 and ERM (ezrin/radixin/moesin). *Phase 2*, NHE1 and ERM activation lead to cortical cytoskeleton anchoring. *Phase 3*, NHE1 activation leads to de-adhesion and Ca^{2+} influx that participates to both de-adhesion and contractility. ECM: extracellular matrix; MMP-9: matrix metalloproteinase 9.

a) NHE1 Regulates Cell Volume during early events of Cell Migration

NHE1-mediated changes of cell volume are of importance for directed movement of *eg.* neutrophil granulocytes. These cells rapidly swell upon stimulation with the chemoattractant N-formylmethionyl-leucyl-phenylalanine (fMLP). This effect can partially be inhibited by NHE1 inhibitors, or by replacement of extracellular Na^+ (Ritter *et al.*, 1998); stimulated migration in neutrophils thus requires a fully active NHE1.

Cell volume is actually increased by the simultaneous action of NHE1, the anion exchanger 2 AE2, and the aquaporin AQP1. As these transporters are located at the front of migrating cells (Klein *et al.*, 2000), they mediate local salt and water uptake at the leading edge of the lamellipodium. The growing lamellipodium and the following gradual cell swelling increase the tension of the plasma membrane, which consequently activates Ca^{2+} permeable mechanosensitive cation channels, an event necessary for triggering the retraction of the rear pole of the cells (Downey *et al.*, 1990), or the release of focal adhesion contacts mediated by integrins (Palecek *et al.*, 1998). After volume loss and retraction of the rear pole, the Ca^{2+} concentration returns to its basic value and cycle can start all over again.

b) NHE1 Regulates Actin Polymerization

Actin filaments play a central role in cell migration. They form a dense meshwork in the lamellipodium. *De novo* actin polymerization at the leading edge of the lamellipodium drives the protrusion of this pole. The retraction of the rear end of the migrating cell is based on the contraction of the cortical actomyosin network underneath the cell membrane and may be regulated by the microtubule network. An increase in pH_i has been speculated to promote *de novo* actin polymerization during cell movement. In mammalian cells, actin polymerization is directly correlated with the generation of free-barbed filament ends by activity of cofilin (Arber *et al.*, 1998), and an increased pH_i is required to stimulate cofilin to the leading edge membrane of migrating cells (Bernstein *et al.*, 2000). In contrast, the activity of gelsolin that nucleates the assembly of filaments is stimulated by low pH (Lagarrigue *et al.*, 2003). All this suggests that NHE1 affects cell migration *via* the regulation of intracellular pH.

c) NHE1 Acts as an Anchor of the Cortical Cytoskeleton

NHE1 assembles complexes of signalling molecules that interact structurally and/or functionally with each other in order to affect cell migration by contributing to dynamic cytoskeletal reorganization. Several NHE1-interacting proteins keep ERM proteins in an activated conformation at the plasma membrane. This then facilitates F-actin binding and maintains the localization of NHE1 at the leading edge of the lamellipodium. The protein kinases ROCK1 and NIK, which target both ERM and NHE1, play an important role in this reorganization. Thus the colocalization of ERM proteins and NHE1 appears to facilitate this reciprocal modulation of cytoskeletal dynamics required for cell migration (Stock and Schwab, 2006).

d) Cell Migration Requires both ion Translocation and Cytoskeletal Anchoring by NHE1

In migrating fibroblasts, a front facing protrusion is transiently stabilized by newly formed focal adhesions, which then dissociate or are displaced reward to facilitate pseudopod retraction and translocation (Lauffenburger and Horwitz, 1996). Using mutant fibroblasts for NHE1, it has been shown that both ion translocation and anchor functions of NHE1 act cooperatively to remodel focal adhesions in migrating cells, with cytoskeletal anchoring being necessary for the assembly of focal complexes and translocation activity promoting de-adhesion (Denker and Barber, 2002).

e) NHE1 Interacts with Signal Molecules at Focal Adhesion

A coordinated formation and release of focal adhesion contacts to the extracellular matrix mediated by integrin receptor molecule is one of the requirements for cell migration (Beningo *et al.*, 2001). NHE1 and adhesion mediating integrin dimers are colocalized at focal adhesion sites. In many cell types, FAK appears to be the initial protein that is phosphorylated in response to adhesion, and launches a signal cascade that promotes cell migration (Sieg *et al.*, 1999). Signals from integrins converge with signals from other cell-surface receptors (Saoncella *et al.*, 1999) such as syndecan-4 or CD44 (Verfaillie *et al.*, 1994), and then affect NHE1 activity *via* the RhoA-ROCK cascade (Tominaga *et al.*, 1998).

f) NHE1 Regulates Gene Expression

As stated above, NHE1 activity is associated with the regulation of diverse genes, notably genes involved in the organization of the cytoskeleton, cell adhesion and ECM assembly (Putney and Barber, 2004). In LAP cells, a complete loss of NHE1 activity led to downregulation of the expression of NF2 (merlin), a tumor suppressor protein and member of the ERM family. Moreover the lack of NHE1 activity was associated with a decreased expression of myosin regulatory light chain, and an increased expression of gelsolin which is the actin-severing and capping protein. A number of microtubule-related genes such as Clip170, KIF4, kinesin light chain and a dynein heavy chain, were also upregulated (Putney and Barber, 2004). Increased expression levels of MMP-9 and lipocalins were also detected; such an observation correlates well with the role NHE1 plays in metastasis, as detailed in the following paragraph.

A.2.3. NHE1 and Metastasis

Tumor invasion and metastasis associated with neoplastic progression are the major causes of deaths by cancer. The invasive process occurs through a complex series of interactions with the host tissue resulting in infiltration and penetration of normal tissue by cancer cell (Chambers, 1999). Metastasis involves alterations in the coordination and regulation of various cellular processes. These include neoangiogenesis, cell adhesion loss (required for cell escape from the primary tumor), degradation and remodelling of the extracellular matrix (ECM) barrier (represented by basal laminae and connective tissue), directed motility, resistance to anoikis while in the circulation, and sustained ectopic growth in the invaded organ (Chambers *et al.*, 2002).

Four biochemical and physiological steps characterize the invasion process: (1) there is a loss of tumor-cell attachment to either basement membranes or ECM; (2) a local degradation of these latter structures, dependent on tumor-cell acid extrusion, occurs; (3) the third step is the secretion of proteases and their activation by acid extrusion from the tumor cell; and finally (4) an increased tumor-cell locomotion into the modified region occurs. The second, third and fourth processes are regulated by both pH_e and pH_i.

a) NHE1 Regulates Protease Secretion

In a regulatory loop, low pH_e resulting from NHE1 activation has been shown to stimulate the release of cathepsin B and to activate MMP-9. Indeed, following the binding of hyaluronan to CD44, NHE1 is activated due to the recruitment of both the transporter and the Rho Kinase into specialized membrane microdomains. NHE1 activation then leads to a localized extracellular acidosis, which would favor activation of low pH-dependent enzymes, including cathepsin B (secreted in the medium), and ultimately ECM degradation and tumor cell invasion. Two lines of evidence are consistent with this loop. First, there is a reduction in tumor cell invasive capacity either by increasing the pH_e, by incubating the cells with a cocktail of inhibitors for various secreted proteases, or by inhibiting NHE1 (Bourguignon *et al.*, 2004). Second, the stable transfection of an ion-translocation defective NHE1 induces a decrease in both the activity of MMP-9 and the expression of the gene that encodes it (Kato *et al.*, 2005; Putney and Barber, 2004). Furthermore, a tight interdependence between pseudopod formation, the process of NHE1 activation and protease secretion is demonstrated by the preferential localization of lysosome production along the pseudopodia and its tip, which is stimulated by low pH_e (Cardone *et al.*, 2005b).

b) NHE1 Regulates Tumor Cell Locomotion

Increased tumor cell invasiveness is the result of two complementary mechanisms: (1) disruption of cell matrix interactions that arise from increased acid secretion and protease activity, and (2) cell motility. Pharmacological inhibition of NHE1 blocks both the formation of this invasive structure and the capacity to invade (Cardone *et al.*, 2005b).

Besides being necessary for the degradation of ECM, low pH_e also induces an increase in the number and length of pseudopodia in metastatic cells. These pseudopodia contain both F-actin and G-actin, suggesting that tumor cell invasion is associated with the reorganization of the actin cytoskeleton and the formation of dynamic actin-rich domains which drive pseudopodial extension (Le *et al.*, 1998). The epithelial transformation and acquisition of motile and invasive capacities are associated with increased expression of NHE1, increased sensitivity to its inhibitor EIPA (Ethyl Iso Propyl Amiloride), and with redistribution of NHE1 to tumor cell pseudopodia (Paradiso et al., 2004). During the formation of pseudopodia, the sorting of NHE1 and regulation of its activity at the pseudopodia's distal tips appears to be due to diverse signalling molecules, *i.e.* PKA (protein kinase A), RhoA, ROCK1 and MAP14 (mitogen-activated protein kinase 14, also known as p38 MAPK) (Cardone *et al.*, 2005a). All these molecules are components of signal transduction systems that are activated in tumor cells (because of the low serum component) to drive NHE1-dependent establishment of the directed cell polarity that is involved in invasion process. This polarity is the result of the leading edge pseudopodia. Interestingly, immunofluorescence analysis of phospho-RhoA and NHE1 have shown that serum deprivation dynamically remodels the cell, forming long, leading edge pseudopodia and that the signal module (PKA/RhoA/ROCK1/MAP14/NHE1) is preferentially compartmentalized in these leading edge pseudopodia, suggesting a tight topographic relation of the signaling module to an invasion-specific cell structure (Cardone *et al.*, 2005a).

B. Role for NHE1 During Apoptotic Processes

Even though the prominent role of NHE1 during proliferation was first identified, recent works underline also its involvement during cell death. Indeed, this membrane protein can influence cell death as a regulator of intracellular pH (pH_i), but also by controlling Na^+_i, cellular volume and assembly of signalling proteins. All these NHE1 functions will be detailed in the apoptotic context.

B.1. Regulation of pH_i

As reported in the Introduction, NHE1 is a major ubiquitous transporter which is essential for maintaining steady-state pH_i in normal cells, by catalyzing H^+ efflux with Na^+ influx. Thus, inhibition or activation of this transporter may lead to an acidification or alkalinization, respectively.

B.1.1. Acidification

Acidification was the first pH_i variation identified as a pro-apoptotic signal and the involvement of NHE1 has been detected both during the two main apoptotic pathways, namely the extrinsic and intrinsic pathways.

a) Extrinsic Pathway

In Jurkat T-lymphocytes, inhibition of NHE1 was triggered during Fas receptor-induced cell death, via the tyrosine kinase $p56^{Lck}$ (Lang *et al.*, 2000). The protein tyrosine phosphatase SHP-1 was also described to inhibit NHE1 following Fas ligation in human breast cancer cells (Thangaraju *et al.*, 1999a). In both cases, inhibitions by $p56^{Lck}$ or SHP1 are not direct, since NHE1 can only be phosphorylated on serine. The hypotheses put forward to explain such a link are the occurrence of ATP depletion or intermediate protein kinase inhibition, respectively. Regarding the latter hypothesis, $p90^{RSK}$ could be a good candidate for SHP1. Indeed, $p90^{RSK}$ is a positive NHE1 regulator and can be activated by phorbol esters like TPA (12-O-tetradecanoylphorbol-13 acetate). In that case, it prevents Fas-induced apoptosis in lymphocytes (Bertolotto *et al.*, 2000). One may speculate that following Fas-receptor aggregation, SHP1 would inactivate $p90^{RSK}$, leading to NHE1 inhibition, cytosolic acidification and apoptosis (Figure 3A).

Other known inducers of apoptosis which interact with membrane receptors also target NHE1. For instance, TGF β 1 decreases NHE1 activity and pH_i, through a mechanism which remains to be determined, and increases rat hepatocyte apoptosis (Benedetti *et al.*, 1995).

Concerning the death receptor pathway, acidification has been described as mediated by the initiator caspase 8 activation upon somatostatin treatment (Liu *et al.*, 2000). However, the idea of a possible caspase-dependent NHE1 inactivation by degradation was not considered in this study, although it might have occurred. Indeed, the cleavage of NHE1 by effector caspases has already been reported (Wu *et al.*, 2003).

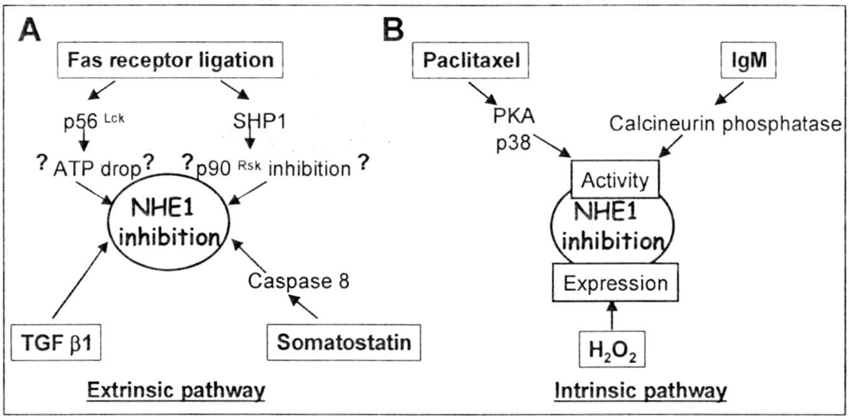

Figure 3. Mechanisms underlying the inhibition of NHE1 during the apoptotic extrinsic *(A)* and intrinsic *(B)* pathways.

b) Intrinsic Pathway

Inhibition of NHE1 during the mitochondrial pathway has been described following exposure to chemicals (Reshkin et al., 2003) or upon treatments with IgM antibodies (Marches *et al.*, 2001). In the former study, activation of PKA and p38 MAPK by paclitaxel inhibited NHE1 and promoted apoptosis, but the mechanism underlying these effects were not determined. In the latter study, the pH_i decrease observed in a human B lymphoma cell line following IgM treatment, was due to a dephosphorylation of NHE1 by calcineurin phosphatase.

The regulation of NHE1 expression has not been much studied in apoptosis. Nevertheless, recent works by Clement and co-workers have demonstrated that exogenous hydrogen peroxide exposure can downregulate NHE1 by acting directly on its promoter; as a result, a sensitization of tumor cells towards anticancer drugs was observed (Akram *et al.*, 2006). However, reactive oxygen species (ROS) treatment can also directly impair NHE1 activity, via the Fenton -reaction dependent ·OH production, thereby orientating cell death to necrosis (Nakamura *et al.*, 2006) (Figure 3B).

In conclusion, inhibition of NHE1 during apoptosis can be triggered by a drop of ATP, known to downregulate the transporter activity (Cassel *et al.*, 1986), or by dephosphorylation of its C-terminal cytosolic domain. However, as NHE1 appears to constitute a potential target of effector caspases 3 and 8, the resulting protein degradation can also partly explain why acidification is often associated with major late apoptotic events (Lagadic-Gossmann *et al.*, 2004), since cell is no longer capable of restoring its basal pH_i.

c) Chemically-induced NHE1 Inhibition

Inhibition of NHE1 in order to induce acidification and apoptosis is one of the strategies proposed to kill cancer cells (Izumi *et al.*, 2003; Marches *et al.*, 2001). Indeed, as reported in Part A -1, NHE1 has been described to be overexpressed and hyperactivated in tumor cells, leading to an alkalinization which favours cell growth and proliferation. That is why the use of amiloride analogs or cariporide, well-known NHE1 inhibitors, has been tested in tumoral cell lines, either alone or in combination with anticancer agents. Thus, the acidification induced by NHE1 inhibitors leads to apoptosis in human breast adenocarcinoma cell lines

(Thangaraju *et al.*, 1999b), human lymphoma cell lines (Marches *et al.*, 2001), human leukemic cells (Rich *et al.*, 2000). Moreover, NHE1 inhibitors can potentialize apoptosis induced by Fas L (Lang *et al.*, 2000), staurosporine (Wu *et al.*, 2003), TRAIL (Cho *et al.*, 2005) or paclitaxel (Reshkin *et al.*, 2003) (Figure 4). However, it has been demonstrated that such NHE1 impairment can also lead to another form of apoptosis, namely paraptosis, in cerebellar granule neurons (Schneider *et al.*, 2004).

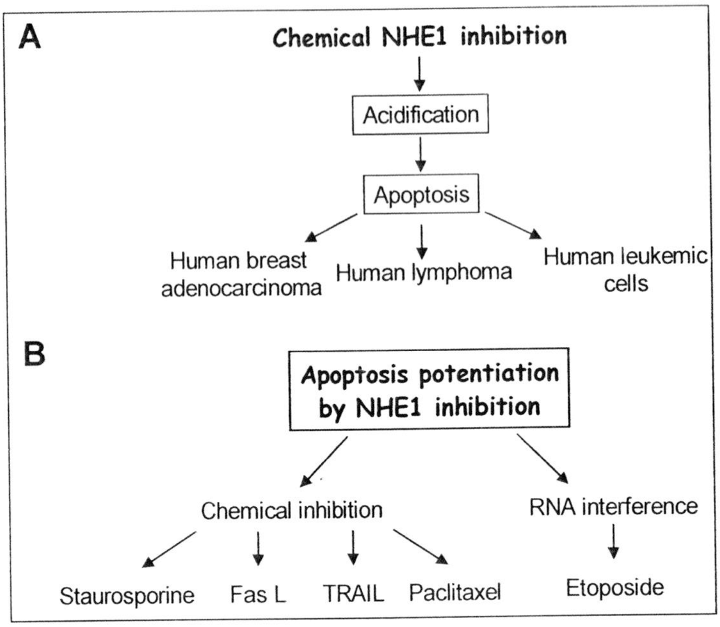

Figure 4. NHE1 as a putative anti-tumoral target. *(A)* Effects of chemical inhibition of NHE1 in tumor cells; *(B)* Effects of NHE1 inhibition as a potentiator of anticancer treatments.

The majority of reported studies have tested NHE1 inhibition by chemical products. In this context, one might argue that these methods may have the disadvantage of being little specific (Chen and O'Brien, 2003; Sparfel *et al.*, 2004). However, it is worth noting that the use of Si RNA in order to silence NHE1 expression has been already employed and has demonstrated that downregulation of NHE1 also sensitizes tumor cells to etoposide-induced apoptosis (Akram *et al.*, 2006).

d) Targets of NHE1 Inhibition

The precise roles of NHE1 inhibition-dependent acidification during apoptosis are not well determined. Experiments demonstrate that pH-clamping to values close to resting pH prevent apoptosis, whereas pre-treatments with NHE1-inhibitors accelerate death processes (Marches *et al.*, 2001), which underlines the importance of such an inhibition for apoptotic transduction. One of the molecular targets of NHE1 inhibition is SHP1: this phosphatase triggers NHE1 inhibition and, in return, consecutive acidification allows the membrane translocation of SHP1, which is necessary for apoptotic induction by somatostatin (Thangaraju *et al.*, 1999b). The consequences of this translocation are not defined, but one can hypothesize that it allows the phosphatase to be closer to its putative substrates (Thangaraju *et al.*, 1999a).

Several studies have analyzed the signalling apoptotic pathways triggered by NHE1 inhibition (Figure 5). DNA degradation during apoptosis is mediated by endonucleases, which are well-known pH-sensitive enzymes (Perez-Sala *et al.*, 1995). Works by Torriglia and co-workers have thus demonstrated the activation of LEI (leukocyte elastase inhibitor)/L-DNase II following NHE1 inhibition-dependent acidification (Altairac *et al.*, 2003). Indeed, LEI is the precursor form of the DNase II and acidification promotes the cleavage of LEI into L-DNase II, independently of caspase activation. Such a cleavage might depend on acidification-elicited cathepsin activation (Huc *et al.*, 2006). Note also that caspase activation has been shown to be sensitive to a decrease of pH (reviewed in (Lagadic-Gossmann *et al.*, 2004)). Therefore, NHE1-inhibition and consequent acidification would promote apoptosis by activating major apoptotic effectors.

Besides these effects, amiloride-induced acidification could modify prosurvival signalling cascade, such as Akt (Cho *et al.*, 2005). In that case, NHE1 inhibition would favour the extinction of growth and survival pathways.

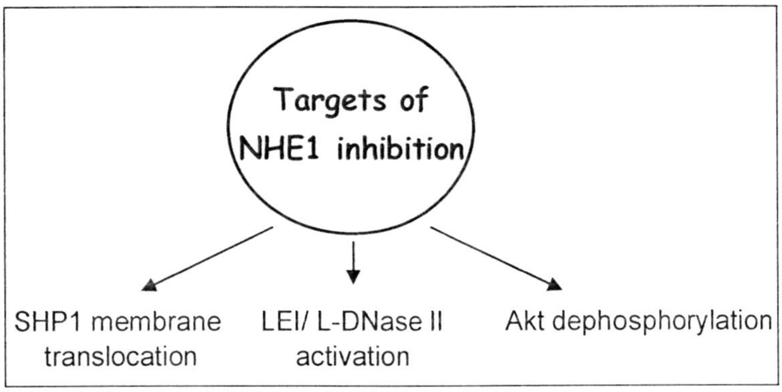

Figure 5. Effects of NHE1 inhibition on apoptotic signalling.

B. 1.2. Alkalinization

The pH_i increase during apoptosis was poorly identified, probably because when it occurs, the phase is early and most often transient (Khaled *et al.*, 1999; Lecureur *et al.*, 2002), followed by an acidification (Fujita *et al.*, 1999; Huc *et al.*, 2004). In this context, it is then possible that some works have missed this phase because of too rapid apoptotic kinetics or of massive cell death due to high concentration of stimulus. The examples of NHE1 activation during apoptosis are rare, but underline the importance of such a transporter in the modulation of cell death transduction.

NHE1-dependent alkalinization was notably identified upon cytokine deprivation (interleukin 7 IL7 and IL3) (Khaled *et al.*, 2001a) or following exposure to xenobiotics, such as Potassium Antimonyl tartrate (PAT) (Lecureur *et al.*, 2002) or Polycyclic Aromatic Hydrocarbons (PAHs) (Huc *et al.*, 2004) (Figure 6).

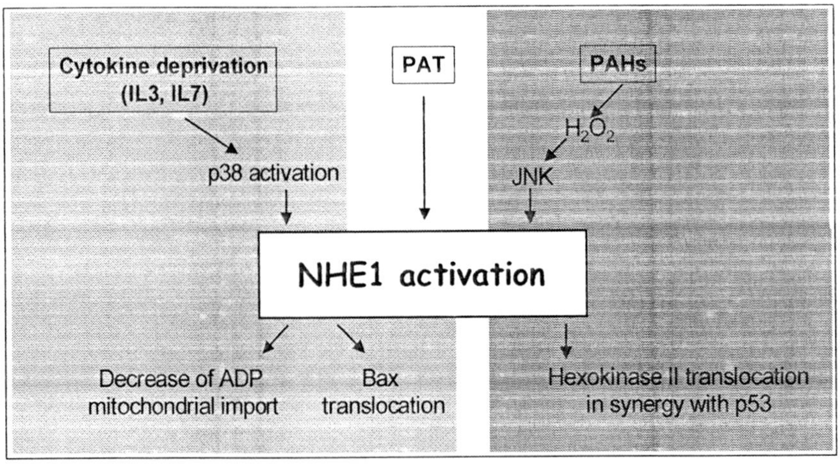

Figure 6. Activation of NHE1 as an early apoptotic event: modalities of regulation and roles. PAT: Potassium antimonyl tartrate; PAHs: polycyclic aromatic hydrocarbons.

After cytokine withdrawal, NHE1 activation in hematopoietic cells is dependent on p38 MAPK phosphorylation. More precisely, alkalinization is associated with an increase of the NHE1 phosphorylation rate, and both events are prevented by chemical inhibition of p38 MAPK or by transfection of dominant negative p38 (Khaled *et al.*, 2001a). Moreover, these authors have demonstrated that NHE1 phosphorylation by p38 MAPK was direct.

In our laboratory, we have focused on the origins of NHE1 activation following PAH exposure and more particularly upon exposure of liver epithelial cells to benzo[*a*]pyrene (BP). This lipophilic xenobiotic is metabolized by cytochromes P450 (CYP), in order to be eliminated. However this metabolism is associated with the production of highly reactive metabolites which bind to DNA (Pelkonen and Nebert, 1982). This genotoxicity was shown to be the trigger of carcinogenic and apoptogenic power of BP (Ethier and Ullrich, 1982; Solhaug *et al.*, 2004). However, during CYP-mediated metabolism, ROS can also be generated; and this is this process that leads to NHE1 activation (Huc *et al.*, 2004). Indeed, early H_2O_2 production activates c-jun kinase JNK, which, in turn, increases NHE1 activity (Huc and Lagadic-Gossmann, unpublished results). The identification of JNK as a modulator of NHE1 is surprising, according to the fact that there is no phosphorylation site for JNK on the cytosolic domain of NHE1. Therefore JNK would not be alone responsible for NHE1 activation and our on-going works are currently dealing with this question.

Up to now, there are only a few data concerning the role of NHE1 activation in apoptotic transduction. However, Khaled and coworkers have identified the Bcl2 protein pro-apoptotic member Bax as a potential target of NHE1-dependent alkalinization (Khaled *et al.*, 1999). Indeed, pH_i increase would favor charge changes in the termini domains of the protein, leading to the exposure of membrane seeking regions of Bax. Consequently, this step would promote Bax translocation from the cytosol to the mitochondria membrane, thereby activating the well-described intrinsic pathway.

pH_i increase may also act on energy metabolism, as suggested by different teams including ours. Firstly, it has been demonstrated that upon IL3 deprivation, mitochondrial ADP import into mitochondria was inhibited by the alkalinization, this event leading to ATP consumption (Khaled *et al.*, 2001b). Secondly, upon PAH treatment, we have associated

NHE1 activation with the late mitochondria ROS production. Indeed, the mitochondrial ROS level is tightly linked to the energy coupling between glycolysis and respiratory chain. The glycolytic enzyme Hexokinase II (HKII) is bound to the external mitochondrial membrane and favours an energy shuttle between glycolytic ADP and mitochondrial ATP. When HKII interaction with mitochondria is disrupted, electron transport along respiratory complex chain is altered and the consecutive electron leak, combined with oxygen, promotes anion superoxide $O_2^{\cdot -}$ production (da-Silva *et al.*, 2004). In BP-induced apoptosis, our recent works have demonstrated that NHE1 activation triggered HKII translocation from the mitochondrial membrane, leading to the fatal ROS production (unpublished results).

Another important question was: does NHE1 activation affect "classical" apoptotic cascade? Under some conditions, the Bax pathway seems to be promoted by the transporter (Khaled *et al.*, 1999). On the contrary, our studies on NHE1 and p53 pathways have underlined the absence of crosstalk, but suggested that they can act in synergy on mitochondrial dysfunction (Huc *et al.*, 2003).

In conclusion, the principal targets of NHE1 activation appear to be related to the mitochondrial function (Bax, HKII, p53), especially at two levels: the external membrane permeability level and the energy level. NHE1 activation during apoptosis has been mainly identified during intrinsic apoptotic pathways, which suggests that this event acts as a permissive parameter for the classical cascades that involve mitochondrial dysfunction.

B.2. Regulation of Na^+_i

Ionic gradients are important for cellular functions. Maintenance of these gradients requires energy; that is why, *eg*. during ischemia, dissipation of ion gradients lead to an entry of Na^+ and water. The consecutive swelling triggers cell rupture, ultimately leading to necrosis. In this context, loss of ion homeostasis has been very often linked to necrosis. However, current works now demonstrate that moderate alterations of ion repartition can induce apoptosis (Figure 7).

NHE1 activation catalyzes both the efflux of H^+ and the influx of Na^+. The low intracellular level of this latter ion is maintained by the pump Na^+/K^+ ATPase. Activated NHE1 in cardiac myocyte is responsible for up to 50% of the basal membrane permeability to Na^+ (Frelin *et al.*, 1984).

Reduction of functional cardiomyocytes by apoptosis and necrosis contributes to ischemia-reperfusion (I/R) injury in the heart. Evidence of apoptosis was detected following 10 min of ischemia and was maximal at 30 min. During I/R sequence in heart (or in brain), NHE1 is activated not only in response to acidosis, but also in response to ischemic metabolites, such as hydrogen peroxide (Sabri *et al.*, 1998), lysophosphatidylcholine (Hoque *et al.*, 1997), or paracrine and autocrine factors which act on NHE1 phosphorylation (Karmazyn, 1996). Recent works by Maekawa and coworkers, using transgenic mice overexpressing the dominant negative of $p90^{RSK}$ and *in vitro* methods, have demonstrated the primordial role of this kinase in the activation of NHE1 during cardiac I/R (Maekawa *et al.*, 2006). Moreover, acute stimulation of NHE1 is not the only factor involved in I/R. Indeed, ischemia as well as cardiotoxic metabolites can up-regulate NHE1 expression, which then also contributes to the upregulation of its activity (Gan *et al.*, 1999).

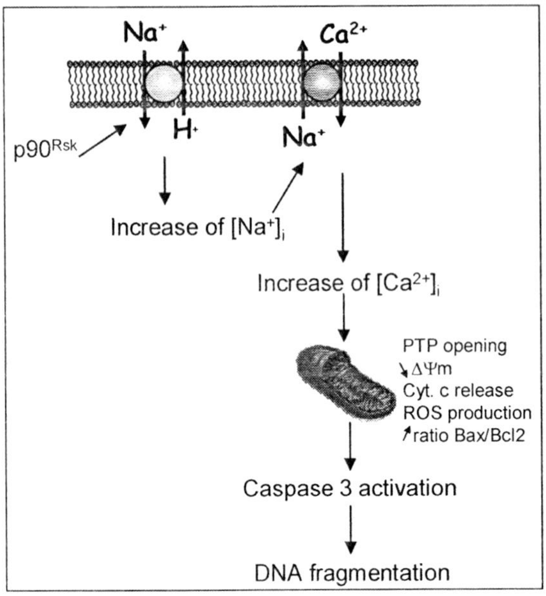

Figure 7. NHE1-mediated Na^+ overloading during apoptosis induced by ischemia/reperfusion. PTP: Permeability Transition Pore; Cyt. c: cytochrome c: $\Delta\psi_m$, miochondrial membrane potential.

NHE1 activation leads to Na^+ overload, which cannot be dissipated by depressed Na^+/K^+ ATPase activity. Consequently, it triggers reverse activation of Na^+/Ca^{2+} exchange. Therefore, NHE1 activation is directly coupled to elevations of $[Ca^{2+}]_i$, thus resulting in cell injury. As a consequence, blocking NHE1 by specific inhibitors allows the reduction of both cytosolic Na^+ and Ca^{2+} accumulations (Levitsky *et al.*, 1998), but also limits ATP depletion, by preventing the ATP consumption necessary for active ion extrusion (Pedersen *et al.*, 2006).

Downstream events of NHE1 activation are a mitochondrial overload of Ca^{2+} which leads to the opening of the mitochondrial permeability transition pore (PTP), associated with the loss of ΔΨm. Release of cytochrome c, ROS production and the increase Bax/Bcl2 ratio have also been detected, followed by the activation of caspase-3, nuclear fragmentation and apoptotic cell death (Jung *et al.*, 2006; Sun *et al.*, 2004) (Figure 7).

Finally, NHE1 inhibition produces cardioprotective effects (Karmazyn, 2001; Sun *et al.*, 2004). Pretreatment with cariporide for 15 min prior to onset of ischemia has been shown to result in a strong reduction in the number of apoptotic cells and an enhanced recovery of contractile function following reperfusion (Chakrabarti *et al.*, 1997). However, the major disadvantage of using chemical inhibitors is that NHE1 prosurvival properties are altered. Therefore, the discovery of $p90^{RSK}$ as a major NHE1 activating protein may offer new therapeutic approach; indeed, inhibition of NHE1 activation by direct targeting of this kinase seems to allow the maintenance of the exchanger basal activity (Maekawa *et al.*, 2006).

B.3. Regulation of Cell Volume

The regulation of volume is necessary for cell homeostasis. All metabolic processes continuously generate osmotically active products, which, if not regulated, would tend to drive water into the cell, causing swelling. Cells possess diverse mechanisms to counteract osmotic perturbations, among them NHE1. Indeed, directed influx of extracellular Na^+ served to compensate a decreased cell volume (for example under hypertonic conditions) through the subsequent osmotically coupled entry of water. This process is called "regulatory volume increase" (RVI). Since Na^+/H^+ exchange is electroneutral, NHE1 does not seem to be osmotically interesting. However, extruded protons are quickly replaced by cellular intrinsic buffers; that is why its activity results in a net gain of Na^+. Moreover, in the case of pH increase, $[HCO_3^-]_i$ rises and allows the activation of anion exchangers (AEs) which catalyze efflux of HCO_3^- coupled to the osmotically active Cl^- influx (Grinstein *et al.*, 1985). Actually, NHE1 is the primary transporter driving RVI.

RVI process is an important mechanism counteracting shrinkage-induced cell death. That is for this reason that cells incapable of RVI have been found to be sensitive to apoptosis (Bortner and Cidlowski, 1996; Friis *et al.*, 2005). Moreover, cell shrinkage is an early event, which suggests its importance in apoptotic signalling. Little is known about NHE1-dependent volume regulatory mechanisms during apoptosis, but several works give us some information. If NHE1 is involved in RVI, its inhibition may facilitate apoptosis (Figure 8). This idea is supported by works by Lang and coworkers; indeed, in Jurkat lymphocytes, stimulation of Fas is followed by a slow inhibition of NHE1, related to the decline of cytosolic ATP (Lang *et al.*, 2000). Thus, this event would favor cell shrinkage (Lang *et al.*, 2004). Moreover, inhibition of NHE1 accelerates cell death, suggesting the importance of NHE1 in the regulation of apoptosis related-cell shrinkage. These results are strengthened by those performed on renal tubular epithelial cells (RTEC): indeed, RTEC, which possess a great capacity to regulate RVI (Lang *et al.*, 1998), exhibit a resistance towards apoptosis following exposure to hypertonic conditions (Bortner and Cidlowski, 1996). Moreover, under particular conditions where hypertonicity can trigger apoptosis in RTEC, NHE1 expression is downregulated by its specific cleavage by caspase 3. Furthermore, when RTEC are exposed to hypertonic media in presence of NHE1 inhibitors, apoptosis is enhanced (Wu *et al.*, 2003). However, these data do not exclude that other NHE1 functions (survival, pH, protein assembling) might also be targeted by its degradation. With respect to that, it is noteworthy that using several NHE1 mutants, it has been demonstrated that Na^+/H^+ exchange activity is essential for resistance towards hypertonic shock-induced apoptosis, and it seems that the ERM-binding domain would be only partially involved (Wu *et al.*, 2003).

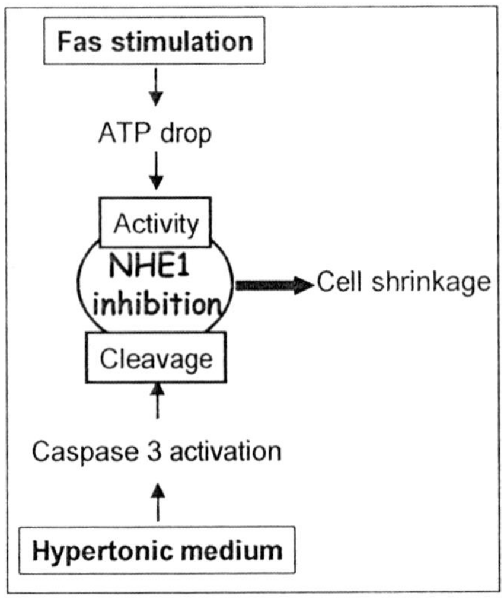

Figure 8. NHE1 inhibition as a cell shrinkage promoter during apoptosis.

B.4. Regulation of Protein Assembling

While cell volume, pH and $[Na^+]_i$ changes are often central events in the physiological effects of NHE1 activation, some effects can also be mediated by physical interactions between NHE1 and other proteins involved in signalling or in cytoskeleton anchoring.

NHE1 interacts directly with actin-binding proteins ERM (ezrin/radixin/moesin), independently of its ion translocation function. Thus NHE1, whose localization is not homogenous in cells (at the leading edge of fibroblasts, into lamellipodiae for example), participates for the maintenance of cell shape (Denker *et al.*, 2000). Thus, NHE1 may have a potential role in anoikis, that is, detachment-induced cell death, but to date no data concerns this topic. However, as reported above, NHE1 can be regulated by integrin stimulation (Tominaga and Barber, 1998) and its role in migration (Patel and Barber, 2005) allows us to think that future studies could be promising and could be useful for the better understanding of the metastasis occurrence (Cardone *et al.*, 2005a and b).

As described previously, degradation of NHE1 by caspases could affect cytoskeleton scaffolding (Wu *et al.*, 2003), and promote ERM membrane dissociation, an event which has been associated with apoptotic induction (Knepper-Nicolai *et al.*, 1998). However, a recent study reveals that NHE1-ERM interaction with PI3K stimulates the Akt/PKB survival pathway during apoptosis (Wu *et al.*, 2004). Indeed, activation and physical association of NHE1 and ezrin rescues cells from apoptosis induced by staurosporine exposure. By using several ezrin mutants, it has been further demonstrated that PI3K interaction with ERM is necessary for such a protection. Finally, Akt is a downstream target of PI3K and promote anti-apoptotic signals, such as Bad phosphorylation.

Therefore, NHE1 appears to act as an important assembling signalling complex. The existing data clearly indicate that this protein represents a major actor in the formation of a pro-survival complex, but one might also suppose that, depending on cell context, NHE1 could also help for the formation of a pro-apoptotic signalling complex.

CONCLUSION

In total, it seems that, besides its well-known physiological role in pH_i regulation, the Na^+/H^+ exchanger represents an important node protein, which integrates changes in cell microenvironment and intracellular signals. Indeed, as described above, NHE1 appears to be modulated by numerous endogenous or exogenous molecules and during diverse pathophysiological processes. This modulation then exerts various intracellular effects that largely seem to be determinant for the cell and hence organ fate, leading, for example, to cell death or proliferation. A better knowledge of how NHE1 regulates the balance between survival/proliferation *versus* apoptosis may help defining new therapeutic strategies for cancer treatment.

REFERENCES

Akram, S., Teong, H. F., Fliegel, L., Pervaiz, S., and Clement, M. V. (2006). Reactive oxygen species-mediated regulation of the Na^+-H^+ exchanger 1 gene expression connects intracellular redox status with cells' sensitivity to death triggers. *Cell Death Differ 13*, 628-641.

Altairac, S., Zeggai, S., Perani, P., Courtois, Y., and Torriglia, A. (2003). Apoptosis induced by Na^+/H^+ antiport inhibition activates the LEI/L-DNase II pathway. *Cell Death Differ 10*, 548-557.

Arber, S., Barbayannis, F. A., Hanser, H., Schneider, C., Stanyon, C. A., Bernard, O., and Caroni, P. (1998). Regulation of actin dynamics through phosphorylation of cofilin by LIM-kinase. *Nature 393*, 805-809.

Avkiran, M., and Marber, M. S. (2002). $Na(^+)/H(^+)$ exchange inhibitors for cardioprotective therapy: progress, problems and prospects. *J Am Coll Cardiol 39*, 747-753.

Baumgartner, M., Patel, H., and Barber, D. L. (2004). $Na(^+)/H(^+)$ exchanger NHE1 as plasma membrane scaffold in the assembly of signaling complexes. *Am J Physiol Cell Physiol 287*, C844-850.

Benedetti, A., Di Sario, A., Casini, A., Ridolfi, F., Bendia, E., Pigini, P., Tonnini, C., D'Ambrosio, L., Feliciangeli, G., Macarri, G., and Svegliati-Baroni, G. (2001). Inhibition of the $Na(^+)/H(^+)$ exchanger reduces rat hepatic stellate cell activity and liver fibrosis: an in vitro and in vivo study. *Gastroenterology 120*, 545-556.

Benedetti, A., Di Sario, A., Svegliati Baroni, G., and Jezequel, A. M. (1995). Transforming growth factor beta 1 increases the number of apoptotic bodies and decreases intracellular pH in isolated periportal and perivenular rat hepatocytes. *Hepatology 22*, 1488-1498.

Beningo, K. A., Dembo, M., Kaverina, I., Small, J. V., and Wang, Y. L. (2001). Nascent focal adhesions are responsible for the generation of strong propulsive forces in migrating fibroblasts. *J Cell Biol 153*, 881-888.

Bernstein, B. W., Painter, W. B., Chen, H., Minamide, L. S., Abe, H., and Bamburg, J. R. (2000). Intracellular pH modulation of ADF/cofilin proteins. *Cell Motil Cytoskeleton 47*, 319-336.

Bertolotto, C., Maulon, L., Filippa, N., Baier, G., and Auberger, P. (2000). Protein kinase C theta and epsilon promote T-cell survival by a rsk-dependent phosphorylation and inactivation of BAD. *J Biol Chem 275*, 37246-37250.

Besson, P., Fernandez-Rachubinski, F., Yang, W., and Fliegel, L. (1998). Regulation of Na^+/H^+ exchanger gene expression: mitogenic stimulation increases NHE1 promoter activity. *Am J Physiol 274*, C831-839.

Bortner, C. D., and Cidlowski, J. A. (1996). Absence of volume regulatory mechanisms contributes to the rapid activation of apoptosis in thymocytes. *Am J Physiol 271*, C950-961.

Bourguignon, L. Y., Singleton, P. A., Diedrich, F., Stern, R., and Gilad, E. (2004). CD44 interaction with Na^+-H^+ exchanger (NHE1) creates acidic microenvironments leading to hyaluronidase-2 and cathepsin B activation and breast tumor cell invasion. *J Biol Chem 279*, 26991-27007.

Brakebusch, C., and Fassler, R. (2003). The integrin-actin connection, an eternal love affair. *Embo J 22*, 2324-2333.

Bretscher, M. S., and Aguado-Velasco, C. (1998). Membrane traffic during cell locomotion. *Curr Opin Cell Biol 10*, 537-541.

Brett, C. L., Donowitz, M., and Rao, R. (2005). Evolutionary origins of eukaryotic sodium/proton exchangers. *Am J Physiol Cell Physiol 288*, C223-239.

Brunet, A., Datta, S. R., and Greenberg, M. E. (2001). Transcription-dependent and -independent control of neuronal survival by the PI3K-Akt signaling pathway. *Curr Opin Neurobiol 11*, 297-305.

Bumke, M. A., Neri, D., and Elia, G. (2003). Modulation of gene expression by extracellular pH variations in human fibroblasts: a transcriptomic and proteomic study. *Proteomics 3*, 675-688.

Burridge, K., and Chrzanowska-Wodnicka, M. (1996). Focal adhesions, contractility, and signaling. *Annu Rev Cell Dev Biol 12*, 463-518.

Bussolino, F., Wang, J. M., Turrini, F., Alessi, D., Ghigo, D., Costamagna, C., Pescarmona, G., Mantovani, A., and Bosia, A. (1989). Stimulation of the Na^+/H^+ exchanger in human endothelial cells activated by granulocyte- and granulocyte-macrophage-colony-stimulating factor. Evidence for a role in proliferation and migration. *J Biol Chem 264*, 18284-18287.

Cantley, L. C. (2002). The phosphoinositide 3-kinase pathway. Science *296*, 1655-1657.

Cardone, R. A., Bagorda, A., Bellizzi, A., Busco, G., Guerra, L., Paradiso, A., Casavola, V., Zaccolo, M., and Reshkin, S. J. (2005a). Protein kinase A gating of a pseudopodial-located RhoA/ROCK/p38/NHE1 signal module regulates invasion in breast cancer cell lines. *Mol Biol Cell 16*, 3117-3127.

Cardone, R. A., Casavola, V., and Reshkin, S. J. (2005b). The role of disturbed pH dynamics and the Na^+/H^+ exchanger in metastasis. *Nat Rev Cancer 5*, 786-795.

Cassel, D., Katz, M., and Rotman, M. (1986). Depletion of cellular ATP inhibits Na^+/H^+ antiport in cultured human cells. Modulation of the regulatory effect of intracellular protons on the antiporter activity. *J Biol Chem 261*, 5460-5466.

Chakrabarti, S., Hoque, A. N., and Karmazyn, M. (1997). A rapid ischemia-induced apoptosis in isolated rat hearts and its attenuation by the sodium-hydrogen exchange inhibitor HOE 642 (cariporide). *J Mol Cell Cardiol 29*, 3169-3174.

Chambers, A. F. (1999). The metastatic process: basic research and clinical implications. *Oncol Res 11*, 161-168.

Chambers, A. F., Groom, A. C., and MacDonald, I. C. (2002). Dissemination and growth of cancer cells in metastatic sites. *Nat Rev Cancer 2*, 563-572.

Chang, F., Lee, J. T., Navolanic, P. M., Steelman, L. S., Shelton, J. G., Blalock, W. L., Franklin, R. A., and McCubrey, J. A. (2003). Involvement of PI3K/Akt pathway in cell cycle progression, apoptosis, and neoplastic transformation: a target for cancer chemotherapy. *Leukemia 17*, 590-603.

Chen, L., Chen, C. X., Gan, X. T., Beier, N., Scholz, W., and Karmazyn, M. (2004). Inhibition and reversal of myocardial infarction-induced hypertrophy and heart failure by NHE-1 inhibition. *Am J Physiol Heart Circ Physiol 286*, H381-387.

Chen, Y. X., and O'Brien, E. R. (2003). Ethyl isopropyl amiloride inhibits smooth muscle cell proliferation and migration by inducing apoptosis and antagonizing urokinase plasminogen activator activity. *Can J Physiol Pharmacol 81*, 730-739.

Cho, Y. L., Lee, K. S., Lee, S. J., Namkoong, S., Kim, Y. M., Lee, H., Ha, K. S., Han, J. A., and Kwon, Y. G. (2005). Amiloride potentiates TRAIL-induced tumor cell apoptosis by intracellular acidification-dependent Akt inactivation. *Biochem Biophys Res Commun 326*, 752-758.

Clark, E. A., and Brugge, J. S. (1995). Integrins and signal transduction pathways: the road taken. *Science 268*, 233-239.

Counillon, L., and Pouyssegur, J. (2000). The expanding family of eucaryotic $Na(^+)/H(^+)$ exchangers. *J Biol Chem 275*, 1-4.

Craig, S. W., and Johnson, R. P. (1996). Assembly of focal adhesions: progress, paradigms, and portents. *Curr Opin Cell Biol 8*, 74-85.

Cramer, L., and Mitchison, T. J. (1993). Moving and stationary actin filaments are involved in spreading of postmitotic PtK2 cells. *J Cell Biol 122*, 833-843.

Dallenbach, A., Marti, U., and Renner, E. L. (1994). Hepatocellular Na^+/H^+ exchange is activated early, transiently and at a posttranscriptional level during rat liver regeneration. *Hepatology 19*, 1290-1301.

da-Silva, W. S., Gomez-Puyou, A., de Gomez-Puyou, M. T., Moreno-Sanchez, R., De Felice, F. G., de Meis, L., Oliveira, M. F., and Galina, A. (2004). Mitochondrial bound hexokinase activity as a preventive antioxidant defense: steady-state ADP formation as a regulatory mechanism of membrane potential and reactive oxygen species generation in mitochondria. *J Biol Chem 279*, 39846-39855.

Denker, S. P., and Barber, D. L. (2002). Cell migration requires both ion translocation and cytoskeletal anchoring by the Na-H exchanger NHE1. *J Cell Biol 159*, 1087-1096.

Denker, S. P., Huang, D. C., Orlowski, J., Furthmayr, H., and Barber, D. L. (2000). Direct binding of the Na-H exchanger NHE1 to ERM proteins regulates the cortical cytoskeleton and cell shape independently of $H(^+)$ translocation. *Mol Cell 6*, 1425-1436.

Di Sario, A., Bendia, E., Taffetani, S., Marzioni, M., Candelaresi, C., Pigini, P., Schindler, U., Kleemann, H. W., Trozzi, L., Macarri, G., and Benedetti, A. (2003). Selective Na^+/H^+

exchange inhibition by cariporide reduces liver fibrosis in the rat. *Hepatology 37*, 256-266.

Downey, G. P., Chan, C. K., Trudel, S., and Grinstein, S. (1990). Actin assembly in electropermeabilized neutrophils: role of intracellular calcium. *J Cell Biol 110*, 1975-1982.

Duggan, S. P., Gallagher, W. M., Fox, E. J., Abdel-Latif, M. M., Reynolds, J. V., and Kelleher, D. (2006). Low pH induces co-ordinate regulation of gene expression in oesophageal cells. *Carcinogenesis 27*, 319-327.

Ethier, S. P., and Ullrich, R. L. (1982). Induction of mammary tumors in virgin female BALB/c mice by single low doses of 7,12-dimethylbenz[a]anthracene. *J Natl Cancer Inst 69*, 1199-1203.

Fausto, N., Campbell, J. S., and Riehle, K. J. (2006). Liver regeneration. *Hepatology 43*, S45-53.

Frelin, C., Vigne, P., and Lazdunski, M. (1984). The role of the Na^+/H^+ exchange system in cardiac cells in relation to the control of the internal Na^+ concentration. A molecular basis for the antagonistic effect of ouabain and amiloride on the heart. *J Biol Chem 259*, 8880-8885.

Friedl, P., and Wolf, K. (2003). Proteolytic and non-proteolytic migration of tumour cells and leucocytes. *Biochem Soc Symp*, 277-285.

Friis, M. B., Friborg, C. R., Schneider, L., Nielsen, M. B., Lambert, I. H., Christensen, S. T., and Hoffmann, E. K. (2005). Cell shrinkage as a signal to apoptosis in NIH 3T3 fibroblasts. *J Physiol 567*, 427-443.

Fujita, H., Ishizaki, Y., Yanagisawa, A., Morita, I., Murota, S. I., and Ishikawa, K. (1999). Possible involvement of a chloride-bicarbonate exchanger in apoptosis of endothelial cells and cardiomyocytes. *Cell Biol Int 23*, 241-249.

Gan, X. T., Chakrabarti, S., and Karmazyn, M. (1999). Modulation of Na+/H+ exchange isoform 1 mRNA expression in isolated rat hearts. *Am J Physiol 277*, H993-998.

Gary, R., and Bretscher, A. (1995). Ezrin self-association involves binding of an N-terminal domain to a normally masked C-terminal domain that includes the F-actin binding site. *Mol Biol Cell 6*, 1061-1075.

Gatenby, R. A., and Gawlinski, E. T. (2003). The glycolytic phenotype in carcinogenesis and tumor invasion: insights through mathematical models. *Cancer Res 63*, 3847-3854.

Gatenby, R. A., and Gillies, R. J. (2004). Why do cancers have high aerobic glycolysis? *Nat Rev Cancer 4*, 891-899.

Gerweck, L. E., Vijayappa, S., and Kozin, S. (2006). Tumor pH controls the in vivo efficacy of weak acid and base chemotherapeutics. *Mol Cancer Ther 5*, 1275-1279.

Gillies, R. J., Raghunand, N., Karczmar, G. S., and Bhujwalla, Z. M. (2002). MRI of the tumor microenvironment. *J Magn Reson Imaging 16*, 430-450.

Grinstein, S., Cohen, S., Goetz, J. D., and Rothstein, A. (1985). Na+/H+ exchange in volume regulation and cytoplasmic pH homeostasis in lymphocytes. *Fed Proc 44*, 2508-2512.

Harguindey, S., Orive, G., Luis Pedraz, J., Paradiso, A., and Reshkin, S. J. (2005). The role of pH dynamics and the Na^+/H^+ antiporter in the etiopathogenesis and treatment of cancer. Two faces of the same coin--one single nature. *Biochim Biophys Acta 1756*, 1-24.

Harguindey, S., Pedraz, J. L., Garcia Canero, R., Perez de Diego, J., and Cragoe, E. J., Jr. (1995). Hydrogen ion-dependent oncogenesis and parallel new avenues to cancer prevention and treatment using a $H(^+)$-mediated unifying approach: pH-related and pH-unrelated mechanisms. *Crit Rev Oncog 6*, 1-33.

Haussinger, D. (2001). Liver fibrosis and $Na(^+)/H(^+)$ exchange. *Gastroenterology 120*, 572-575.

Hoque, A. N., Haist, J. V., and Karmazyn, M. (1997). Na^+-H^+ exchange inhibition protects against mechanical, ultrastructural, and biochemical impairment induced by low concentrations of lysophosphatidylcholine in isolated rat hearts. *Circ Res 80*, 95-102.

Huc, L., Gilot, D., Gardyn, C., Rissel, M., Dimanche-Boitrel, M. T., Guillouzo, A., Fardel, O., and Lagadic-Gossmann, D. (2003). Apoptotic mitochondrial dysfunction induced by benzo(a)pyrene in liver epithelial cells: role of p53 and pH_i changes. *Ann N Y Acad Sci 1010*, 167-170.

Huc, L., Rissel, M., Solhaug, A., Tekpli, X., Gorria, M., Torriglia, A., Holme, J. A., Dimanche-Boitrel, M. T., and Lagadic-Gossmann, D. (2006). Multiple apoptotic pathways induced by p53-dependent acidification in benzo[a]pyrene-exposed hepatic F258 cells. *J Cell Physiol. 208*, 527-537.

Huc, L., Sparfel, L., Rissel, M., Dimanche-Boitrel, M. T., Guillouzo, A., Fardel, O., and Lagadic-Gossmann, D. (2004). Identification of Na^+/H^+ exchange as a new target for toxic polycyclic aromatic hydrocarbons. *Faseb J 18*, 344-346.

Izumi, H., Torigoe, T., Ishiguchi, H., Uramoto, H., Yoshida, Y., Tanabe, M., Ise, T., Murakami, T., Yoshida, T., Nomoto, M., and Kohno, K. (2003). Cellular pH regulators: potentially promising molecular targets for cancer chemotherapy. *Cancer Treat Rev 29*, 541-549.

Jons, T., and Drenckhahn, D. (1992). Identification of the binding interface involved in linkage of cytoskeletal protein 4.1 to the erythrocyte anion exchanger. *Embo J 11*, 2863-2867.

Jung, Y. S., Kim, M. Y., Kim, M. J., Oh, K. S., Yi, K. Y., Lee, S., Yoo, S. E., and Lee, B. H. (2006). Pharmacological profile of KR-33028, a highly selective inhibitor of Na^+/H^+ exchanger. *Eur J Pharmacol 535*, 220-227.

Karmazyn, M. (1996). The sodium-hydrogen exchange system in the heart: its role in ischemic and reperfusion injury and therapeutic implications. *Can J Cardiol 12*, 1074-1082.

Karmazyn, M. (2001). Role of sodium-hydrogen exchange in cardiac hypertrophy and heart failure: a novel and promising therapeutic target. *Basic Res Cardiol 96*, 325-328.

Karmazyn M, Avkiran M, Fliegel L. The sodium-hydrogen exchanger. From Molecule to its role in disease. Dordrecht : Kluwer Academic Publishers; 2003.

Kato, Y., Lambert, C. A., Colige, A. C., Mineur, P., Noel, A., Frankenne, F., Foidart, J. M., Baba, M., Hata, R., Miyazaki, K., and Tsukuda, M. (2005). Acidic extracellular pH induces matrix metalloproteinase-9 expression in mouse metastatic melanoma cells through the phospholipase D-mitogen-activated protein kinase signaling. *J Biol Chem 280*, 10938-10944.

Khaled, A. R., Kim, K., Hofmeister, R., Muegge, K., and Durum, S. K. (1999). Withdrawal of IL-7 induces Bax translocation from cytosol to mitochondria through a rise in intracellular pH. *Proc Natl Acad Sci U S A 96*, 14476-14481.

Khaled, A. R., Moor, A. N., Li, A., Kim, K., Ferris, D. K., Muegge, K., Fisher, R. J., Fliegel, L., and Durum, S. K. (2001a). Trophic factor withdrawal: p38 mitogen-activated protein kinase activates NHE1, which induces intracellular alkalinization. *Mol Cell Biol 21*, 7545-7557.

Khaled, A. R., Reynolds, D. A., Young, H. A., Thompson, C. B., Muegge, K., and Durum, S. K. (2001b). Interleukin-3 withdrawal induces an early increase in mitochondrial membrane potential unrelated to the Bcl-2 family. Roles of intracellular pH, ADP transport, and $F_{(0)}F_{(1)}$-ATPase. *J Biol Chem 276*, 6453-6462.

Klein, M., Seeger, P., Schuricht, B., Alper, S. L., and Schwab, A. (2000). Polarization of $Na(^{+})/H(^{+})$ and $Cl(^{-})/HCO_{(3)}(^{-})$ exchangers in migrating renal epithelial cells. *J Gen Physiol 115*, 599-608.

Knepper-Nicolai, B., Savill, J., and Brown, S. B. (1998). Constitutive apoptosis in human neutrophils requires synergy between calpains and the proteasome downstream of caspases. *J Biol Chem 273*, 30530-30536.

Kozin, S. V., Shkarin, P., and Gerweck, L. E. (2001). The cell transmembrane pH gradient in tumors enhances cytotoxicity of specific weak acid chemotherapeutics. *Cancer Res 61*, 4740-4743.

Lacroix, J., Poet, M., Maehrel, C., and Counillon, L. (2004). A mechanism for the activation of the Na/H exchanger NHE-1 by cytoplasmic acidification and mitogens. *EMBO Rep 5*, 91-96.

Lagadic-Gossmann, D., Huc, L., and Lecureur, V. (2004). Alterations of intracellular pH homeostasis in apoptosis: origins and roles. *Cell Death Differ 11*, 953-961.

Lagarrigue, E., Ternent, D., Maciver, S. K., Fattoum, A., Benyamin, Y., and Roustan, C. (2003). The activation of gelsolin by low pH: the calcium latch is sensitive to calcium but not pH. *Eur J Biochem 270*, 4105-4112.

L'Allemain, G., Paris, S., and Pouyssegur, J. (1984). Growth factor action and intracellular pH regulation in fibroblasts. Evidence for a major role of the Na^{+}/H^{+} antiport. *J Biol Chem 259*, 5809-5815.

Lang, F., Busch, G. L., Ritter, M., Volkl, H., Waldegger, S., Gulbins, E., and Haussinger, D. (1998). Functional significance of cell volume regulatory mechanisms. *Physiol Rev 78*, 247-306.

Lang, F., Gulbins, E., Szabo, I., Lepple-Wienhues, A., Huber, S. M., Duranton, C., Lang, K. S., Lang, P. A., and Wieder, T. (2004). Cell volume and the regulation of apoptotic cell death. *J Mol Recognit 17*, 473-480.

Lang, F., Madlung, J., Bock, J., Lukewille, U., Kaltenbach, S., Lang, K. S., Belka, C., Wagner, C. A., Lang, H. J., Gulbins, E., and Lepple-Wienhues, A. (2000). Inhibition of Jurkat-T-lymphocyte Na^{+}/H^{+}-exchanger by CD95(Fas/Apo-1)-receptor stimulation. *Pflugers Arch 440*, 902-907.

Lauffenburger, D. A., and Horwitz, A. F. (1996). Cell migration: a physically integrated molecular process. *Cell 84*, 359-369.

Le, P. U., Nguyen, T. N., Drolet-Savoie, P., Leclerc, N., and Nabi, I. R. (1998). Increased beta-actin expression in an invasive moloney sarcoma virus-transformed MDCK cell variant concentrates to the tips of multiple pseudopodia. *Cancer Res 58*, 1631-1635.

Lecureur, V., Lagadic-Gossmann, D., and Fardel, O. (2002). Potassium antimonyl tartrate induces reactive oxygen species-related apoptosis in human myeloid leukemic HL60 cells. *Int J Oncol 20*, 1071-1076.

Lee, C. H., Cragoe, E. J., Jr., and Edwards, A. M. (2003). Control of hepatocyte DNA synthesis by intracellular pH and its role in the action of tumor promoters. *J Cell Physiol 195*, 61-69.

Leem, C. H., Lagadic-Gossmann, D., and Vaughan-Jones, R. D. (1999). Characterization of intracellular pH regulation in the guinea-pig ventricular myocyte. *J Physiol 517 (Pt 1)*, 159-180.

Legg, J. W., and Isacke, C. M. (1998). Identification and functional analysis of the ezrin-binding site in the hyaluronan receptor, CD44. *Curr Biol 8*, 705-708.

Levitsky, J., Gurell, D., and Frishman, W. H. (1998). Sodium ion/hydrogen ion exchange inhibition: a new pharmacologic approach to myocardial ischemia and reperfusion injury. *J Clin Pharmacol 38*, 887-897.

Linz, W. J., and Busch, A. E. (2003). NHE-1 inhibition: from protection during acute ischaemia/reperfusion to prevention/reversal of myocardial remodelling. *Naunyn Schmiedebergs Arch Pharmacol 368*, 239-246.

Liu, D., Martino, G., Thangaraju, M., Sharma, M., Halwani, F., Shen, S. H., Patel, Y. C., and Srikant, C. B. (2000). Caspase-8-mediated intracellular acidification precedes mitochondrial dysfunction in somatostatin-induced apoptosis. *J Biol Chem 275*, 9244-9250.

Maekawa, N., Abe, J., Shishido, T., Itoh, S., Ding, B., Sharma, V. K., Sheu, S. S., Blaxall, B. C., and Berk, B. C. (2006). Inhibiting p90 ribosomal S6 kinase prevents (Na+)-H+ exchanger-mediated cardiac ischemia-reperfusion injury. *Circulation 113*, 2516-2523.

Marches, R., Vitetta, E. S., and Uhr, J. W. (2001). A role for intracellular pH in membrane IgM-mediated cell death of human B lymphomas. *Proc Natl Acad Sci U S A 98*, 3434-3439.

Matsui, T., Maeda, M., Doi, Y., Yonemura, S., Amano, M., Kaibuchi, K., and Tsukita, S. (1998). Rho-kinase phosphorylates COOH-terminal threonines of ezrin/radixin/moesin (ERM) proteins and regulates their head-to-tail association. *J Cell Biol 140*, 647-657.

Matsuyama, S., and Reed, J. C. (2000). Mitochondria-dependent apoptosis and cellular pH regulation. *Cell Death Differ 7*, 1155-1165.

Meurette, O., Huc, L., Rebillard, A., G, L. E. M., Lagadic-Gossmann, D., and Dimanche-Boitrel, M. T. (2005). TRAIL (TNF-Related Apoptosis-Inducing Ligand) Induces Necrosis-Like Cell Death in Tumor Cells at Acidic Extracellular pH. *Ann N Y Acad Sci 1056*, 379-387.

Montcourrier, P., Silver, I., Farnoud, R., Bird, I., and Rochefort, H. (1997). Breast cancer cells have a high capacity to acidify extracellular milieu by a dual mechanism. *Clin Exp Metastasis 15*, 382-392.

Mukhin, Y. V., Garnovskaya, M. N., Ullian, M. E., and Raymond, J. R. (2004). ERK is regulated by sodium-proton exchanger in rat aortic vascular smooth muscle cells. *J Biol Chem 279*, 1845-1852.

Nakamura, U., Iwase, M., Uchizono, Y., Sonoki, K., Sasaki, N., Imoto, H., Goto, D., and Iida, M. (2006). Rapid intracellular acidification and cell death by H2O2 and alloxan in pancreatic beta cells. *Free Radic Biol Med 40*, 2047-2055.

Orlowski, J., and Grinstein, S. (2004). Diversity of the mammalian sodium/proton exchanger SLC9 gene family. *Pflugers Arch 447*, 549-565.

Palecek, S. P., Huttenlocher, A., Horwitz, A. F., and Lauffenburger, D. A. (1998). Physical and biochemical regulation of integrin release during rear detachment of migrating cells. *J Cell Sci 111 (Pt 7)*, 929-940.

Paradiso, A., Cardone, R. A., Bellizzi, A., Bagorda, A., Guerra, L., Tommasino, M., Casavola, V., and Reshkin, S. J. (2004). The Na^+-H^+ exchanger-1 induces cytoskeletal changes involving reciprocal RhoA and Rac1 signaling, resulting in motility and invasion in MDA-MB-435 cells. *Breast Cancer Res 6*, R616-628.

Parkins, C. S., Stratford, M. R., Dennis, M. F., Stubbs, M., and Chaplin, D. J. (1997). The relationship between extracellular lactate and tumour pH in a murine tumour model of ischaemia-reperfusion. *Br J Cancer 75*, 319-323.

Patel, H., and Barber, D. L. (2005). A developmentally regulated Na-H exchanger in Dictyostelium discoideum is necessary for cell polarity during chemotaxis. *J Cell Biol 169*, 321-329.

Pedersen, S. F., O'Donnell, M. E., Anderson, S. E., and Cala, P. M. (2006). Physiology and pathophysiology of Na^+/H^+ exchange and Na^+ -K^+ -2Cl^- cotransport in the heart, brain, and blood. *Am J Physiol Regul Integr Comp Physiol 291*, R1-25.

Pelkonen, O., and Nebert, D. W. (1982). Metabolism of polycyclic aromatic hydrocarbons: etiologic role in carcinogenesis. *Pharmacol Rev 34*, 189-222.

Perez-Sala, D., Collado-Escobar, D., and Mollinedo, F. (1995). Intracellular alkalinization suppresses lovastatin-induced apoptosis in HL-60 cells through the inactivation of a pH-dependent endonuclease. *J Biol Chem 270*, 6235-6242.

Pouyssegur, J., Chambard, J. C., Franchi, A., Paris, S., and Van Obberghen-Schilling, E. (1982). Growth factor activation of an amiloride-sensitive Na^+/H^+ exchange system in quiescent fibroblasts: coupling to ribosomal protein S6 phosphorylation. *Proc Natl Acad Sci U S A 79*, 3935-3939.

Pouyssegur, J., Dayan, F., and Mazure, N. M. (2006). Hypoxia signalling in cancer and approaches to enforce tumour regression. *Nature 441*, 437-443.

Pouyssegur, J., Franchi, A., and Pages, G. (2001). pH_i, aerobic glycolysis and vascular endothelial growth factor in tumour growth. *Novartis Found Symp 240*, 186-196; discussion 196-188.

Putney, L. K., and Barber, D. L. (2003). Na-H exchange-dependent increase in intracellular pH times G2/M entry and transition. *J Biol Chem 278*, 44645-44649.

Putney, L. K., and Barber, D. L. (2004). Expression profile of genes regulated by activity of the Na-H exchanger NHE1. *BMC Genomics 5*, 46.

Reichert, M., Steinbach, J. P., Supra, P., and Weller, M. (2002). Modulation of growth and radiochemosensitivity of human malignant glioma cells by acidosis. *Cancer 95*, 1113-1119.

Reshkin, S. J., Bellizzi, A., Albarani, V., Guerra, L., Tommasino, M., Paradiso, A., and Casavola, V. (2000a). Phosphoinositide 3-kinase is involved in the tumor-specific activation of human breast cancer cell $Na(^+)/H(^+)$ exchange, motility, and invasion induced by serum deprivation. *J Biol Chem 275*, 5361-5369.

Reshkin, S. J., Bellizzi, A., Caldeira, S., Albarani, V., Malanchi, I., Poignee, M., Alunni-Fabbroni, M., Casavola, V., and Tommasino, M. (2000b). Na^+/H^+ exchanger-dependent intracellular alkalinization is an early event in malignant transformation and plays an essential role in the development of subsequent transformation-associated phenotypes. *Faseb J 14*, 2185-2197.

Reshkin, S. J., Bellizzi, A., Cardone, R. A., Tommasino, M., Casavola, V., and Paradiso, A. (2003). Paclitaxel induces apoptosis via protein kinase A- and p38 mitogen-activated protein-dependent inhibition of the Na^+/H^+ exchanger (NHE) NHE isoform 1 in human breast cancer cells. *Clin Cancer Res 9*, 2366-2373.

Rich, I. N., Worthington-White, D., Garden, O. A., and Musk, P. (2000). Apoptosis of leukemic cells accompanies reduction in intracellular pH after targeted inhibition of the $Na(^+)/H(^+)$ exchanger. *Blood 95*, 1427-1434.

Ritter, M., Schratzberger, P., Rossmann, H., Woll, E., Seiler, K., Seidler, U., Reinisch, N., Kahler, C. M., Zwierzina, H., Lang, H. J., *et al.* (1998). Effect of inhibitors of Na^+/H^+-exchange and gastric H^+/K^+ ATPase on cell volume, intracellular pH and migration of human polymorphonuclear leucocytes. *Br J Pharmacol 124*, 627-638.

Rosengren, S., Henson, P. M., and Worthen, G. S. (1994). Migration-associated volume changes in neutrophils facilitate the migratory process in vitro. *Am J Physiol 267*, C1623-1632.

Sabri, A., Byron, K. L., Samarel, A. M., Bell, J., and Lucchesi, P. A. (1998). Hydrogen peroxide activates mitogen-activated protein kinases and Na^+-H^+ exchange in neonatal rat cardiac myocytes. *Circ Res 82*, 1053-1062.

Saoncella, S., Echtermeyer, F., Denhez, F., Nowlen, J. K., Mosher, D. F., Robinson, S. D., Hynes, R. O., and Goetinck, P. F. (1999). Syndecan-4 signals cooperatively with integrins in a Rho-dependent manner in the assembly of focal adhesions and actin stress fibers. *Proc Natl Acad Sci U S A 96*, 2805-2810.

Sardet, C., Franchi, A., and Pouyssegur, J. (1989). Molecular cloning, primary structure, and expression of the human growth factor-activatable Na^+/H^+ antiporter. *Cell 56*, 271-280.

Schneider, D., Gerhardt, E., Bock, J., Muller, M. M., Wolburg, H., Lang, F., and Schulz, J. B. (2004). Intracellular acidification by inhibition of the Na^+/H^+-exchanger leads to caspase-independent death of cerebellar granule neurons resembling paraptosis. *Cell Death Differ 11*, 760-770.

Schwab, A., Rossmann, H., Klein, M., Dieterich, P., Gassner, B., Neff, C., Stock, C., and Seidler, U. (2005). Functional role of Na^+-HCO_3^- cotransport in migration of transformed renal epithelial cells. *J Physiol 568*, 445-458.

Shimoda, L. A., Fallon, M., Pisarcik, S., Wang, J., and Semenza, G. L. (2006). HIF-1 regulates hypoxic induction of NHE1 expression and alkalinization of intracellular pH in pulmonary arterial myocytes. *Am J Physiol Lung Cell Mol Physiol. In press*

Shimokawa, N., Kumaki, I., Qiu, C. H., Ohmiya, Y., Takayama, K., and Koibuchi, N. (2005). Extracellular acidification enhances DNA binding activity of MafG-FosB heterodimer. *J Cell Physiol 205*, 77-85.

Sieg, D. J., Hauck, C. R., and Schlaepfer, D. D. (1999). Required role of focal adhesion kinase (FAK) for integrin-stimulated cell migration. *J Cell Sci 112 (Pt 16)*, 2677-2691.

Solhaug, A., Refsnes, M., and Holme, J. A. (2004). Role of cell signalling involved in induction of apoptosis by benzo[a]pyrene and cyclopenta[c,d]pyrene in Hepa1c1c7 cells. *J Cell Biochem 93*, 1143-1154.

Sparfel, L., Huc, L., Le Vee, M., Desille, M., Lagadic-Gossmann, D., and Fardel, O. (2004). Inhibition of carcinogen-bioactivating cytochrome P450 1 isoforms by amiloride derivatives. *Biochem Pharmacol 67*, 1711-1719.

Stock, C., and Schwab, A. (2006). Role of the Na/H exchanger NHE1 in cell migration. *Acta Physiol (Oxf) 187*, 149-157.

Sun, H. Y., Wang, N. P., Halkos, M. E., Kerendi, F., Kin, H., Wang, R. X., Guyton, R. A., and Zhao, Z. Q. (2004). Involvement of Na^+/H^+ exchanger in hypoxia/re-oxygenation-induced neonatal rat cardiomyocyte apoptosis. *Eur J Pharmacol 486*, 121-131.

Suzuki, T., and Tsukamoto, I. (2004). Apoptosis induced by 5-(N,N-hexamethylene)-amiloride in regenerating liver after partial hepatectomy. *Eur J Pharmacol 503*, 1-7.

Svegliati-Baroni, G., Di Sario, A., Casini, A., Ferretti, G., D'Ambrosio, L., Ridolfi, F., Bolognini, L., Salzano, R., Orlandi, F., and Benedetti, A. (1999). The Na^+/H^+ exchanger modulates the fibrogenic effect of oxidative stress in rat hepatic stellate cells. *J Hepatol 30*, 868-875.

Takeuchi, K., Kawashima, A., Nagafuchi, A., and Tsukita, S. (1994). Structural diversity of band 4.1 superfamily members. *J Cell Sci 107 (Pt 7)*, 1921-1928.

Thangaraju, M., Sharma, K., Leber, B., Andrews, D. W., Shen, S. H., and Srikant, C. B. (1999a). Regulation of acidification and apoptosis by SHP-1 and Bcl-2. *J Biol Chem 274*, 29549-29557.

Thangaraju, M., Sharma, K., Liu, D., Shen, S. H., and Srikant, C. B. (1999b). Interdependent regulation of intracellular acidification and SHP-1 in apoptosis. *Cancer Res 59*, 1649-1654.

Tominaga, T., and Barber, D. L. (1998). Na-H exchange acts downstream of RhoA to regulate integrin-induced cell adhesion and spreading. *Mol Biol Cell 9*, 2287-2303.

Tominaga, T., Ishizaki, T., Narumiya, S., and Barber, D. L. (1998). p160ROCK mediates RhoA activation of Na-H exchange. *Embo J 17*, 4712-4722.

Tsukita, S., and Yonemura, S. (1997). ERM proteins: head-to-tail regulation of actin-plasma membrane interaction. *Trends Biochem Sci 22*, 53-58.

Turunen, O., Wahlstrom, T., and Vaheri, A. (1994). Ezrin has a COOH-terminal actin-binding site that is conserved in the ezrin protein family. *J Cell Biol 126*, 1445-1453.

Ulrich-Baker, M. G., Wang, P., Fitzpatrick, L., and Johnson, L. R. (1988). Amiloride inhibits rat mucosal ornithine decarboxylase activity and DNA synthesis. *Am J Physiol 254*, G408-415.

Ungewickell, E., Bennett, P. M., Calvert, R., Ohanian, V., and Gratzer, W. B. (1979). In vitro formation of a complex between cytoskeletal proteins of the human erythrocyte. *Nature 280*, 811-814.

Vairo, G., Cocks, B. G., Cragoe, E. J., Jr., and Hamilton, J. A. (1992). Selective suppression of growth factor-induced cell cycle gene expression by Na^+/H^+ antiport inhibitors. *J Biol Chem 267*, 19043-19046.

Verfaillie, C. M., Benis, A., Iida, J., McGlave, P. B., and McCarthy, J. B. (1994). Adhesion of committed human hematopoietic progenitors to synthetic peptides from the C-terminal heparin-binding domain of fibronectin: cooperation between the integrin alpha 4 beta 1 and the CD44 adhesion receptor. *Blood 84*, 1802-1811.

Wu, K. L., Khan, S., Lakhe-Reddy, S., Jarad, G., Mukherjee, A., Obejero-Paz, C. A., Konieczkowski, M., Sedor, J. R., and Schelling, J. R. (2004). The NHE1 Na^+/H^+ exchanger recruits ezrin/radixin/moesin proteins to regulate Akt-dependent cell survival. *J Biol Chem 279*, 26280-26286.

Wu, K. L., Khan, S., Lakhe-Reddy, S., Wang, L., Jarad, G., Miller, R. T., Konieczkowski, M., Brown, A. M., Sedor, J. R., and Schelling, J. R. (2003). Renal tubular epithelial cell apoptosis is associated with caspase cleavage of the NHE1 Na^+/H^+ exchanger. *Am J Physiol Renal Physiol 284*, F829-839.

Zhou, D., Xue, J., Gavrialov, O., and Haddad, G. G. (2004). Na^+/H^+ exchanger 1 deficiency alters gene expression in mouse brain. *Physiol Genomics 18*, 331-339.

In: Cell Apoptotic Signaling Pathways
Editor: Charles O. Pickens, pp. 115-132

ISBN: 978-1-60021-507-0

Chapter IV

SIGNALING OF APOPTOTIC CELLS TOWARDS ALTERNATIVE MACROPHAGE ACTIVATION

Bernhard Brüne*[*], *Andreas Weigert and Axel M. Johann
Institute of Biochemistry I, Faculty of Medicine, Johann Wolfgang Goethe-University, 60590 Frankfurt, Germany.

ABSTRACT

Cells entering the route of programmed cell death are recognized and removed by professional phagocytes. The uptake of apoptotic cells prior to cell membrane disintegration, averts an inflammatory response in the surrounding tissue by preventing contact of neighboring cells with cellular constituents derived from secondary necrosis. During the last decade it became apparent that prevention/termination of inflammation not only is due to apoptotic cell clearance, but also results from apoptotic cell-derived signals, which provoke a macrophage phenotype shift. Along this phenotype switch, a proinflammatory mediator profile characterized by the formation of e.g. TNFα (tumor necrosis factor-α) or nitric oxide changes to an immunosuppressive one, dominated by the release of e.g. interleukin-10, TGF-β (transforming growth factor-β) or PGE_2 (prostaglandin E_2).

Cell clearance is tightly controlled by a variety of signaling molecules and receptors. Thus, phagocyte attraction is facilitated, among other signals by bioactive lipids secreted from apoptotic cells. At the same time, macrophages primed by contact with apoptotic cells produce antiinflammatory cytokines and upregulate antiapoptotic proteins which guarantee their survival in an otherwise hostile environment. Recognition, binding and subsequent uptake of apoptotic cells by macrophages involves so-called 'eat-me flag' on the surface of apoptotic cells but also the formation of soluble factors by dying cells ('remember me signals'). These alterations initiated during apoptosis orchestrate a process in macrophages known as 'alternative activation' to affect cell viability, cytokine

[*] Correspondence concerning this article should be addressed to Bernhard Brüne, Johann Wolfgang Goethe-University, Biochemistry I, Theodor-Stern-Kai 7, 60590 Frankfurt, Germany. Phone: +49-69-6301 7423; Fax: +49-69-6301 4203; E-mail: bruene@zbc.kgu.de.

production as well as NO and/or O_2^- formation. We discuss the impact of apoptotic cells on macrophage function and focus on apoptosis-derived signals, which determine apoptosis and inflammatory responses.

Keywords: Macrophages, M2 phenotype, phagocytosis, oxygen species, lipid mediators, S1P, nitric oxide, antiinflammatory, alternative activation

Apoptotic Cell Death – More than a Waste Disposal

Apoptosis, or programmed cell death, is a major form of cell demise characterized by a series of stereotypic morphological and biochemical features [1,2] and is important in normal cell development as well as different types of disease [3,4]. The process of apoptosis occurs as an initial commitment phase followed by an execution phase. The latter involves condensation and fragmentation of nuclear chromatin as well as alterations at the cell membrane such as phosphatidylserine exposure which guarantees recognition and phagocytosis of apoptotic cell debris. Briefly, pathways of cell death are categorized by either involving mitochondria or death receptors. In the death receptor pathway, receptors such as tumor necrosis factor-α (TNF-α) receptor-1 or Fas/CD95 are activated by their cognate ligands, which recruits and activates procaspase-8 through interactions between death effector domains found in the receptor as well as procaspase-8. Active caspase-8 by cleaving e.g. Bid generates signals to connect the death receptor with the intrinsic death pathway. In the mitochondrial pathway, cytochrome c is released from the intermembrane space to associate in an ATP or dATP-dependent manner with apoptosis protease activating factor 1 (Apaf-1) to form a multimeric complex, known as the apoptosome that recruits and activates procaspase-9. This is followed by activation of executioner caspases such as caspase-3 or -7. Not surprisingly, apoptosis is highly regulated, among others by the Bcl-2 family of proteins [5] that control the release of mitochondrial components, i.e. cytochrome c. Final regulation of cell death pathways is assured by a delicate balance between pro- versus antiapoptotic signals which will affect the outcome to direct cell survival and cell death. For example, the tumor suppressor p53 is just one important proapoptotic factor [6]. p53 may serve as a sensor of cellular stress and once activated can initiate apoptosis in a transcription-dependent and/or -independent manner [6].

Apoptosis and removal of cell debris occurs in many tissues of our body and is vital for life. The quick and efficient removal of dying cells prevents secondary necrosis and is perhaps the ultimate objective of the apoptotic program. While enormous progress has been achieved in understanding molecular details of the dying program, details on clearing apoptotic material as well as the meaning of this process have lagged behind [7]. It is now appreciated that failure to remove apoptotic debris has serious consequences for inflammation and autoimmunity [8,9]. Interestingly, the phagocytosis of apoptotic cells does not provoke inflammation, which is in contrast to other types of phagocytosis, e.g. Fc receptor mediated which provokes the production of proinflammatory cytokines such as TNFα or interleukin-1 (IL-1). In fact, phagocytosis of apoptotic material is antiinflammatory and stimulates the release of TGF-β, IL-10, or PGE_2 (prostaglandin E_2), which may contribute to immune tolerance. Moreover, an inflammatory response induced by potent proinflammatory stimuli such as lipopolysaccharide (LPS) is dampened by introducing apoptotic cells into inflamed tissues [10].

It was the surgeon John Hunter who in 1794 noticed that inflammation in itself is not considered to be a disease but a salutary operation consequent to some violence or some disease. These considerations emphasized that the usual outcome of acute inflammation, besides showing cardinal systems such as calor, rubor, dolor and tumor, is successful resolution and repair rather than persistent, i.e. chronic inflammation, which can lead to loss of tissue and organ function. Acute inflammation is characterized by emigration of granulocytes, typically neutrophils followed by monocytes that mature into macrophages to provoke proinflammatory reactions that are directed at destroying the pathogen. Macrophages release, among others, reactive oxygen species, nitric oxide, cytokines, arachidonate metabolites, proteases or coagulation factors [11]. Resolution of inflammation can occur if granulocytes are eliminated and the tissue mononuclear cell population returns to physiological numbers and function. During resolution of inflammation, macrophages apparently actively contribute to healing by releasing growth factors, antiinflammatory cytokines, angiogenetic factors, matrix-remodeling activities, and suppress superoxide as well as NO formation [11]. The phenotype switch from a pro- towards an antiinflammatory operating macrophage needs a coordinated 'resolution program' involving a switch to local production of specialized intra- as well as extracellular messengers, programmed leukocyte cell death and subsequent clearance of dying cells by phagocytes that then disappear from the inflamed tissue through lymphatics.

Mononuclear phagocytes are versatile cells, which express different functional programs in response to microenvironmental signals. Extremes of a continuum of functional behaviors are fully polarized M1 and M2 (also termed alternatively activated) macrophages [12]. Polarized macrophages differ in terms of receptor expression, cytokine production or NO as well as O_2^- formation. Classical proinflammatory macrophage stimulation with LPS and/or interferon-γ (IFNγ) produces the M1 phenotype, which is characterized among other signals by releasing NO, O_2^-, TNFα, IL-1β, IL-6 or IFNγ. Stimulation of macrophages with IL-4, IL-10 or IL-13 shifts them towards the M2 phenotype by dampening O_2^- release, blocking NO production via arginase expression, enhancing IL-10 generation or expressing scavenger receptors. At the same time, M2 macrophages produce little TNFα, IL-1β or IL-12. Macrophages are a major cellular constituent of the leukocyte infiltrate of tumors. Tumor associated macrophages (TAM) acquire complex functions in their interactions with neoplastic cells and multiple evidence suggests that they are part of the program which promotes tumor growth and progression [13]. TAMs have a phenotype and function similar to M2 macrophages. They show little cytotoxicity for tumor cells and their cytokine as well as mediator profile helps to promote tumor cell proliferation including the release of angiogenetic factors.

Early studies followed the idea that phagocytosis of apoptotic cells follows the doctrine of being silently and immunologically neutral. However, later on Voll and coworkers recognized that apoptotic lymphocytes co-cultured with LPS-stimulated monocytes inhibited TNFα production whereas antiinflammatory mediator formation of TGF-β and IL-10 was increased [14]. Experiments were reproduced with apoptotic neutrophils or activated platelets with indications that TGF-β, IL-10, PGE_2 as well as omega-3 polyunsaturated fatty acid components such as resolvins or protectins contribute towards antiinflammatory and reparative capacities of macrophages following the ingestion of apoptotic cells [15]. Moreover, extreme proinflammatory macrophage phenotypes such as the capacity to induce apoptosis of tumor or glumerular cells are suppressed by the uptake of apoptotic debris [16].

The removal of apoptotic cells can now be regarded as an integral component of the program to terminate inflammation. Clearance of apoptotic cells initiates an interacting program of chemical mediators by altering membrane topology of apoptotic cells or by releasing a mediator profile, which resembles that of a M2 macrophage. 'Find me signals' generated by apoptotic cells attract macrophages towards dying cells, 'eat me signals' provoke their removal by phagocytosis and 'remember me signals' are released by dying cells which directly affect macrophage polarization.

Here we briefly summarize current knowledge on signals released by apoptotic cells to affect macrophage function. Particularly, we focus on superoxide and nitric oxide formation, the pro- vs. antiinflammatory cytokine profile and the ability of apoptotic cells to interfere with execution of cell death in macrophages.

Eat me and Recognition Signals

During the live span of a multicellular organism cells die to assure tissue homeostasis and are silently removed by professional phagocytes. This 'quiescent meal' is tightly regulated, highly conserved between the different species [14,17] and requires a multiplicity of soluble signals as well as bridging/recognition signals or receptors. Moreover, it allows phagocytes to strictly differentiate between apoptotic and necrotic cell debris [18,19] and thus to respond accordingly with the production of either anti- or proinflammatory signals. Here we will mainly focus on apoptotic cell derived signals.

The release of different mediators by apoptotic cells e.g. IL-10 [9], TGF-β [20] or the bioactive lipid sphingosine-1-phosphate (S1P) [21] (which will be discussed later) has been described. Moreover, it became apparent that cells undergoing apoptosis are also able to release lipid attraction signals such as lysophosphatidylcholine (LPC) via a caspase-3-mediated mechanism which initiates migration of phagocytes into inflamed tissues [22]. While single dying cells may be engulfed by surrounding resident tissue macrophages the chemoattractant LPC may be of importance to handle the removal of large numbers of dying cells.

During apoptosis changes at the plasma membrane serve as 'tether' or 'eat me signals'. CD31 (also known as platelet-endothelial cell adhesion molecule-1, PECAM-1) of leukocytes normally interacts with endothelial CD31 and regulates the motility and active movement across endothelial cell surfaces, directing leukocytes towards intercellular junctions [23]. Brown *et al.* studied human leukocyte binding to macrophage monolayers and found that CD31 on apoptotic cells was modified and did not represent any longer a 'detachment' signal as known for CD31 on viable cells which prevents macrophage ingestion [24]. Rather, modified CD31 mediates homophilic phagocyte 'tethering' of apoptotic cells under conditions of flow, which is important in allowing phagocytes to engage other molecular 'eat me' signals displayed by apoptotic cells. Furthermore, immunoprecipitation showed that CD31-mediated signaling in apoptotic but not viable leucocytes is disabled [24] because CD31 failed to associate with the cytoplasmic signaling molecule SHP-1 and/or SHP-2 [25].

Another important and general feature of apoptotic cell surface alterations is phosphatidylserine (PS) exposure [26-38]. In vital cells PS is actively transported from the outer to the inner leaflet by a P-Type ATPase [29], known as aminophospholipid translocase [30,31]. The resulting asymmetric distribution of PS can be disturbed by activation of scramblase, which catalyzes bidirectional transbilayer movement of all classes of phospholipids across the membrane [32,33]. In apoptotic cells the coordinated inhibition of the translocase but activation of the scramblase provokes PS exposure [28,34] which then serves as a recognition signal for phagocytes. More recent studies have shown that apoptosis induced by various stimuli in a number of different cell types is accompanied by almost simultaneous PS oxidation and externalisation [35]. Oxidized PS (PSox) has been shown to exist in the outer and inner surface of apoptotic cells and may represent a general feature of apoptotic cells [36]. Although described in several publications, recognitions and uptake of dying cells via PS seem not to be apoptosis specific. It was demonstrated that PS externalisation contributes to the uptake of necrotic cells as well [37], suggesting that PS is a more general rather than apoptotic marker. Nevertheless, PS is recognised by a variety of bridging molecules which link apoptotic cells to phagocytes, such as β_2-glycoprotein I [38], milk-fat globule epidermal growth factor 8 [39] or GAS6 [40]. Moreover, the existence of a specific PS receptor (PSR) has been suggested and was linked to the phagocytosis of apoptotic cells [10,41,42]. However, studies with PSR knock out mice demonstrated an immunomodulatory effect of PSR rather than being involved in apoptotic cell phagocytosis [43] and reported on an exclusive nuclear localisation of PSR rather than in the plasma membrane [44]. Recently, Mitchell *et al.* suggested that PSR neither plays a role in proinflammatory transcription nor in the antiinflammatory modulation of the transcriptional response triggered by apoptotic cells, thus making an involvement of the PSR in phagocytosis of apoptotic cells unlikely [45].

Despite the uncertainties on PS and its receptor, other oxidized phospholipids may play a role in apoptotic cell (AC) recognition and/or phagocytosis. Activation of NADPH oxidase during apoptosis provokes phospholipid oxidation not only of PS but also of phosphatidylcholin (PC) and phosphatidylethaolamine (PE) [46]. Interestingly, using the pancaspase inhibitor z-VAD-fmk prevented PS oxidation during apoptosis in differentiated HL-60 cells, whereas modification of PC and PE remained unaffected by z-VAD-fmk [47]. Experiments using a monoclonal antibody directed towards oxidized PC effectively blocked the uptake of AC by macrophages [48], suggesting that oxidized phospholipids others than PS may play a role in the removal of cell corpses as well. Phagocyte receptors such as CD36 [49], SRA [50] or LOX1 [51] may recognize oxidized lipoproteins (oxLDL)-like sites and promote AC recognition. Along that line, carbohydrate structures, which undergo poorly identified alterations at the cell surface of AC [52-56] or other membrane molecules (e.g. ABC1 [57]) or transglutaminase type II [58]) as well as large number of receptors on the macrophage surface (e.g. CD14 [59], CD91-calreticulin [60], the vitronectin receptor [61] or Mer receptor kinase [62]) are also involved in recognition and uptake of apoptotic cells. A schematic outline of essential signaling pathways initiated by apoptotic cells is presented in figure 1. This includes activation of potential receptors by apoptotic cells and/or derived mediators as well as some of the alterations in macrophages that are reminiscent of the M2 polarized macrophage phenotype.

For more information on macrophage receptors being involved in the clearance process of apoptotic cells we refer to detailed reviews [8,63]. Essentially, the likely interplay between different signaling pathways assures a save and tightly regulated corpse removal without proinflammatory consequences.

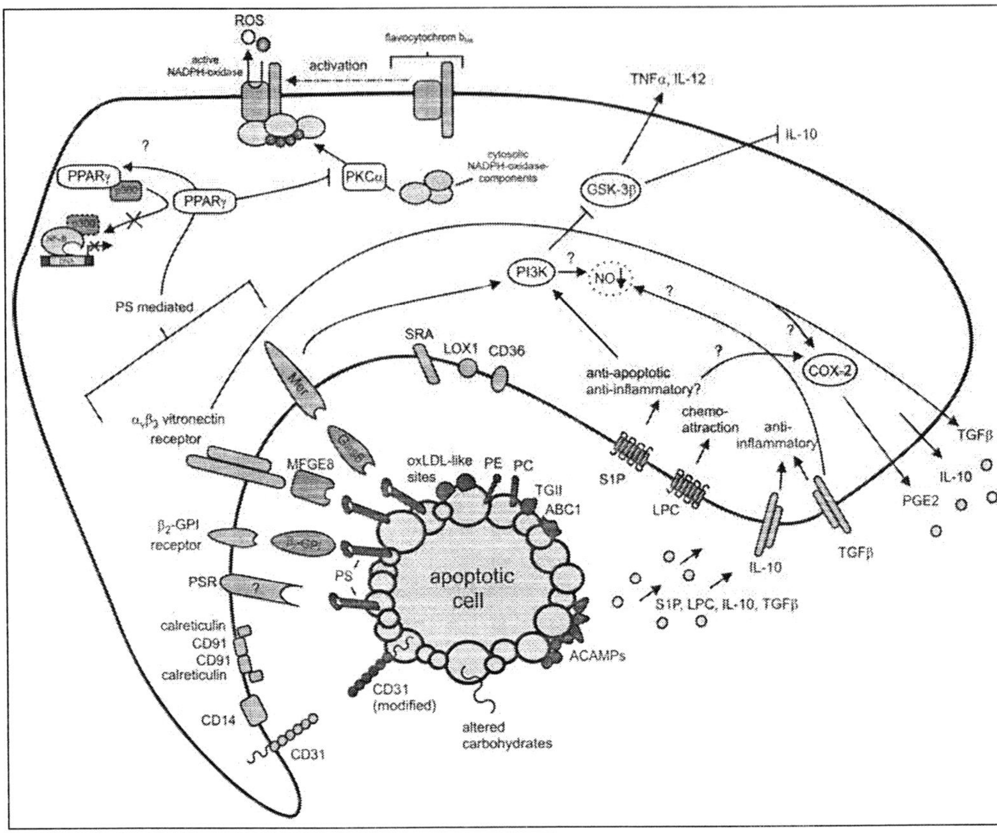

Figure 1. Recognition of apoptotic cells and downstream signaling. A schematic view of apoptotic cell derived signals, bridging molecules, phagocyte receptors and macrophage M2-type characteristic intracellular signaling pathways. Question marks point to pathways and/or receptors being controversially discussed in literature. ABC1, ATP binding-cassette-transporter 1; ACAMPs, apoptotic cell-associated molecular patterns; β2-GPI, β2 glycoprotein I; CD14, lipopolysaccharide receptor CD14; CD31, platelet endothelial cell adhesion molecule-1; CD36, scavenger receptor CD36; CD91, calreticulin/heat shock protein receptor; COX-2, cyclooxygenase 2; Gas-6, growth arrest specific gene-6; GSK-3β, glycogen synthase kinase 3β; IL-10, interleukin 10; IL-12, interleukin 12; Lox-1, oxidised low density lipoprotein receptor 1; LPC, lysophosphatidylcholine; MER, myeloid epithelial reproductive tyrosine kinase; MFG-E8, milk fat globule epidermal growth factor-8; NF-κB, nuclear factor κB; NO, nitric oxide; oxLDL oxidized low densitiy lipoprotein; PE, phosphatidylethanolamine; PGE_2, prostaglandin E_2; PI3K phosphoinositide 3-kinase; PKCα, protein kinase Cα; PPAR-γ, peroxysome proliferator-activated receptor-γ; PS, phosphatidylserine; PSR, PS-receptor; ROS, reactive oxygen species; S1P, sphingosine-1-phosphate; SRA, scavenger receptor A; TNFα, tumor necrosis factor α.

SUPPRESSION OF INFLAMMATION

During the last decade, it became apparent that suppression of inflammation is an intrinsic feature of apoptotic cell death. A first hint indicating a role for apoptosis in the resolution of inflammation came from studies where apoptotic neutrophil granulocytes were cleared by macrophages [64]. Later it became apparent that other inflammatory cell types were cleared from sites of inflammation as well [65,66] and it was generally accepted that the process of cell clearance represents a major step in the prevention and resolution of inflammation [67].

In 1997 Voll et. al. provided evidence that apoptotic cells modulate immune responses aside from simply being removed by phagocytosis [14]. In cocultures of human monocytes and apoptotic cells the authors noticed downregulation of TNFα and IL-1β, as well as upregulation of IL-10 following stimulation with LPS. These effects were mimicked by CD36 activation and supressed by blocking CD36 dependent phagocytosis. One year later, Fadok and colleagues supported the idea that phagocytes may actively contribute to suppressing inflammation [68]. Human macrophages incubated with apoptotic neutrophils or thymocytes displayed an antiinflammatory or suppressive phenotype. This phenotype was characterized by production of TGF-β, PGE_2 or platelet-activating factor (PAF), which prevented LPS-induced upregulation of proinflammatory cytokines such as TNFα, IL-8 or IL-1β Supporting argumentation came from studies in CD14 knockout mice showing phagocytosis-independent effects on inflammation [69]. Knockdown of CD14 impaired phagocytosis of apoptotic cells in vivo, but neither affected TGF-β production nor downregulation of TNFα and although apoptotic cells persisted, no proinflammatory answer occurred. This suggests that termination of inflammation occurred independently of the clearing process.

Although active immunosuppression by the interaction of macrophages with apoptotic cells is now widely accepted, the type of macrophage/apoptotic cell interaction as well as underlying molecular signaling circuits are ill defined.

It has been suggested that recognition of externalized PS contributes to immunosuppression. Masking PS on irradiated tumor cells (ITC) by annexin V prevented cytokine alterations such as upregulation of TGF-β and TNFα as well as downregulation of IL-1β in macrophages [70] and injecting annexin V coated ITC into mice hindered tumor development. Moreover, secretion of TGF-β was directly linked to activation of PSR, as its inhibition attenuated TGF-β production in mouse macrophages by apoptotic cells [10]. Although the importance of the PSR in suppression of inflammation was supported by PSR-knockout studies [42,71], recent evidence suggests that PSR may be dispensable for recognition of apoptotic cells as well as prevention of inflammation [43-45]. Since this topic remains contradictory, other PS receptors such as the vitronectin receptor (VnR) or the Mer tyrosine kinase might be of relevance.

Three independent studies in Mer-deficient mice suggest a role for Mer in the clearance of apoptotic cells [62,72,73]. Although a direct link between Mer-deficiency and an antiinflammatory response elicited by apoptotic cells is missing, one may propose that Mer attenuates Toll-like receptor signaling. Macrophages expressing a kinase-dead Mer produced increased TNFα following LPS administration due to enhanced NF-κB activity [74].

Furthermore, Mer may signal via crosstalk with VnR [75]. In fact, a mutant of milk fat globule-EGF-factor 8 (MGF-E8), which normally links PS to VnR, inhibited IL-10 production by macrophages [76]. Along that line, inhibition of VnR on murine macrophages reduced TGF-β and PGE_2 production in response to apoptotic cells [77]. As PGE_2 production was diminished by pharmacological inhibition of cyclooxygenase-2 (COX-2) [68], one may suggest COX-2 expression by apoptotic cells via VnR. The authors went on and linked VnR-dependent TGF-β production to reduced NO generation in LPS-stimulated macrophages, thus providing evidence that apoptotic cells interfere with macrophage cytotoxicity. Unfortunately, in another study, PS-enriched Jurkat cells did not prevented LPS-induced NO generation, although apoptotic Jurkat cells actively did. Therefore, one must conclude that PS exposure alone was insufficient in down-regulating NO-production [78]. Taking further into consideration that PS-enriched Jurkat cells upregulated TGF-β suggests that factors other than TGF-β inhibited macrophage NO-production.

Another feature of macrophage cytotoxicity resides in the production of reactive oxygen species (ROS). ROS generation was inhibited by PS-enriched Jurkat cells [78] as well as apoptotic cells in LPS-stimulated murine macrophages [79]. In macrophages, apoptotic cells provoked rapid activation of the peroxisome proliferator-activated receptor (PPAR)-γ, which was linked to inhibition of protein kinase C (PKC)- α-mediated NAPDH-oxidase activation [79]. For this process binding but not phagocytosis of apoptotic cells was necessary. Although a potential role of PS in affecting PPAR-γ remains unclear it is intriguing to speculate on the involvement of PPAR-γ in suppressing inflammation [80].

Recently, the group of Ucker showed that LPS-induced NF-κB activation was impaired in macrophages that had recognized apoptotic cells [18]. NF-κB inhibition may be the most rational explanation of a molecular program to shift macrophages towards an antiinflammatory phenotype [81]. Considering that contact of macrophages with apoptotic cells provokes alternative activation rather than deactivation does not allow explaining this phenotype transition with modulation of NF-κB alone. Indeed, antiinflammatory signaling per se was unrelated to NF-κB, thus arguing for additional transcriptional modulator systems under the control of apoptotic cells [82]. Supporting evidence for NF-κB-independent effects was provided in studies with microglial cells, where PS liposomes did not impair LPS-induced Iκ-Bα degradation [83]. As an explanation one may envision depletion of transcriptional coactivators for e.g. NF-κB [18]. Interestingly, this is an established function of PPARs [80], thus putting PPAR-γ into focus once more. Alternatively, the zinc finger GC binding protein (GC-BP) is activated by apoptotic cells [84] and may provide further information to understand immunosuppression by apoptotic cells. Transglutaminase 2 (TG2) may arise as an alternative target because TG2 knockout mice are defective in upregulating TGF-ß, clearing apoptotic cells and reveal compromised autoimmunity [58,85]. TG2 assembles polymeric protein structures during apoptosis [86] and therefore might be needed for efficient linkage of apoptotic cells to macrophages and subsequent phagocytosis.

The direct interaction between macrophage and apoptotic cell appears to be multifaceted and yet another level of complexity is added by findings that apoptotic cells secrete soluble, potentially immunomodulatory factors. Chen et al. observed TGF-β release from apoptotic T-cells, which repressed LPS-mediated TNFα production in murine macrophages [20]. Similar observations have been reported for apoptotic epithelial cells [87]. In addition, evidence is provided that IL-10 is released by apoptotic cells [88].

Besides IL-10 and TGF-β as protein factors being released from apoptotic cells also lipid mediators are produced. Arachidonic acid (AA) and lysophosphatidylcholine (LPC) are generated by the cleavage of phosphatidylcholine (PC) due to the action of a cytosolic calcium-independent phospholipase A_2 (iPLA_2) [22,89]. iPLA_2 is activated during apoptosis via cleavage of ankyrin repeats by caspase-3. Although immune-modulatory properties of arachidonate as well as LPC are poorly defined, at least LPC may provoke proinflammatory responses [90]. Discussing the role of caspase-3 in LPC production it is of interest that immunosuppressive properties of apoptotic cells rather appear caspase-3 independent, since caspase-3 deficient MCF-7 cells effectively acted immunosuppressive [91].

Recently, the release of another lipid mediator, sphingosine-1-phosphate (S1P), from apoptotic cells by the action of sphingosine kinase 2 (SK2) was reported [21]. There is evidence that S1P binds to G-protein coupled receptors and FTY720, an agonist of these receptors, is known to increase the immunosuppressive activity of regulatory T-cells [92]. Furthermore, S1P inhibited the secretion of TNFα and IL-12, but enhanced the release of IL-10 from human dendritic cells [93] and intriguingly, S1P mimics at least in part antiinflammatory signaling properties of TGF-β [94]. In this context, induction of PGE_2 may also be mediated by S1P, since S1P has been correlated to COX-2 induction [95]. It is further established that S1P-receptors target PI3K [96], another feature which might be relevant for immunomodulation. In this respect, knockout of Src homology2-containing inositol-5'-phosphatase (SHIP), a potent negative regulator of PI3K, induced on itself a M2-like phenotype in mice, associated with impaired NO-production and enhanced tumor growth [97]. Moreover, phosphorylation-dependent inhibition of glycogen synthase kinase 3β (GSK3β), a downstream target of PI3K, reduced TNFα and IL-12 but increased IL-10 from monocytic cells [98], thereby adding to the immune regulatory role of PI3K.

It is an intrinsic feature of apoptosis not only to ensure the silent removal of cellular debris, but also to avoid harmful inflammatory responses in the environment. Although an immune regulatory function of apoptotic cells is now generally accepted, underlying molecular pathways remain heterogeneous. Several reports favoured PS exposure to account for immunosuppression. However, taking into consideration that PS is also exposed on necrotic or even viable cells [99] implies that additional signals are needed to reprogram macrophages. Further studies need to precisely define the type of apoptotic cell / macrophage interaction and to characterize the contribution of soluble messengers in affecting the (anti)inflammatory responses elicited during cell demise. There might be no single molecular switch which controls the pro- vs. antiinflammatory phenotype because physiological cell death associated with macrophage polarization represents a continuum of signaling patterns.

Control of Apoptosis by Dying Cells

Apoptotic cells show the ability to reprogram macrophages from 'killers' to 'healers'. The aim of immunoregulation during apoptosis should help to dampen cytotoxicity. To achieve this goal it makes sense to reprogram macrophages towards an antiinflammatory phenotype and to protect these macrophages from entering the route of cell demise.

Reddy et. al. observed protection of murine macrophages from apoptosis induced by cytokine withdrawal which demanded phagocytosis of apoptotic material and activation of PI3K-signaling as suggested by the use of colchicine and LY29002, respectively [100]. Unfortunately, colchicine is rather unspecific and besides blocking phagocytosis it depletes calcium from internal stores [101]. Therefore, this study does not allow to discriminate whether uptake of apoptotic material and/or PI3K activation are distinct or uniform cell responses. However, protection from macrophage apoptosis induced by apoptotic cells was corroborated in another study and in addition was proven for non-professional phagocytic epithelial cells [102]. Specifically, the authors achieved survival effects with supernatants from phagocytes co-cultured with apoptotic cells. This implies paracrine signaling, which is in contrast to above-mentioned phagocytosis. Survival promotion correlated with VEGF secretion from epithelial cells and upregulation of Bcl-2. Although VEGF is known to act para- and autocrine in regulating epithelial cell survival, in part by PI3K/Akt-induced Bcl-2 expression [103], a direct connection between VEGF secretion and Bcl-2 was not investigated [102].

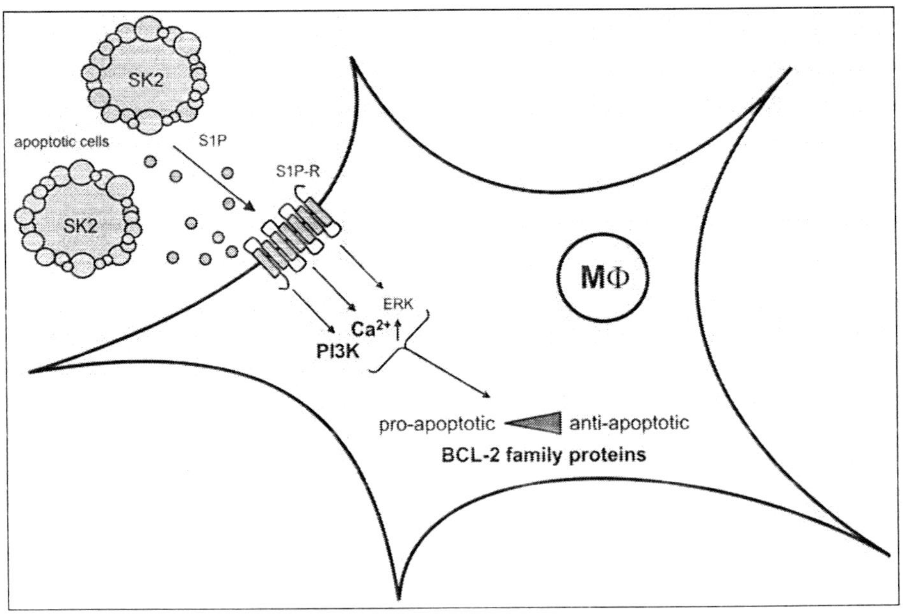

Figure 2. S1P released from apoptotic cells protects macrophages from apoptosis. Induction of apoptosis provokes the release of S1P (sphingosine-1-phosphate) via SK2 (sphingosine kinase 2). S1P binds to its specific G-protein-coupled receptors (S1P-R, S1P-receptors) on macrophages and induces activation of PI3K-, ERK1/2- and Ca^{2+}-dependent signaling pathways. Activation of these pathways alters the expression of Bcl-2 family members from a pro- towards an antiapoptotic profile. PI3K, phosphoinositide 3-kinase; ERK1/2, extracellular signal-regulated kinase 1/2.

More recently it was observed that S1P, being released from apoptotic cells, promotes survival in macrophages. Supernatants from apoptotic cells as well as authentic S1P averted macrophage cell death induced by different proapoptotic stimuli such as etoposide or

TNFα/CHX (cycloheximide). Protection correlated with PI3K activation and calcium-dependent phosphorylation of Bad at Ser136 as well as ERK1/2-dependent upregulation of Bcl-2 or Bcl-X_L [21]. Knockdown of sphingosine kinase-2 (SK2) in apoptotic cells eliminated protection and reestablished cell death in macrophages upon agonist challenge, thus making this isoenzyme responsible for S1P production. Although activation of SK2 in apoptotic cells has been observed previously, but so far only been related to promotion of apoptosis [104], one might speculate that programmed mechanisms exist in apoptotic cells that normally assure a vital outcome. Figure 2 briefly summarizes pathways attributed to S1P signaling in assuring survival of macrophages in an otherwise hostile environment.

Besides those reports, indicating attenuated apoptosis in phagocytes following the uptake of apoptotic material, opposing information exists. Distler et al. isolated microparticles released from apoptotic cells, which induced macrophage cell death as determined by caspase activation or PS exposure [105]. Furthermore, shedding of LPC from dying cells [22] is reported to induce apoptosis [106]. Further investigations will be needed to define conditions when pro- vs. antiapoptotic signals derive from dying cells, which transducing pathways generate signals in the process of cell demise and which signal transmitting steps activate or block cell survival decisions in macrophages that have recognized apoptotic material.

Conclusions

In the past years, we changed our view in looking at dying cells. Rather considering their removal by phagocytosis as a simple waste disposal, we now attribute sophisticated signal initiation towards this process. In close association with the recognition and/or uptake of apoptotic debris macrophages become polarized towards a M2 phenotype, an operationally useful however simplified description of the functional repertoire of alternatively activated macrophages. Functional macrophages responses including gene and cytokine signatures should allow a better understanding of their diverse role in human pathologies and help to depicter their role in inflammation, immunomodulation, wound healing or cancer progression. In the future, it will be challenging to explore the significance of macrophage polarization in clinical conditions and to approach the option whether therapeutic interventions can be built on our new understanding of apoptotic cell recognition/removal as part of a new meaning of cell death.

Acknowledgements

We apologize to those researchers whose immanent contributions to this field have only been cited indirectly. During the last years we grateful thank the German Research Association (DFG), German Cancer League (Deutsche Krebshilfe), Sander Foundation, and European Community for support of individual research projects.

REFERENCES

[1] Ellis, HM and HR Horvitz (1986). Genetic control of programmed cell death in the nematode C. elegans. *Cell*, *44*, 817-829.

[2] Hengartner, MO (2000). The biochemistry of apoptosis. *Nature*, *407*, 770-776.

[3] Fadeel, B, S Orrenius, et al. (1999). Apoptosis in human disease: a new skin for the old ceremony? *Biochem Biophys Res Commun*, *266*, 699-717.

[4] Meier, P, A Finch, et al. (2000). Apoptosis in development. *Nature*, *407*, 796-801.

[5] Huang, Z (2000). Bcl-2 family proteins as targets for anticancer drug design. *Oncogene*, *19*, 6627-6631.

[6] Vousden, KH and X Lu (2002). Live or let die: the cell's response to p53. *Nat Rev Cancer*, *2*, 594-604.

[7] Stuart, LM and RA Ezekowitz (2005). Phagocytosis: elegant complexity. *Immunity*, *22*, 539-550.

[8] Savill, J, I Dransfield, et al. (2002). A blast from the past: clearance of apoptotic cells regulates immune responses. *Nat Rev Immunol*, *2*, 965-975.

[9] Albert, ML (2004). Death-defying immunity: do apoptotic cells influence antigen processing and presentation? *Nat Rev Immunol*, *4*, 223-231.

[10] Fadok, VA, DL Bratton, et al. (2000). A receptor for phosphatidylserine-specific clearance of apoptotic cells. *Nature*, *405*, 85-90.

[11] Duffield, JS (2003). The inflammatory macrophage: a story of Jekyll and Hyde. *Clin Sci (Lond)*, *104*, 27-38.

[12] Mantovani, A, S Sozzani, et al. (2002). Macrophage polarization: tumor-associated macrophages as a paradigm for polarized M2 mononuclear phagocytes. *Trends Immunol*, *23*, 549-555.

[13] Lewis, CE and JW Pollard (2006). Distinct role of macrophages in different tumor microenvironments. *Cancer Res*, *66*, 605-612.

[14] Voll, RE, M Herrmann, et al. (1997). Immunosuppressive effects of apoptotic cells. *Nature*, *390*, 350-351.

[15] Serhan, CN and J Savill (2005). Resolution of inflammation: the beginning programs the end. *Nat Immunol*, *6*, 1191-1197.

[16] Duffield, JS, CF Ware, et al. (2001). Suppression by apoptotic cells defines tumor necrosis factor-mediated induction of glomerular mesangial cell apoptosis by activated macrophages. *Am J Pathol*, *159*, 1397-1404.

[17] McDonald, PP, VA Fadok, et al. (1999). Transcriptional and translational regulation of inflammatory mediator production by endogenous TGF-beta in macrophages that have ingested apoptotic cells. *J Immunol*, *163*, 6164-6172.

[18] Cvetanovic, M and DS Ucker (2004). Innate immune discrimination of apoptotic cells: repression of proinflammatory macrophage transcription is coupled directly to specific recognition. *J Immunol*, *172*, 880-889.

[19] Barker, RN, LP Erwig, et al. (2002). Antigen presentation by macrophages is enhanced by the uptake of necrotic, but not apoptotic, cells. *Clin Exp Immunol*, *127*, 220-225.

[20] Chen, W, ME Frank, et al. (2001). TGF-beta released by apoptotic T cells contributes to an immunosuppressive milieu. *Immunity*, *14*, 715-725.

[21] Weigert, A, AM Johann, et al. (2006). Apoptotic cells promote macrophage survival by releasing the anti-apoptotic mediator sphingosine-1-phosphate. *Blood*, (ahead of print).

[22] Lauber, K, E Bohn, et al. (2003). Apoptotic cells induce migration of phagocytes via caspase-3-mediated release of a lipid attraction signal. *Cell*, *113*, 717-730.

[23] Muller, WA, SA Weigl, et al. (1993). PECAM-1 is required for transendothelial migration of leukocytes. *J Exp Med*, *178*, 449-460.

[24] Brown, S, I Heinisch, et al. (2002). Apoptosis disables CD31-mediated cell detachment from phagocytes promoting binding and engulfment. *Nature*, *418*, 200-203.

[25] Pumphrey, NJ, V Taylor, et al. (1999). Differential association of cytoplasmic signalling molecules SHP-1, SHP-2, SHIP and phospholipase C-gamma1 with PECAM-1/CD31. *FEBS Lett*, *450*, 77-83.

[26] Martin, SJ, CP Reutelingsperger, et al. (1995). Early redistribution of plasma membrane phosphatidylserine is a general feature of apoptosis regardless of the initiating stimulus: inhibition by overexpression of Bcl-2 and Abl. *J Exp Med*, *182*, 1545-1556.

[27] Fadok, VA, DR Voelker, et al. (1992). Exposure of phosphatidylserine on the surface of apoptotic lymphocytes triggers specific recognition and removal by macrophages. *J Immunol*, *148*, 2207-2216.

[28] Verhoven, B, RA Schlegel, et al. (1995). Mechanisms of phosphatidylserine exposure, a phagocyte recognition signal, on apoptotic T lymphocytes. *J Exp Med*, *182*, 1597-1601.

[29] Tang, X, MS Halleck, et al. (1996). A subfamily of P-type ATPases with aminophospholipid transporting activity. *Science*, *272*, 1495-1497.

[30] Seigneuret, M and PF Devaux (1984). ATP-dependent asymmetric distribution of spin-labeled phospholipids in the erythrocyte membrane: relation to shape changes. *Proc Natl Acad Sci U S A*, *81*, 3751-3755.

[31] Zachowski, A, A Herrmann, et al. (1987). Phospholipid outside-inside translocation in lymphocyte plasma membranes is a protein-mediated phenomenon. *Biochim Biophys Acta*, *897*, 197-200.

[32] Verhoven, B, RA Schlegel, et al. (1992). Rapid loss and restoration of lipid asymmetry by different pathways in resealed erythrocyte ghosts. *Biochim Biophys Acta*, *1104*, 15-23.

[33] Williamson, P, A Kulick, et al. (1992). Ca2+ induces transbilayer redistribution of all major phospholipids in human erythrocytes. *Biochemistry*, *31*, 6355-6360.

[34] Bratton, DL, VA Fadok, et al. (1997). Appearance of phosphatidylserine on apoptotic cells requires calcium-mediated nonspecific flip-flop and is enhanced by loss of the aminophospholipid translocase. *J Biol Chem*, *272*, 26159-26165.

[35] Kagan, VE, GG Borisenko, et al. (2003). Appetizing rancidity of apoptotic cells for macrophages: oxidation, externalization, and recognition of phosphatidylserine. *Am J Physiol Lung Cell Mol Physiol*, *285*, L1-17.

[36] Matsura, T, A Togawa, et al. (2005). The presence of oxidized phosphatidylserine on Fas-mediated apoptotic cell surface. *Biochim Biophys Acta*, *1736*, 181-188.

[37] Brouckaert, G, M Kalai, et al. (2004). Phagocytosis of necrotic cells by macrophages is phosphatidylserine dependent and does not induce inflammatory cytokine production. *Mol Biol Cell, 15*, 1089-1100.

[38] Balasubramanian, K, J Chandra, et al. (1997). Immune clearance of phosphatidylserine-expressing cells by phagocytes. The role of beta2-glycoprotein I in macrophage recognition. *J Biol Chem, 272*, 31113-31117.

[39] Hanayama, R, M Tanaka, et al. (2002). Identification of a factor that links apoptotic cells to phagocytes. *Nature, 417*, 182-187.

[40] Wu, Y, S Singh, et al. (2005). A role for Mer tyrosine kinase in alphavbeta5 integrin-mediated phagocytosis of apoptotic cells. *J Cell Sci, 118*, 539-553.

[41] Huynh, ML, VA Fadok, et al. (2002). Phosphatidylserine-dependent ingestion of apoptotic cells promotes TGF-beta1 secretion and the resolution of inflammation. *J Clin Invest, 109*, 41-50.

[42] Li, MO, MR Sarkisian, et al. (2003). Phosphatidylserine receptor is required for clearance of apoptotic cells. *Science, 302*, 1560-1563.

[43] Bose, J, AD Gruber, et al. (2004). The phosphatidylserine receptor has essential functions during embryogenesis but not in apoptotic cell removal. *J Biol, 3*, 15.

[44] Cikala, M, O Alexandrova, et al. (2004). The phosphatidylserine receptor from Hydra is a nuclear protein with potential Fe(II) dependent oxygenase activity. *BMC Cell Biol, 5*, 26.

[45] Mitchell, JE, M Cvetanovic, et al. (2006). The presumptive phosphatidylserine receptor is dispensable for innate anti-inflammatory recognition and clearance of apoptotic cells. *J Biol Chem, 281*, 5718-5725.

[46] Kadl, A, VN Bochkov, et al. (2004). Apoptotic cells as sources for biologically active oxidized phospholipids. *Antioxid Redox Signal, 6*, 311-320.

[47] Arroyo, A, M Modriansky, et al. (2002). NADPH oxidase-dependent oxidation and externalization of phosphatidylserine during apoptosis in Me2SO-differentiated HL-60 cells. Role in phagocytic clearance. *J Biol Chem, 277*, 49965-49975.

[48] Chang, MK, C Bergmark, et al. (1999). Monoclonal antibodies against oxidized low-density lipoprotein bind to apoptotic cells and inhibit their phagocytosis by elicited macrophages: evidence that oxidation-specific epitopes mediate macrophage recognition. *Proc Natl Acad Sci U S A, 96*, 6353-6358.

[49] Savill, J, N Hogg, et al. (1992). Thrombospondin cooperates with CD36 and the vitronectin receptor in macrophage recognition of neutrophils undergoing apoptosis. *J Clin Invest, 90*, 1513-1522.

[50] Platt, N, H Suzuki, et al. (1996). Role for the class A macrophage scavenger receptor in the phagocytosis of apoptotic thymocytes in vitro. *Proc Natl Acad Sci U S A, 93*, 12456-12460.

[51] Murphy, JE, D Tacon, et al. (2006). LOX-1 scavenger receptor mediates calcium-dependent recognition of phosphatidylserine and apoptotic cells. *Biochem J, 393*, 107-115.

[52] Morris, RG, AD Hargreaves, et al. (1984). Hormone-induced cell death. 2. Surface changes in thymocytes undergoing apoptosis. *Am J Pathol, 115*, 426-436.

[53] Duvall, E, AH Wyllie, et al. (1985). Macrophage recognition of cells undergoing programmed cell death (apoptosis). *Immunology, 56*, 351-358.

[54] Hall, SE, JS Savill, et al. (1994). Apoptotic neutrophils are phagocytosed by fibroblasts with participation of the fibroblast vitronectin receptor and involvement of a mannose/fucose-specific lectin. *J Immunol, 153*, 3218-3227.

[55] Dini, L, F Autuori, et al. (1992). The clearance of apoptotic cells in the liver is mediated by the asialoglycoprotein receptor. *FEBS Lett, 296*, 174-178.

[56] Yamanaka, M, S Eda, et al. (2005). Carbohydrate chains and phosphatidylserine successively work as signals for apoptotic cell removal. *Biochem Biophys Res Commun, 328*, 273-280.

[57] Marguet, D, MF Luciani, et al. (1999). Engulfment of apoptotic cells involves the redistribution of membrane phosphatidylserine on phagocyte and prey. *Nat Cell Biol, 1*, 454-456.

[58] Falasca, L, V Iadevaia, et al. (2005). Transglutaminase type II is a key element in the regulation of the anti-inflammatory response elicited by apoptotic cell engulfment. *J Immunol, 174*, 7330-7340.

[59] Devitt, A, OD Moffatt, et al. (1998). Human CD14 mediates recognition and phagocytosis of apoptotic cells. *Nature, 392*, 505-509.

[60] Ogden, CA, A deCathelineau, et al. (2001). C1q and mannose binding lectin engagement of cell surface calreticulin and CD91 initiates macropinocytosis and uptake of apoptotic cells. *J Exp Med, 194*, 781-795.

[61] Savill, J, I Dransfield, et al. (1990). Vitronectin receptor-mediated phagocytosis of cells undergoing apoptosis. *Nature, 343*, 170-173.

[62] Scott, RS, EJ McMahon, et al. (2001). Phagocytosis and clearance of apoptotic cells is mediated by MER. *Nature, 411*, 207-211.

[63] Gregory, CD and A Devitt (2004). The macrophage and the apoptotic cell: an innate immune interaction viewed simplistically? *Immunology, 113*, 1-14.

[64] Savill, JS, AH Wyllie, et al. (1989). Macrophage phagocytosis of aging neutrophils in inflammation. Programmed cell death in the neutrophil leads to its recognition by macrophages. *J Clin Invest, 83*, 865-875.

[65] Surh, CD and J Sprent (1994). T-cell apoptosis detected in situ during positive and negative selection in the thymus. *Nature, 372*, 100-103.

[66] Stern, M, L Meagher, et al. (1992). Apoptosis in human eosinophils. Programmed cell death in the eosinophil leads to phagocytosis by macrophages and is modulated by IL-5. *J Immunol, 148*, 3543-3549.

[67] Cohen, JJ, RC Duke, et al. (1992). Apoptosis and programmed cell death in immunity. *Annu Rev Immunol, 10*, 267-293.

[68] Fadok, VA, DL Bratton, et al. (1998). Macrophages that have ingested apoptotic cells in vitro inhibit proinflammatory cytokine production through autocrine/paracrine mechanisms involving TGF-beta, PGE2, and PAF. *J Clin Invest, 101*, 890-898.

[69] Devitt, A, KG Parker, et al. (2004). Persistence of apoptotic cells without autoimmune disease or inflammation in CD14-/- mice. *J Cell Biol, 167*, 1161-1170.

[70] Bondanza, A, VS Zimmermann, et al. (2004). Inhibition of phosphatidylserine recognition heightens the immunogenicity of irradiated lymphoma cells in vivo. *J Exp Med*, *200*, 1157-1165.

[71] Kunisaki, Y, S Masuko, et al. (2004). Defective fetal liver erythropoiesis and T lymphopoiesis in mice lacking the phosphatidylserine receptor. *Blood*, *103*, 3362-3364.

[72] D'Cruz, PM, D Yasumura, et al. (2000). Mutation of the receptor tyrosine kinase gene Mertk in the retinal dystrophic RCS rat. *Hum Mol Genet*, *9*, 645-651.

[73] Duncan, JL, H Yang, et al. (2003). Inherited retinal dystrophy in Mer knockout mice. *Adv Exp Med Biol*, *533*, 165-172.

[74] Camenisch, TD, BH Koller, et al. (1999). A novel receptor tyrosine kinase, Mer, inhibits TNF-alpha production and lipopolysaccharide-induced endotoxic shock. *J Immunol*, *162*, 3498-3503.

[75] Wu, Y, N Tibrewal, et al. (2006). Phosphatidylserine recognition by phagocytes: a view to a kill. *Trends Cell Biol*, *16*, 189-197.

[76] Asano, K, M Miwa, et al. (2004). Masking of phosphatidylserine inhibits apoptotic cell engulfment and induces autoantibody production in mice. *J Exp Med*, *200*, 459-467.

[77] Freire-de-Lima, CG, DO Nascimento, et al. (2000). Uptake of apoptotic cells drives the growth of a pathogenic trypanosome in macrophages. *Nature*, *403*, 199-203.

[78] Serinkan, BF, F Gambelli, et al. (2005). Apoptotic cells quench reactive oxygen and nitrogen species and modulate TNF-alpha/TGF-beta1 balance in activated macrophages: involvement of phosphatidylserine-dependent and -independent pathways. *Cell Death Differ*, *12*, 1141-1144.

[79] Johann, AM, A von Knethen, et al. (2005). Recognition of apoptotic cells by macrophages activates the peroxisome proliferator-activated receptor-gamma and attenuates the oxidative burst. *Cell Death Differ*, (ahead of print).

[80] Daynes, RA and DC Jones (2002). Emerging roles of PPARs in inflammation and immunity. *Nat Rev Immunol*, *2*, 748-759.

[81] Lawrence, T, M Bebien, et al. (2005). IKKalpha limits macrophage NF-kappaB activation and contributes to the resolution of inflammation. *Nature*, *434*, 1138-1143.

[82] Cvetanovic, M, JE Mitchell, et al. (2006). Specific recognition of apoptotic cells reveals a ubiquitous and unconventional innate immunity. *J Biol Chem*, (ahead of print).

[83] Ajmone-Cat, MA, R De Simone, et al. (2003). Effects of phosphatidylserine on p38 mitogen activated protein kinase, cyclic AMP responding element binding protein and nuclear factor-kappaB activation in resting and activated microglial cells. *J Neurochem*, *84*, 413-416.

[84] Kim, S, KB Elkon, et al. (2004). Transcriptional suppression of interleukin-12 gene expression following phagocytosis of apoptotic cells. *Immunity*, *21*, 643-653.

[85] Szondy, Z, Z Sarang, et al. (2003). Transglutaminase 2-/- mice reveal a phagocytosis-associated crosstalk between macrophages and apoptotic cells. *Proc Natl Acad Sci U S A*, *100*, 7812-7817.

[86] Fesus, L and M Piacentini (2002). Transglutaminase 2: an enigmatic enzyme with diverse functions. *Trends Biochem Sci*, *27*, 534-539.

[87] Hodge, S, G Hodge, et al. (2002). Up-regulation of production of TGF-beta and IL-4 and down-regulation of IL-6 by apoptotic human bronchial epithelial cells. *Immunol Cell Biol*, *80*, 537-543.

[88] Tomimori, Y, Y Ikawa, et al. (2000). Ultraviolet-irradiated apoptotic lymphocytes produce interleukin-10 by themselves. *Immunol Lett*, *71*, 49-54.

[89] Atsumi, G, M Tajima, et al. (1998). Fas-induced arachidonic acid release is mediated by Ca2+-independent phospholipase A2 but not cytosolic phospholipase A2, which undergoes proteolytic inactivation. *J Biol Chem*, *273*, 13870-13877.

[90] Fadok, VA (2003). The sirens' call. *Nat Cell Biol*, *5*, 697-699.

[91] Turner, C, A Devitt, et al. (2003). Macrophage-mediated clearance of cells undergoing caspase-3-independent death. *Cell Death Differ*, *10*, 302-312.

[92] Sawicka, E, G Dubois, et al. (2005). The sphingosine 1-phosphate receptor agonist FTY720 differentially affects the sequestration of CD4+/CD25+ T-regulatory cells and enhances their functional activity. *J Immunol*, *175*, 7973-7980.

[93] Idzko, M, E Panther, et al. (2002). Sphingosine 1-phosphate induces chemotaxis of immature and modulates cytokine-release in mature human dendritic cells for emergence of Th2 immune responses. *Faseb J*, *16*, 625-627.

[94] Xin, C, S Ren, et al. (2004). Sphingosine 1-phosphate cross-activates the Smad signaling cascade and mimics transforming growth factor-beta-induced cell responses. *J Biol Chem*, *279*, 35255-35262.

[95] Pettus, BJ, K Kitatani, et al. (2005). The coordination of prostaglandin E2 production by sphingosine-1-phosphate and ceramide-1-phosphate. *Mol Pharmacol*, *68*, 330-335.

[96] Taha, TA, KM Argraves, et al. (2004). Sphingosine-1-phosphate receptors: receptor specificity versus functional redundancy. *Biochim Biophys Acta*, *1682*, 48-55.

[97] Rauh, MJ, V Ho, et al. (2005). SHIP represses the generation of alternatively activated macrophages. *Immunity*, *23*, 361-374.

[98] Martin, M, K Rehani, et al. (2005). Toll-like receptor-mediated cytokine production is differentially regulated by glycogen synthase kinase 3. *Nat Immunol*, *6*, 777-784.

[99] Appelt, U, A Sheriff, et al. (2005). Viable, apoptotic and necrotic monocytes expose phosphatidylserine: cooperative binding of the ligand Annexin V to dying but not viable cells and implications for PS-dependent clearance. *Cell Death Differ*, *12*, 194-196.

[100] Reddy, SM, KH Hsiao, et al. (2002). Phagocytosis of apoptotic cells by macrophages induces novel signaling events leading to cytokine-independent survival and inhibition of proliferation: activation of Akt and inhibition of extracellular signal-regulated kinases 1 and 2. *J Immunol*, *169*, 702-713.

[101] Oka, T, M Hori, et al. (2005). Microtubule disruption suppresses allergic response through the inhibition of calcium influx in the mast cell degranulation pathway. *J Immunol*, *174*, 4584-4589.

[102] Golpon, HA, VA Fadok, et al. (2004). Life after corpse engulfment: phagocytosis of apoptotic cells leads to VEGF secretion and cell growth. *Faseb J*, *18*, 1716-1718.

[103] Gerber, HP, V Dixit, et al. (1998). Vascular endothelial growth factor induces expression of the antiapoptotic proteins Bcl-2 and A1 in vascular endothelial cells. *J Biol Chem*, *273*, 13313-13316.

[104] Maceyka, M, H Sankala, et al. (2005). SphK1 and SphK2, sphingosine kinase isoenzymes with opposing functions in sphingolipid metabolism. *J Biol Chem*, *280*, 37118-37129.

[105] Distler, JH, LC Huber, et al. (2005). The release of microparticles by apoptotic cells and their effects on macrophages. *Apoptosis*, *10*, 731-741.

[106] Lin, P and RD Ye (2003). The lysophospholipid receptor G2A activates a specific combination of G proteins and promotes apoptosis. *J Biol Chem*, *278*, 14379-14386.

In: Cell Apoptotic Signaling Pathways
Editor: Charles O. Pickens, pp. 133-162

ISBN: 978-1-60021-507-0

Chapter V

CELL DEATH MECHANISMS IN PANCREATITIS

Madhav Bhatia**, *Sharmila Adhikari, Tamizhselvi Ramasamy and Yang Cao
Department of Pharmacology, National University of Singapore, Singapore.

ABSTRACT

Acute pancreatitis has numerous causes, culminating in a common pathway of changes within the pancreatic acinar cell. The pancreatic acinar cell being the functional unit of the exocrine pancreas, it synthesizes, stores, and secretes digestive enzymes. Evidence has accumulated of the significance of apoptotic cell death in the systemic manifestation associated with acute pancreatitis and studies suggest that apoptotic cell death may play a considerable role in affecting mortality and morbidity in severe acute pancreatitis. Apoptosis in pancreatitis is mediated by caspase activation and mitochondrial dysfunction in acinar cells. It is suggested that caspases play an important role in apoptosis. The expression of caspases is a frequent event in chronic pancreatitis and its distribution pattern may reflect the functions of this protease in its role in the apoptotic pathway in atrophic acinar. Also acute pancreatitis modifies mRNA level of both pro and anti-apoptotic genes simultaneously. Numerous studies suggest that pancreatic bcl2, bax and p53 mRNA levels increased extensively in the mild acute pancreatitis and the levels were unchanged during severe acute pancreatitis. This article presents a brief overview of the apoptotic machinery in pancreatitis, including specific and indirect death signals and the underlying molecular events in cell.

* Correspondence concerning this article should be addressed to Madhav Bhatia, Ph.D. Department of Pharmacology, National University of Singapore, Yong Loo Lin School of Medicine, Bldg MD2, 18 Medical Drive, SINGAPORE 117597. Tel. (65)-6516-8256; Fax. (65)-6873-7690; email. mbhatia@nus.edu.sg.

INTRODUCTION

Acute pancreatitis is a severe disease with a significant morbidity and mortality [1]. The overall mortality rate of acute pancreatitis is around 10%. However, among severe cases in which the functions of multiple organs are compromised, the mortality rate can reach 20-30% [2]. Clinically, the severity of acute pancreatitis varies considerably. Edematous pancreatitis is the mildest form of the disease, characterized by interstitial edema and in this form the acinar cell structure remains intact. On the contrary, severe acute pancreatitis is characterized by massive destruction of pancreatic parenchyma and infiltration of leukocytes [2]. It has been hypothesized that milder forms of experimental pancreatitis are associated with more apoptosis and the relatively severe forms, with more necrosis [3,4]. Most of the reports support the idea that pancreatic acinar cell apoptosis acts as a self-defense mechanism against the development of pancreatic necrosis, extensively reviewed elsewhere [5,6].

In pancreatitis, acinar cell death occurs by both apoptosis and necrosis [7-10]. Apoptosis plays a key role in a variety of biological events, including cell differentiation, organogenesis and various disease conditions [11,12]. "Apoptosis" is one of the most frequently used words in biology and medicine today. It was introduced by Kerr, and colleagues in 1972 to describe a form of cell death distinct from necrosis [13]. Apoptosis was initially recognized as "programmed cell death," a type of cell death that serves to eliminate excessive or unwanted cells, in the course of organ development. Apoptotic cells are phagocytosed by neighbouring cells, while necrotic cells cause local and general inflammatory reactions. Several reports demonstrate that a chemical at a low concentration may cause apoptosis, while at a higher (toxic) concentration it may cause necrosis [14-18]. Since apoptotic processes require energy, depletion of intracellular ATP results in the switch from apoptosis to necrosis [19-21]. It is well established that reactive oxygen species (ROS) are intimately related to apoptosis and a burst in the intracellular level of ROS triggers the switch from apoptosis to necrosis [16,22,23]. Necrosis has been considered a passive degenerative phenomenon induced by direct toxic or physical injuries, which most often occurs accidentally. Nuclear DNA is randomly cleaved as a consequence of cellular degeneration. Leakage of the cytoplasm through plasma membrane disruption induces cellular inflammatory responses [24,25].

There are two major signaling pathways of apoptosis, the death receptor-mediated extrinsic pathway and the mitochondrial intrinsic pathway. In the death receptor signaling pathway, signal is provided by the interaction between the ligand and death receptor. Initiated by caspase 8 or 10 followed by the activation of effector caspases such as caspase 3 [26-28]. In the mitochondrial pathway, permeabilization of the mitochondrial outer membrane, results in the release of apoptotic factors such as cytochrome c followed by the activation of caspase cascade. The regulatory role of Bcl-2 family proteins in mitochondrial pathway of apoptosis is well-established [29,30]. Although the two pathways appear to be distinct, death receptor signaling may mediate apoptosis via mitochondrial pathway via tBid [31]. This review outlines the mechanisms of apoptosis and necrosis and its involvement in pancreatitis.

Mechanism of Cell Death

Two Alternative Forms of Cell Death

The concept of apoptosis or programmed cell death (PCD), with unique and dynamic morphological and biochemical features is distinguishable from necrosis (Table 1). Apoptosis is a type of cell death that does not involve an inflammatory response and occurs in a tightly controlled manner. In contrast to necrosis, apoptosis involves activation of catabolic and enzymes prior to cytolysis. Many cellular changes that take place during this form of death have an important role in recognition and elimination of those cells by neighboring phagocyte cells [32,33].

Table 1. Morphological and biochemical features

Apoptosis	Necrosis
• Shrinkage of the cell and its organelles and condensation of chromatin	• Cell swelling and rupture of the plasma membrane
• No leakage of cellular contents	• Leakage of cellular constituents
• Apoptotic bodies engulfed by phagocytes	• Cellular and nuclear lysis
• Activation of specific cysteine proteases, the caspases, and internucleosomal DNA fragmentation	• Absent in necrotic cells
• Regulated signaling mechanisms	• Poorly regulated
• No inflammatory response	• Acute inflammatory response

The Phases of Death

The death programme can be divided into four different phases. The `initiation' phase, during which cells receive signals which result in the activation of the death programme. The `commitment' phase, the point after which death signals become irreversible. The `amplification' phase, where multiple caspases are recruited to cooperate in the destruction of the cell, and finally, the `demolition' phase, where a panoply of active caspases dismantle cellular structures, either directly, or via activation of other enzymes such as CAD/DFF45 [34].

Caspase Activation Mechanisms

Caspases

A group of cysteine proteases - caspases [35,36], are essential for programmed cell death in a variety of species [37]. Cysteine protease activity can be detected in all cells undergoing apoptosis, regardless of their origin or the death stimulus. Caspases, the interleukin-1β-converting enzyme family proteases, are highly homologous to *Caenorhabditis elegans* cell death gene *CED-3*. Fourteen caspases have been identified so far, all of which share some common properties: they are all aspartate-specific cysteine proteases; they all have a conservative pentapeptide active site 'QACXG' (X can be R, Q or D); their precursors are all zymogens known as procaspases. The N-terminal of the prodomain in procaspases contains a highly diverse structure required for caspase activation; and they are all capable of autoactivating as well as activating other caspases, to produce a heterodimer with a big and a small subunit, and two heterodimers form an enzymatic active heterotetramer [38]. Based on their homology in amino acid sequences, caspases are divided into three subfamilies:

Subfamily I: Apoptosis activator/initiator- Caspase-2, 8, 9 and 10
Subfamily II: Apoptosis executioner/effector- Caspase-3, 6 and 7
Subfamily III: Inflammatory mediator- Caspase- 1, 4, 5, 11, 12, 13, and 14

Generally, there are two pathways through which the caspase family proteases can be activated: one is the death receptor-mediated pathway; the other is the mitochondrion-mediated pathway.

Death Receptor-Mediated Apoptotic Pathway

Cell death signals, such as Fas ligand (FasL) and tumor necrosis factor (TNF)-2, can be specifically recognized by their corresponding death receptors, such as Fas or TNF receptor (TNFR)-1, in the plasma membrane. Their binding will in turn activate the death receptors. Fas can bind to the Fas-associated death domain (FADD) (or TNFR-associated death domain, TRADD) and cause FADD aggregation and the emergence of death effector domains (DEDs) [39]. The death receptors (CD95/Fas/APO-1, TNFR1, DR3/WSL-1/ TRAMP, DR4/TRAIL-R1, DR5/TRAIL-R2, DR6) are a subset of the TNF/NGF receptor family of cell surface molecules that possess a common motif within their cytoplasmic tails, called the death domain [28,40]. The death domains of these receptors are responsible for recruiting adaptor molecules that, in turn, recruit caspases to the receptor complex. This simple strategy greatly increases the probability of caspase activation by facilitating cross processing of caspases within the cluster. This proximity-induced processing is thought to drive apical caspase activation events that are initiated by all of the death receptors. Thus, death receptors can drive caspase activation in a very direct manner [41,42]. The exposed DEDs interact with the DEDs in the prodomain of procaspase-8, which will induce the oligomerization of procaspase-8 localized on the cytosolic side of the plasma membrane. Then a massive molecule complex known as the death-inducing signal complex (DISC) is formed. In DISC, two linear subunits of procaspase-8 compact to each other followed by procaspase-8 autoactivation to caspase-8 [43]. The activation of the downstream pathways of caspase-8 varies with different cell types. In Type I cells (cells of some lymphoid cell lines), caspase-8 is vigorously activated and can directly activate the downstream procaspases (e.g. procaspase-3). In Type II cells (other than Type I cells), caspase-8 is only mildly activated and unable to activate procaspase-3 directly. However, it can activate the mitochondrion-mediated pathway by truncating Bid (a pro-apoptotic Bcl-2 family member), a kind of proapoptotic protein in the cytosol, into its active form, tBid. tBid will trigger the activation of the mitochondrion pathway: cytochrome c, apoptosis-inducing factor (AIF) and other molecules are released from mitochondria, leading to the initiation of apoptosis [44] (Figure

1). Caspase-10 can function independently of caspase-8 in initiating Fas- and TNF-related apoptosis in lymphoid cells [45].

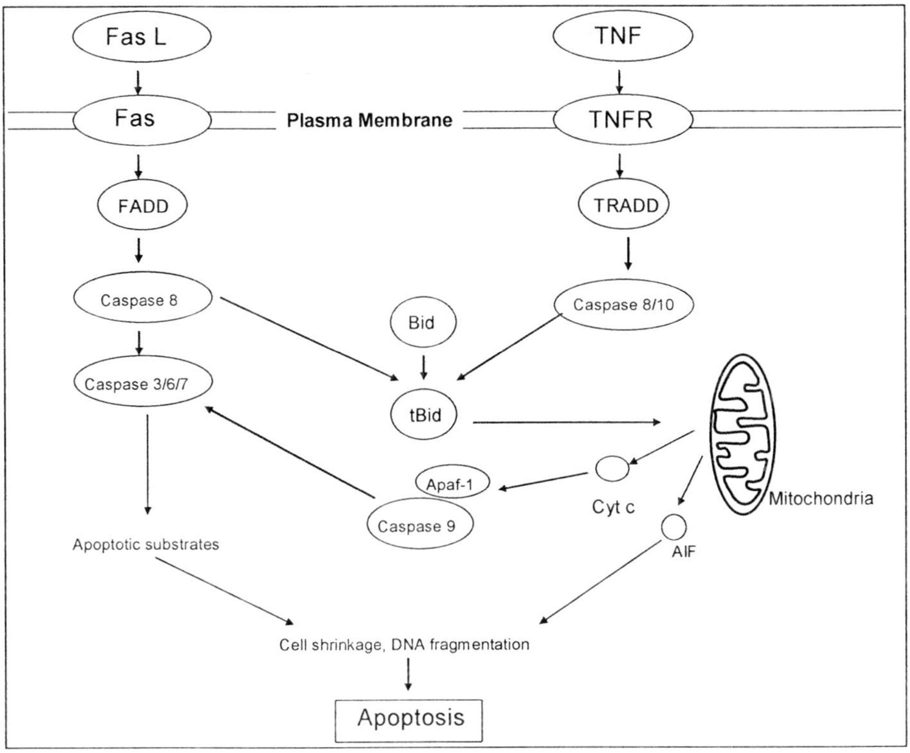

Figure 1. Death receptor-dependent procaspase-activation pathway. AIF, apoptosis-inducing factor; Apaf-1, apoptotic protease activation factor-1; Cyto c, cytochrome c; FADD, Fas-associated death domain; TNF, tumor necrosis factor; TNFR, TNF receptor; TRADD, TNFR-associated death domain.

Mitochondrion-Mediated Apoptotic Pathway

The mitochondrial pathway of apoptosis is dependent on the process of mitochondrial outer membrane permeabilization, which leads to the release of proteins from the mitochondrial intermembrane space into the cytosol. Mitochondria play a major role in apoptosis triggered by many stimuli. They integrate death signals through Bcl-2 family members and coordinate caspase activation through the release of cytochrome c as a result of the outer mitochondrial membrane becoming permeable [46]. In general terms, antiapoptotic members of the Bcl-2 family tend to inhibit the release of mitochondrial proteins, whereas proapoptotic members (such as BID, BAX and BAK) favor release [47]. Proteins of the intermembrane space are released to the cytosol [48]. One is holocytochrome c, which binds to cytosolic, monomeric apoptotic protease activating factor–1 (APAF-1). The interaction with cytochrome c permits a 2′-deoxyadenosine 5′-triphosphate (dATP)–mediated conformational change in APAF-1, promoting APAF-1 oligomerization to form an "apoptosome". The apoptosome then binds the proform of a protease, caspase-9. The oligomerization of caspase-9 on the apoptosome activates the protease; initiator caspases such as caspase-9 can only be activated by dimerization [49].

The active caspase-9 cleaves two other caspases, caspase-3 and caspase-7. These are "executioner" caspases, and, unlike initiator caspases, these can only be activated by cleavage at specific sites, which are targeted by initiator caspases. Once activated, the executioner caspases orchestrate apoptosis through the cleavage of key substrates within the cell, resulting in demolition and packaging of the dying cell for removal by phagocytic cells (Figure 2). The process of apoptosis in mitochondrial outer membrane permeabilization is regulated at the apoptosome and each of the caspases. The latter are controlled by inhibitor of apoptosis proteins (IAPs) that bind and ubiquitinate the caspases for proteasomal degradation. The function of the IAPs, in turn, is blocked by IAP inhibitors. These include Smac (also known as DIABLO) and Omi (also known as HtrA2), which compete with caspases for IAP binding. Like cytochrome c, Smac and Omi are sequestered in the mitochondrial intermembrane space and are only available to regulate the cytosolic IAPs upon mitochondrial outer membrane permeabilization (Bcl-2, IAP will be discussed in detail below) [48].

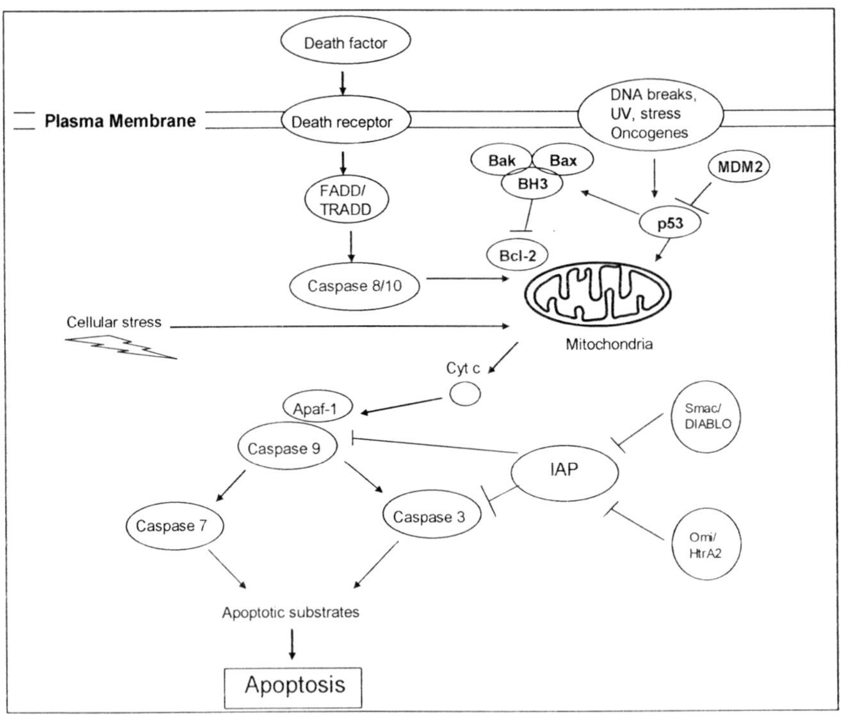

Figure 2. Mitochondrion-mediated caspase-dependent pathway. Apaf-1, apoptotic protease activation factor-1; Cyto c, cytochrome c; FADD,Fas-associated death domain; TRADD, tumour necrosis factor receptor-associated death domain. IAP, inhibitor of apoptosis proteins.

Upon activation by DNA damage-induced or oncogene-induced signaling pathways, p53 promotes the expression of a number of genes that are involved in apoptosis, including those encoding death receptors [50,51] and proapoptotic members of the Bcl-2 family [52-54]. In most cases, p53-induced apoptosis proceeds through mitochondrial release of cytochrome c, which leads to caspase activation [55]. Furthermore, p53 modifications can influence the ability of p53 to bind its negative regulator Mdm2 (a p53-inducible E3 ubiquitin ligase capable of targeting p53 for degradation) [56].

Caspase-Independent Apoptosis

Characterization of *C. elegans* ced-3 mutants revealed that PCD underwent even in the absence of *ced-3* [57] which indicated that cell death can occur in the absence of caspase activation. Subsequent findings were consistent with the existence of *ced-3*-independent PCD [58]. Some studies have suggested caspase-independent PCD in mammals. Genetic studies on PCD induced by BH3-domain only proteins, such as tBID, BIM and BAD, showed that these proteins, which have been shown to promote caspase activation and apoptosis, can also kill cells independently of Apaf-1 and downstream caspases. Specifically, Apaf-1 $^{-/-}$ mouse embryonic fibroblasts (MEFs) could still die in response to the overexpression of BH3-domain-only proteins [59]. Caspase activation was not detected in the dying cells when assayed with fluorogenic substrates for caspase-2, -3, -6 or -7, nor could cell death be blocked by the pan-caspase inhibitor zVAD.fmk [59].

Also other mammalian mediators of caspase-independent PCD have been identified as apoptosis-inducing factor (AIF), a conserved mitochondrial oxidoreductase [60] and endonuclease G [61]. In the early mouse embryo, PCD occurs in the ectoderm to create the pro-amniotic cavity. An in vitro model of this event uses an aggregate of embryonic stem (ES) cells in which the inner ectodermal cells undergo cell death [60]. Both have been suggested to be capable of inducing apoptotic changes in the nucleus (i.e., DNA fragmentation), but neither have been demonstrated to be responsible for other stereotypical changes that occur in apoptosis (e.g., exposure of phosphatidylserine and proteolysis of vital substrates [60,61].

Caspases are proteases, it is possible that caspase-independent cell death might require other proteases. Indeed, there is weak, yet suggestive, evidence that might support roles for cathepsins, calpains and granzyme B in caspase-independent PCD. Cathepsins are lysosomal proteases, and mice lacking cystatin B, an inhibitor of cathepsins B, H, L and S, show cerebellar cell death. This suggests that cathepsins might mediate cell death, although it is also possible that the phenotype was caused by a cathepsin-independent effect of cystatin B [62]. Calpains are a family of calcium-dependent proteases that have also been described as possible facilitators of cell death. In breast cancer cells, vitamin-D-induced cell death appeared to be caspase-independent [63] and promoted by calpains [64].

Necrotic Signaling Pathway

The signaling mechanisms mediating necrotic cell death are, in general, poorly understood. Necrosis-like programmed cell death is mediated by a coordinated series of signaling events that can be triggered via death receptors [65,66]. Fas kills activated primary T cells efficiently in the absence of active caspases, which results in necrotic morphological changes and late mitochondrial damage but no cytochrome c release. This Fas ligand–

induced caspase-independent death is absent in T cells that are deficient in either Fas-associated death domain (FADD) or receptor-interacting protein (RIP). RIP is also required for necrotic death induced by tumor necrosis factor (TNF) and TNF-related apoptosis-inducing ligand (TRAIL). Thus, Fas, TRAIL and TNF receptors can initiate cell death by two alternative pathways, one relying on caspase-8 and the other dependent on the kinase RIP [67-69]. Evidence suggests that Fas may also cause necrosis. It was found that, when overexpressed in a highly sensitive fibrosarcoma cell line, Fas triggers cell death in a caspase-independent necrotic manner [70]. In addition, artificial aggregation of FADD initiates cell death with necrotic morphology in caspase-8–deficient Jurkat cells [91]. Using genetically deficient Jurkat T cells, the kinase receptor-interacting protein (RIP) was identified as a crucial component not only of Fas-mediated, but also of TNF and TNF-related apoptosis-inducing ligand (TRAIL)-triggered caspase-independent cell death.

Apart from caspase-8 and caspase-10, FLIP is currently the only protein known to bind to the DED domain of FADD. FLIP was shown to enhance RIP recruitment to Fas [72]. Overexpression of FLIP is toxic to cells [73] and thus its implication in the necrotic pathway cannot be excluded. FADD has an opposite observed effect in the TNF-triggered necrotic pathway where a FADD-interacting protein has an anti-necrotic effect. A likely candidate is caspase-8, the inhibition of which can sensitize cells to TNF-induced necrosis [66]. Necrosis occurs as a consequence of acute ATP depletion. Several ATP-dependent ion channels become ineffective leading to disruption of actin cytoskeleton, cell swelling, membrane blebbing and eventual collapse of the cell [74,75]. Also Na^+ overloading Ca^{2+} accumulation and changes in mitochondrial permeability, may also participate in necrotic process [76-78]. Opening of high conductance permeability transition (PT) pores in the mitochondrial inner membrane initiates the mitochondrial permeability transition (MPT). Ca^{2+}, inorganic phosphate, alkaline pH, and reactive oxygen species (ROS) are a few of the many agents that promote the MPT, whereas the immunosuppressive drug, cyclosporin A (CsA), Mg^{2+}, acidic pH, and phospholipase inhibitors, including trifluoperazine, dibucaine, and quinacrine, block opening of the PT pores [79,80] (Figure 3).

Interestingly, cathepsin and calpain proteases have been described as key regulators of necrotic cell death in C. elegans [81].

Regulating Factors

The activation and inactivation of caspases are regulated by various proteins, ions and other factors, such as IAP, Bcl-2 family proteins, calpain, Ca^{2+}, FLICE-Inhibitory Protein (FLIP), Gran B, cytokine response modifier A (Crm A), p35 and NF-κB.

IAP

In the mid 1990s, a screen for viral-encoded suppressers of apoptosis led to the identification of two death-suppressing genes from *Cydai pomonella granulovirus* (CpGV) and *Orgyia pseudotsugata nucleopolyhedrovirus* (OpNPV), each with a high degree of similarity to the other [82,83]. These viral proteins, Cp-IAP and Op-IAP, defined the first members of a family of inhibitors of apoptosis proteins (IAP) containing one or more baculoviral IAP repeats (BIR). This BIR domain consists of approximately 70 amino acids with a conserved spacing of cysteine and histidine residues, which suggests that this structure represents a novel zinc binding fold [84]. Homology to this BIR domain has been used to isolate a number of cellular proteins that inhibit apoptosis. Other members of the IAP family include XIAP (X-linked mammalian inhibitor of apoptosis protein) c-IAP1, c-IAP2, NAIP (neuronal apoptosis inhibitory protein), livin and Survivin in mammals; DIAP1, DIAP2, and Deterin in Drosophila, and CeBIR1 and CeBIR2 in *C. elegans*; TnIAP and SfIAP in Lepidopteran; and SpIAP and ScIAP in yeast [84,85]. Various members of the family of inhibitory proteins contain one to three N- terminal baculovirus BIR domains and one conservative C-terminal RING (really interesting new gene) domain. The IAP family (X-IAP, c-IAP-1, c-IAP-2, N-AIP) appear to negatively regulate caspase activities by acting as pseudosubstrates for the caspases, binding to active caspases and neutralizing their activity [86,87]. The activity of mammalian IAP can be inhibited by Smac/DIABLO released from mitochondria. As the four residue sequence (Ala1-Val2-Pro3-Ile4) in the newly formed N-terminus of Smac/DIABLO can recognize and bind to the caspase-9-binding site of XIAP, so that XIAP will be inactivated, its inhibiting effect on caspase-9 will in turn be relieved [88]. IAP family proteins may also have other functions besides caspase inhibition. For example, IAP family members in yeast can neither unite caspases nor induce apoptosis [85].

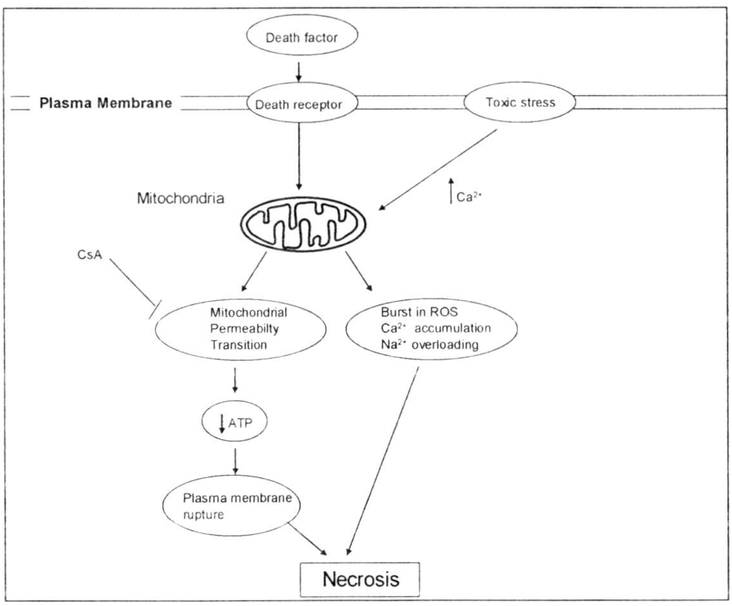

Figure 3. Cell death by necrosis. CsA, Cyclosporin A; ROS, Reactive oxygen species.

Bcl-2 Family Proteins

The mitochondria-dependent pathway for apoptosis is governed by Bcl-2 family proteins. Mitochondria play a major role in apoptosis triggered by many stimuli. They integrate death signals through Bcl-2 family members and coordinate caspase activation through the release of cytochrome c as a result of the outer mitochondrial membrane becoming permeable [46,89]. Bcl-2 belongs to a growing family of proteins that regulate apoptosis [89,90].

According to functional and structural criteria, the members can be divided into two groups. Group I proteins are all anti-apoptotic proteins, including A1/Bfl1, Bcl-2, Bcl-w, Bcl-xL, Boo/Diva, Mcl-1, NR-13 and Nrf3 in mammals, BHRF-1, E1B19K, Ks-Bcl-2, LMW5-HL and ORF16 in bacteria, and Ced-9 in *C. elegans* [91-93]. They all have four short Bcl-2 homology (BH) domains: BH1, BH2, BH3 and BH4. Group II proteins are all pro-apoptotic proteins, including Bad, Bak, Bax, Bcl-rambo, Bcl-xS, Bid, Bik, Bim, Blk, BNIP3, Bok/Mtd, Hrk and Nip3 in mammals, and Egl-1 in *C. elegans* [92].

Bcl-2 family proteins contain at least one of the four conserved regions called Bcl-2 homology domains (BH1–BH4). These motifs are formed by helices and enable the different members of the family to form either homo- or heterodimers and to regulate each other [94,95]. The Bcl-2-related proteins display either anti-apoptotic or pro-apoptotic function. The balance between Bcl-2 family members defines whether a cell will live or die. The members that inhibit apoptosis, such as Bcl-2 and Bcl-X_L, harbor at least three BH domains. Over-expression of Bcl-2 and Bcl-X_L inhibits apoptosis induced by a variety of stimuli, in particular those that utilize the mitochondria-dependent pathway [93,96]. This is often associated with protection against the release of cytochrome *c* as well as collapse of membrane potential [97,98]. Moreover, co-expression of Bcl-2, Bcl-X_L or other anti-apoptotic Bcl-2 family members can rescue yeast from the cytotoxic effects of Bax. Studies of the effects of Bax on mitochondria in intact yeast have provided evidence of hyper-polarization (based on cationic dyes) and matrix alkalinization (based on pH-GFP), with concomitant cytosol acidification [47,99-101]. Furthermore, Bcl-2 or Bcl-X_L prevents these effects of Bax on mitochondria and pH regulation. Moreover, channel formation by pore forming Bcl-2 family proteins in synthetic membranes (including anti-apoptotic Bcl-2, Bcl-X_L and pro-apoptotic Bax and Bid) is also markedly enhanced by low pH again with optima in the pH 4-5 range [102]. Among the death promoters, some proteins (e.g. Bax, Bak) contain BH1, BH2 and BH3, and closely resemble Bcl-2. Increased Bax expression is often correlated with apoptotic cell death. Bax and Bak, originally localized in the cytoplasm, can translocate to the mitochondrial outer membrane after an apoptotic program starts. Following the translocation, they will undergo conformation changes, oligomerization and insertion into the mitochondrial outer membrane to elevate the mitochondrion permeability transition pores [39]. The Bcl-2 mediated protection of mitochondria is most likely due to the prevention of pore opening caused by Bax- or Bak- dependent mechanisms [103]. Up-regulation and/or translocation of Bax to the mitochondria is a second major pathway for mitochondrial perturbation preceding the onset of apoptosis. At the mitochondria, Bax invokes cytochrome *c* release and loss of membrane potential [104], possibly via direct pore formation [105] or by association with the voltage-dependent anion channel [106].

Other Bcl-2 family proteins (e.g. Bid, Bad) possess only the BH3 domain, which is essential for both their interaction with other family members and their death-promoting activity [107] these are called 'BH3-only proteins'. The three-dimensional structures of Bid [60] and Bcl-X_L [108] are similar, suggesting that Bid may also form ion channels. Bid requires the cooperation of relatives such as Bax for its effect [109]. Unlike most other members of the Bcl-2 family, BID is normally localized to the cytoplasm. However, caspase-8 can catalyze the cleavage of BID, the C-terminal portion of which then translocates to the mitochondria where it integrates into the outer membrane and provokes the release of cytochrome c - an important accessory molecule for downstream caspase activation events [34]. This caspase-8 mediated activation of Bid represents an important mechanism accounting for cross-talk between the death receptor (extrinsic) and mitochondrial (intrinsic) pathways [110]. Most Bcl-2 family proteins also contain a C-terminal ~20-residue hydrophobic domain that targets them to intracellular membranes, predominantly to the outer membrane of mitochondria [111]. In contrast to Bcl-2, Bcl-X_L, and many of the Bcl-2 family proteins, the pro-apoptotic proteins Bid, Bim, and BAD lack C-terminal transmembrane domains. These proteins are normally found in the cytosol but can be induced to target mitochondria. The principal mechanism by which Bcl-2 family proteins regulate apoptosis is probably by controlling cytochrome-*c* release [112]. In addition to cytochrome c, Bcl-2 family proteins have been reported to control the release of other proteins from mitochondria. These proteins include: caspases [113-115], apoptosis inducing factor [116] and Smac/Diablo, the inhibitor of IAP family proteins [117,118].

Calpain and Ca^{2+} Ion

Calpain is a kind of Ca^{2+}-dependent cysteine protease of the papainase family. It is generally believed that calpain is activated in both necrosis and apoptosis. Calpain and caspase-3 share many common substrates, including fodrin, Ca^{2+}-dependent protein kinase and ADP ribosyltransferase/PARP [119]. In apoptosis induced by ER stress, calpain's functions are particularly salient because of the perturbed Ca^{2+} homeostasis. In the brain cells of rats suffering from unilateral hypoxia-ischemia, m-calpain first cleaved procaspase-3 into 29 kDa fragments to facilitate its further cleavage and activation [120]. Cisplatin, a kind of anticancer agent, can cause ER stress and apoptosis. During this process, the activation of procaspase-12 by cisplatin is dependent on Ca^{2+} and calpain [121]. In addition, calpain can also cleave Bcl-xL in its loop region, which will convert Bcl-xL to a proapoptotic molecule from an anti-apoptotic one [122]. Bcl-2 family proteins emerge as major regulators of cellular calcium handling. In addition, apoptotic factors released from the mitochondria and activated caspase enzymes may also control Ca^{2+} transport proteins [123].

FLICE-Inhibitory Protein (FLIP)

A subgroup of membrane receptors of the TNF superfamily, collectively known as the death receptors, induces apoptosis upon ligand binding [124]. CD95 (Fas) is the most extensively studied member of the death receptor family. Upon ligation of the Fas receptor with its ligand (FasL), receptor clustering at the cell membrane results in conformational changes, enabling recruitment of an adaptor molecule FADD, and procaspase-8 (also known as FLICE) to the membrane-bound receptor in a complex termed the death inducing signaling complex (DISC). Recruitment of procaspase-8 to this complex permits its activation, initiating a caspase cascade that ultimately leads to cell death. Therefore, a level of control can also occur at the level of this receptor complex. A family of FLICE-inhibitory proteins (FLIPs) has recently been identified, with both viral and mammalian members [73,125-128]. The viral proteins (v-FLIPs) contain two tandem death effector domains (DED), while the human homolog (c-FLIP$_L$) contains two DED regions along with an inactive caspase-8–like catalytic domain at the C-terminus. The short form of this protein (c-FLIP$_S$) is similar in structure to the viral FLIPs. FLIPs are thought to be recruited to the DISC through their DED domains, thus preventing procaspase-8 recruitment and activation. Therefore, receptor-mediated cell death can be regulated at the level of the most proximal caspase, caspase-8. Inhibition of apoptosis by FLIPs through the inactivation of caspase-8 occurs in several apoptotic model systems, including germinal center B-cells, human peripheral blood T-lymphocytes, and mast cells [129-131].

Gran B and Crm A

Gran B is a kind of serine protease with an important role in apoptosis in cytotoxic T cells. Gran B can activate various procaspases, such as procaspase-3, procaspase-7, procaspase-8, procaspase-9 and procaspase-10, to initiate apoptosis [131,132]. Also it was found that Gran B could cleave Bid to initiate the mitochondrion-mediated activation pathway [132].

CrmA, a 38-kD cytokine response modifier protein of the cowpox virus, is a protease inhibitor of the serpin family initially shown to inhibit interleukin-1 beta converting enzyme (ICE) and cytokine-induced apoptosis [133-135]. CrmA is a unique member of the serpin family in that it can inhibit both cysteine proteases (caspases) and serine proteases, such as granzyme B [136]. CrmA is a potent inhibitor of diverse apoptotic model systems such as Fas- and TNF-induced apoptosis [135], serum withdrawal [137], and hypoxia-mediated apoptosis [138]. While crmA is most effective at inhibiting caspase-1 (ICE) and caspase-8, it can also inhibit with decreasing effectiveness caspase-10, -6, -3, and -7 [139]. In general, crmA rescues cells primarily from apoptotic signals involving death receptors, or other mechanisms in which caspase-8 (or caspase-10) acts as the apical protease [140]. In contrast, crmA is ineffective at preventing cell death owing to stress or genotoxic damage, which suggests that apoptotic stimuli that act directly at the level of mitochondria are relatively resistant to crmA inhibition. Thus, apoptotic stimuli activating caspases, such as caspase-9, as the initial protease are essentially crmA-insensitive.

The activity of Gran B can be inhibited by Crm A, a kind of serpin from the vaccinia virus. Crm A, a strong inhibitor of caspase-1 and caspase-8, and a weak inhibitor of caspase-3 and caspase-6, can prevent the cross-link of Fas and inactivate Gran B [132].

p35

Another caspase inhibitor identified from a novel gene product of the baculovirus *Autographa californica* nuclear polyhedrosis virus, designated p35, prevents cell death in various apoptotic model systems from insects to mammals [141-143]. p35, also blocks cell death induced with a wide variety of apoptotic induction signals including ionizing radiation, Fas, and TNF [144,145]. p35, with the ability of binding to caspases to cleave and inactivate them, is an effective inhibitor of caspases from caspase-1 to caspase-8 [146]. The crystal structure of the caspase-8/p35 complex was recently solved and revealed that caspase inhibition occurs owing to a covalent thioester linkage between p35 and the active site of caspase-8 [147]. It is interesting to note, p35 inhibits the activity of a number of different caspases; however, it was recently reported that p35 does not prevent caspase-9 activity in a cell-free system of mammalian caspase activation [148].

Proteins such as IAPs, FLIPs, crmA, and p35 prevent apoptosis at the level of caspases to inhibit aberrant cell death. This variety of different regulating proteins used to inhibit specific caspase activities depends on the particular stimuli employed to initiate apoptosis. One advantage of this diverse capacity of various caspase inhibitors to prevent apoptosis is the ability to use these proteins to define a specific caspase cascade that a particular stimulus engages to commit the cell to death. Additionally, structural studies examining the interactions between caspase inhibitors and specific initiator or effector caspases offers new insights into the design of synthetic proteins with increased selectivity for their use in therapeutic approaches for controlling apoptosis.

Nuclear Factor NF-κB

The NF-κB transcription factors are critical players in the control of the apoptotic response to a variety of stimuli. They are best known for their pivotal roles in immune, inflammatory and acute phase responses, but they also play important role in apoptosis [149-151]. Whereas NF-κB is most commonly involved in suppressing apoptosis by transactivating the expression of anti-apoptotic genes, it can promote programmed cell death in response to certain death-inducing signals and in certain cell types. In this context, in mature T cells NF-κB triggered the Fas-death cascade by directly activating the expression of Fas ligand (FasL) [152-154]. TNFα is a key cytokine that regulates immune responses, cell differentiation and apoptosis. Activation of TNF receptors TNFR1 or TNFR2 by TNFα represents the essential pathway balancing the induction of cell death with its inhibition through activation of NF-κB [155]. Also, endogenous caspase-8 mediates TNF-α-induced activation of NF-κB via FLASH (FLICE-associated huge protein) [156]. Bcl-2 and Bcl-xL, which block apoptosis, have also been shown to suppress NF-κB activation induced by TNF in endothelial cells [157].

Pancreatic Acinar Cell Death in Acute Pancreatitis

Progressive pancreatic acinar cell death is a hallmark in both clinical and experimental acute pancreatitis, the mechanisms of which remain poorly understood. [6]. Necrosis has been classically considered the major form of cell death in acute pancreatitis, and the degree of pancreatic necrosis is strongly associated with the prognosis of this disease [158,159]. On the other hand, apoptosis was suggested to mediate the destruction of acinar cells and induce atrophy of the organ [160-162]. However, careful biochemical and morphological examination of experimental models of acute pancreatitis has shown that severe acute pancreatitis (e.g. that induced by pancreatic duct ligation in the opossum, by choline-deficient and ethionine supplemented diet in the mouse, and by caerulein-hyperstimulation in the mouse) is associated primarily with necrosis but little apoptosis, whereas mild acute pancreatitis (e.g. that induced by pancreatic duct ligation and by caerulein-hyperstimulation in the rat) is associated primarily with apoptotic cell death and little necrosis [3,10]. In other words, the severity of acute pancreatitis is inversely related to the extent of acinar cell apoptosis. Moreover, induction of apoptosis reduces the severity of experimental pancreatitis, while inhibition of it worsens the disease [4,163-166]. These suggested that apoptosis is a teleologically beneficial form of cell death in acute pancreatitis. Hence, elucidating the molecular mechanisms of pancreatic cell apoptosis as well as how it benefits against acute pancreatitis would provide valuable insights into clinical pancreatic disorders and offer potential pharmacological interventions.

Mechanism of Apoptosis in Acute Pancreatitis

Caspase Activation

Caspase activation has been observed in isolated pancreatic acinar cells *in vitro*, stimulated with apoptotic inducers (e.g. cholecystokinin, crambene, hydrogen sulphide, and menadione), as well as in experimental pancreatitis induced by cerulein *in vivo* [167-171]. Stimulation of isolated rat pancreatic acini with cholecystokinin, which serves as a model for human acute edematous pancreatitis *in vitro*, stimulates death signaling pathways in rat pancreatic acinar cells, including caspase activation, cytochrome c release, and mitochondrial depolarization, leading to apoptosis. The mitochondrial dysfunction is mediated by upstream caspase(s). CCK causes mitochondrial alterations through both PTP (permeability transition pore)-dependent (cytochrome c release) and PTP-independent (mitochondrial depolarization) mechanisms [171]. Some pathogenic factors, such as oxidative stress, a major pathogenic factor in acute pancreatitis, induces apoptosis of acinar cells involving activation of caspase-3, the executioner caspase, which degrades the DNA repair protein Ku70 and Ku80 [172].

In addition to mediate apoptosis, caspases also regulate other processes in the pancreatic acinar cell that play key roles in pancreatitis. The initiator caspase, caspase 8 has been suggested to mediate degradation of plectin, which represents a critical biochemical event during cholecystokinin-induced secretory blockade and cell injury in acute pancreatitis [173]. Caspases also down regulate the intra-acinar cell activation of trypsin, which initiates autodigestion and acute pancreatitis [171]. Particularly, caspases negatively regulate necrosis through degradation of receptor-interacting protein (RIP), a key mediator of "programmed" necrosis and thus protect from necrosis in pancreatitis [171,172,174]. This may explain the inverse correlation between the extent of apoptosis on the one hand and necrosis along with the severity of the disease on the other hand observed in experimental models of pancreatitis.

Apoptotic Associated Gene

In addition to caspase, the Bcl-2 family of apoptotic regulators is another one of the functional components comprised in the apoptosis pathway. Acute pancreatitits modifies mRNA level of both pro and anti-apoptotic genes simultaneously. Pancreatic bclXL, bax, and p53 mRNA levels increased significantly in a temporal fashion during induction of acute pancreatitis [175].

The increase in expression of p53 and Bax, and a decrease in Bcl-2 expression are also found in glucose/glucose oxidase (G/GO), an oxidative stress induced apoptosis in pancreatic acinar AR42J cells (20). However, in the enhanced severe acute pancreatitis model, pancreatic expression of bax and Bcl-2 remains unchanged, which may due to the decreases of the sensitivity of acinar cells to apoptotic stimuli during the deleterious course of acute pancreatitis [166]. Besides Bcl-2 family, SIP gene has recently been identified, whose expression localized in acinar cell, and demonstrated that as a stress-inducible gene, it is overexpressed during pancreatitis and promoted cell apoptosis [176].

Neutrophils

Neutrophils activation is an important determinant of severity of acute pancreatitis. Neutrophil apoptosis is one of resolutions for such inflammatory responses. On the contrary, delays in such apoptotic response are associated with persistent inflammation and subsequent tissue damage [177,178]. However, the infiltration of neutrophils may itself contribute to regulate pancreatic acinar cell death in acute pancreatitis. The depletion of neutrophils by administration of antineutrophil serum or using ICAM-1 neutralizing antibody, results in a significant increase in acinar cells undergoing apoptosis on the one hand, and significantly reduced the extent of necrosis on the other hand [179]. Yet in the experimental model of taurocholate-induced severe acute pancreatitis, neutrophils, via a TNF-α-dependent mechanism, is involved in the development of both apoptotic and necrotic forms of acinar cell death [180].

Cytokines

In addition to their role in inducing inflammation, cytokines, such as TNF-α or interleukin-1β, which are released by neutrophils (and to a lesser extent, by acinar cells) have also been shown to induce apoptosis in pancreas. Pancreatic acinar cells have been reported produce and release TNF-α, which is not only involved in the development of pancreatitis, but also mediated apoptosis in acinar cell suspension *in vitro* as well as *in vivo* in the caerulein model of acute pancreatitis [9,181]. Other cytokines, such as interleukin 6 plays a role in initiating the inflammatory as well as antiapoptotic responses in severe acute pancreatitis. Blocking IL-6 suppresses STAT-3 (a transcription factor which down regulates cell apoptosis) activation in the pancreas, and consequently attenuates the severity of severe acute pancreatitis by promotion of pancreatic acinar cell apoptosis [164].

Agents which Induce Pancreatic Acinar Cell Apoptosis but do not cause acute Pancreatitis

CCK as well as hydrogen sulfide are inducers of pancreatic acinar cell apoptosis. They can, however, also induce or promote inflammation in acute pancreatitis. In addition to CCK and hydrogen sulfide, other inducers of pancreatic acinar cell apoptosis have been used to investigate cell death in relation to acute pancreatitis. Examples of these compounds are menadione and crambene (1-cyano-2-hydroxy-3-butene-CHB). Menadione is a quinone that is metabolized by flavoprotein reductase to semiquinone, which can be oxidized back to quinone in the presence of molecular oxygen. In this redox cycle the superoxide anion radical, hydrogen peroxide and other reactive oxygen species are generated [182]. Menadione can cause elevations in the cytosolic Ca2+ concentration contributing to cell death [182]. In a recent study, menadione has been shown to evoke repetitive cytosolic Ca^{2+} spikes, partial mitochondrial depolarization, cytochrome c release and apoptosis of pancreatic acinar cells in vitro [170].

Crambene is a stable plant nitrile found in many cruciferous vegetables. Crambene is a potent inducer of "phase II" detoxification enzymes, including certain glutathione Stransferases and quinone reductase, which are important enzymes associated with conjugation and elimination of reactive chemical intermediates and carcinogens [183,184]. Also, it has been shown to cause cell cycle arrest of the Hep G2 cells or colonic adenocarcinoma cells in G2/M phase [170]. The intrinsic pathway of apoptosis mediated by loss of mitochondrial membrane potential, release of mitochondrial cytochrome C as well as activation of caspase 9 is suggested to be the primary pathway involved in crambene induced pancreatic acinar cell apoptosis. [168]. Upon intravenous administration to rat as well as mouse, crambene induces apoptosis of the pancreatic acinar cells [185,186]. Moreover, induction of pancreatic acinar cell apoptosis with crambene protects mice against acute pancreatitis induced by caerulein [4]. This is despite the fact that crambene does not alter the interaction of CCK to its receptor on pancreatic acinar cells [4]. The maximal protective effect of crambene against acute pancreatitis is observed when the pancreatic acinar cells have become committed to apoptosis [4]. A possible role of the gap junction communication (GJC) and connexin gene Cx32 has been found in crambene-induced apoptosis [166]. Mice deficient in Cx32 gene were found to be resistant to apoptosis induced by crambene and were more susceptible to acute pancreatitis than their wild-type counterparts [166], which indicate that defective GJC in Cx32-deficient mice prevented acinar cell apoptosis at a step upstream of caspase-3 activation. Besides crambene, an extract of Artemisia asiatica (DA-9601) has also been shown its beneficial effect in caerulein-induced pancreatitis in rats by induction of pancreatic acinar cells apoptosis [163].

CONCLUSION

Cell death plays an essential role in normal development and functioning of multicellular organisms and that abnormalities in this process can cause disease. Research studies on the molecular mechanisms of apoptosis and its regulation has led to the deeper understanding of the pathophysiology of many diseases. Several strategies have emerged for therapeutic intervention in a wide range of ailments, including cancer, autoimmune disorders, immunodeficiency, inflammation, ischemic heart diseases, stroke and neurodegenerative diseases [187]. Numerous novel approaches are currently being followed employing gene therapy and antisense strategies in order to target specific apoptotic regulators [188]. These approaches should facilitate efforts to rationally manipulate the apoptotic machinery for therapeutic gain.

Both apoptotic and necrotic forms of cell death are seen in clinical, as well as experimental acute pancreatitis [5,6]. The mechanisms of pancreatic acinar cell death - apoptosis and necrosis - in acute pancreatitis are just beginning to be identified. The vast array of molecules involved in both the apoptotic and necrotic cell death pathways offers several potential targets for therapeutic intervention. The extent of pancreatic acinar cell apoptosis has been shown to be inversely related to the severity of the disease, suggesting that apoptosis is a teleologically beneficial form of cell death in acute pancreatitis. Indeed, in animal models, induction of apoptosis has been found to have a protective action against acute pancreatitis. It is, therefore, reasonable to speculate that selective induction of pancreatic acinar cells apoptosis may be of value in clinical acute pancreatitis. As a first step, the effect prophylactic induction of apoptosis in individuals with high risk of pancreatitis (such as those undergoing ERCP procedure) should be investigated. More basic research into the mechanisms of apoptosis of pancreatic acinar cells and how it benefits against acute pancreatitis is, however, needed in order for these studies to be taken into the clinic. Active research in this direction will facilitate a transition of our knowledge of pancreatic acinar cell death in the form of necrosis *versus* apoptosis from bench to bedside.

REFERENCES

[1] Banks PA. Medical management of acute pancreatitis and complications. In: Go VW, DiMagno EP, Gardner JD, Lebenthal E, Reber HA, Scheele GA, Eds. *The Pancreas: Biology, Pathobiology, and Disease*. New York: Raven Press.1993 pp593–611.

[2] Leung, P. S. and Ip, S. P. (2006) Pancreatic acinar cell: Its role in acute pancreatitis. *Int J Biochem Cell Biol 38*, 1024-1030

[3] Gukovskaya, A. S., Perkins, P., Zaninovic, V., Sandoval, D., Rutherford, R., Fitzsimmons, T., Pandol, S. J. and Poucell-Hatton, S. (1996) Mechanisms of cell death after pancreatic duct obstruction in the opossum and the rat. *Gastroenterology 110*, 875-884

[4] Bhatia, M., Wallig, M. A., Hofbauer, B., Lee, H. S., Frossard, J. L., Steer, M. L. and Saluja, A. K. (1998) Induction of apoptosis in pancreatic acinar cells reduces the severity of acute pancreatitis. *Biochem Biophys Res Commun 246*, 476-483

[5] Bhatia, M. (2004) Apoptosis of pancreatic acinar cells in acute pancreatitis: is it good or bad? *J Cell Mol Med 8*, 402-409

[6] Bhatia, M. (2004) Apoptosis versus necrosis in acute pancreatitis. *Am J Physiol Gastrointest Liver Physiol 286*, G189-196

[7] Kaiser, A. M., Saluja, A. K., Lu, L., Yamanaka, K., Yamaguchi, Y. and Steer, M. L. (1996) Effects of cycloheximide on pancreatic endonuclease activity, apoptosis, and severity of acute pancreatitis. *Am J Physiol 271*, C982-993

[8] Walker, N. I., Winterford, C. M., Williamson, R. M. and Kerr, J. F. (1993) Ethionine-induced atrophy of rat pancreas involves apoptosis of acinar cells. *Pancreas 8*, 443-449

[9] Gukovskaya, A. S., Gukovsky, I., Zaninovic, V., Song, M., Sandoval, D., Gukovsky, S. and Pandol, S. J. (1997) Pancreatic acinar cells produce, release, and respond to tumor necrosis factor-alpha. Role in regulating cell death and pancreatitis. *J Clin Invest 100*, 1853-1862

[10] Kaiser, A. M., Saluja, A. K., Sengupta, A., Saluja, M. and Steer, M. L. (1995) Relationship between severity, necrosis, and apoptosis in five models of experimental acute pancreatitis. *Am J Physiol 269*, C1295-1304

[11] Steller, H. (1995) Mechanisms and genes of cellular suicide. *Science 267*, 1445-1449

[12] Thompson, C. B. (1995) Apoptosis in the pathogenesis and treatment of disease. *Science 267*, 1456-1462

[13] Kerr, J. F., Wyllie, A. H. and Currie, A. R. (1972) Apoptosis: a basic biological phenomenon with wide-ranging implications in tissue kinetics. *Br J Cancer 26*, 239-257

[14] Dive, C. and Hickman, J. A. (1991) Drug-target interactions: only the first step in the commitment to a programmed cell death? *Br J Cancer 64*, 192-196

[15] Eastman, A. (1993) Apoptosis: a product of programmed and unprogrammed cell death. *Toxicol Appl Pharmacol 121*, 160-164

[16] Bonfoco, E., Krainc, D., Ankarcrona, M., Nicotera, P. and Lipton, S. A. (1995) Apoptosis and necrosis: two distinct events induced, respectively, by mild and intense insults with N-methyl-D-aspartate or nitric oxide/superoxide in cortical cell cultures. *Proc Natl Acad Sci U S A 92*, 7162-7166

[17] Lieberthal, W., Triaca, V. and Levine, J. (1996) Mechanisms of death induced by cisplatin in proximal tubular epithelial cells: apoptosis vs. necrosis. *Am J Physiol 270*, F700-708

[18] Benchoua, A., Guegan, C., Couriaud, C., Hosseini, H., Sampaio, N., Morin, D. and Onteniente, B. (2001) Specific caspase pathways are activated in the two stages of cerebral infarction. *J Neurosci 21*, 7127-7134

[19] Eguchi, Y., Shimizu, S. and Tsujimoto, Y. (1997) Intracellular ATP levels determine cell death fate by apoptosis or necrosis. *Cancer Res 57*, 1835-1840

[20] Leist, M., Single, B., Castoldi, A. F., Kuhnle, S. and Nicotera, P. (1997) Intracellular adenosine triphosphate (ATP) concentration: a switch in the decision between apoptosis and necrosis. *J Exp Med 185*, 1481-1486

[21] Ha, H. C. and Snyder, S. H. (1999) Poly(ADP-ribose) polymerase is a mediator of necrotic cell death by ATP depletion. *Proc Natl Acad Sci U S A 96*, 13978-13982

[22] Nobel, C. S., Burgess, D. H., Zhivotovsky, B., Burkitt, M. J., Orrenius, S. and Slater, A. F. (1997) Mechanism of dithiocarbamate inhibition of apoptosis: thiol oxidation by dithiocarbamate disulfides directly inhibits processing of the caspase-3 proenzyme. *Chem Res Toxicol 10*, 636-643

[23] Hampton, M. B. and Orrenius, S. (1997) Dual regulation of caspase activity by hydrogen peroxide: implications for apoptosis. *FEBS Lett 414*, 552-556

[24] Hawkins, H. K., Ericsson, J. L., Biberfeld, P. and Trump, B. F. (1972) Lysosome and phagosome stability in lethal cell injury. Morphologic tracer studies in cell injury due to inhibition of energy metabolism, immune cytolysis and photosensitization. *Am J Pathol 68*, 255-258

[25] Alison, M. R. and Sarraf, C. E. (1994) Liver cell death: patterns and mechanisms. *Gut 35*, 577-581

[26] Gupta, S. (2000) Molecular steps of cell suicide: an insight into immune senescence. *J Clin Immunol 20*, 229-239

[27] Krammer, P. H. (2000) CD95's deadly mission in the immune system. *Nature 407*, 789-795

[28] Ashkenazi, A. and Dixit, V. M. (1998) Death receptors: signaling and modulation. *Science 281*, 1305-1308

[29] Kroemer, G. and Reed, J. C. (2000) Mitochondrial control of cell death. *Nat Med 6*, 513-519

[30] Martinou, J. C. and Green, D. R. (2001) Breaking the mitochondrial barrier. *Nat Rev Mol Cell Biol 2*, 63-67

[31] Li, H., Zhu, H., Xu, C. J. and Yuan, J. (1998) Cleavage of BID by caspase 8 mediates the mitochondrial damage in the Fas pathway of apoptosis. *Cell 94*, 491-501

[32] Devitt, A., Moffatt, O. D., Raykundalia, C., Capra, J. D., Simmons, D. L. and Gregory, C. D. (1998) Human CD14 mediates recognition and phagocytosis of apoptotic cells. *Nature 392*, 505-509

[33] Fadok, V. A., Bratton, D. L., Rose, D. M., Pearson, A., Ezekewitz, R. A. and Henson, P. M. (2000) A receptor for phosphatidylserine-specific clearance of apoptotic cells. *Nature 405*, 85-90

[34] Slee, E. A., Adrain, C. and Martin, S. J. (1999) Serial killers: ordering caspase activation events in apoptosis. *Cell Death Differ 6*, 1067-1074

[35] Thornberry, N. A. and Lazebnik, Y. (1998) Caspases: enemies within. *Science 281*, 1312-1316

[36] Alnemri, E. S., Livingston, D. J., Nicholson, D. W., Salvesen, G., Thornberry, N. A., Wong, W. W. and Yuan, J. (1996) Human ICE/CED-3 protease nomenclature. *Cell 87*, 171

[37] Nicholson, D. W. and Thornberry, N. A. (1997) Caspases: killer proteases. *Trends Biochem Sci 22*, 299-306

[38] Launay, S., Hermine, O., Fontenay, M., Kroemer, G., Solary, E. and Garrido, C. (2005) Vital functions for lethal caspases. *Oncogene 24*, 5137-5148

[39] Fan, T. J., Han, L. H., Cong, R. S. and Liang, J. (2005) Caspase family proteases and apoptosis. *Acta Biochim Biophys Sin (Shanghai) 37*, 719-727

[40] Pan, G., Bauer, J. H., Haridas, V., Wang, S., Liu, D., Yu, G., Vincenz, C., Aggarwal, B. B., Ni, J. and Dixit, V. M. (1998) Identification and functional characterization of DR6, a novel death domain-containing TNF receptor. *FEBS Lett 431*, 351-356

[41] Yang, X., Chang, H. Y. and Baltimore, D. (1998) Autoproteolytic activation of pro-caspases by oligomerization. *Mol Cell 1*, 319-325

[42] Muzio, M., Stockwell, B. R., Stennicke, H. R., Salvesen, G. S. and Dixit, V. M. (1998) An induced proximity model for caspase-8 activation. *J Biol Chem 273*, 2926-2930

[43] Wang, Z. B., Liu, Y. Q. and Cui, Y. F. (2005) Pathways to caspase activation. *Cell Biol Int 29*, 489-496

[44] Scaffidi, C., Fulda, S., Srinivasan, A., Friesen, C., Li, F., Tomaselli, K. J., Debatin, K. M., Krammer, P. H. and Peter, M. E. (1998) Two CD95 (APO-1/Fas) signaling pathways. *Embo J 17*, 1675-1687

[45] Wang, J., Chun, H. J., Wong, W., Spencer, D. M. and Lenardo, M. J. (2001) Caspase-10 is an initiator caspase in death receptor signaling. *Proc Natl Acad Sci U S A 98*, 13884-13888

[46] Hengartner, M. O. (2000) The biochemistry of apoptosis. *Nature 407*, 770-776

[47] Gross, A., Pilcher, K., Blachly-Dyson, E., Basso, E., Jockel, J., Bassik, M. C., Korsmeyer, S. J. and Forte, M. (2000) Biochemical and genetic analysis of the mitochondrial response of yeast to BAX and BCL-X(L). *Mol Cell Biol 20*, 3125-3136

[48] Danial, N. N. and Korsmeyer, S. J. (2004) Cell death: critical control points. *Cell 116*, 205-219

[49] Boatright, K. M. and Salvesen, G. S. (2003) Mechanisms of caspase activation. *Curr Opin Cell Biol 15*, 725-731

[50] Owen-Schaub, L. B., Zhang, W., Cusack, J. C., Angelo, L. S., Santee, S. M., Fujiwara, T., Roth, J. A., Deisseroth, A. B., Zhang, W. W., Kruzel, E. and et al. (1995) Wild-type human p53 and a temperature-sensitive mutant induce Fas/APO-1 expression. *Mol Cell Biol 15*, 3032-3040

[51] Muller, M., Wilder, S., Bannasch, D., Israeli, D., Lehlbach, K., Li-Weber, M., Friedman, S. L., Galle, P. R., Stremmel, W., Oren, M. and Krammer, P. H. (1998) p53 activates the CD95 (APO-1/Fas) gene in response to DNA damage by anticancer drugs. *J Exp Med 188*, 2033-2045

[52] Miyashita, T. and Reed, J. C. (1995) Tumor suppressor p53 is a direct transcriptional activator of the human bax gene. *Cell 80*, 293-299

[53] Oda, E., Ohki, R., Murasawa, H., Nemoto, J., Shibue, T., Yamashita, T., Tokino, T., Taniguchi, T. and Tanaka, N. (2000) Noxa, a BH3-only member of the Bcl-2 family and candidate mediator of p53-induced apoptosis. *Science 288*, 1053-1058

[54] Chipuk, J. E., Kuwana, T., Bouchier-Hayes, L., Droin, N. M., Newmeyer, D. D., Schuler, M. and Green, D. R. (2004) Direct activation of Bax by p53 mediates mitochondrial membrane permeabilization and apoptosis. *Science 303*, 1010-1014

[55] Schuler, M., Bossy-Wetzel, E., Goldstein, J. C., Fitzgerald, P. and Green, D. R. (2000) p53 induces apoptosis by caspase activation through mitochondrial cytochrome c release. *J Biol Chem 275*, 7337-7342

[56] Shieh, S. Y., Ikeda, M., Taya, Y. and Prives, C. (1997) DNA damage-induced phosphorylation of p53 alleviates inhibition by MDM2. *Cell 91*, 325-334

[57] Ellis, H. M. and Horvitz, H. R. (1986) Genetic control of programmed cell death in the nematode C. elegans. *Cell 44*, 817-829

[58] Shaham, S., Reddien, P. W., Davies, B. and Horvitz, H. R. (1999) Mutational analysis of the Caenorhabditis elegans cell-death gene ced-3. *Genetics 153*, 1655-1671

[59] Cheng, E. H., Wei, M. C., Weiler, S., Flavell, R. A., Mak, T. W., Lindsten, T. and Korsmeyer, S. J. (2001) BCL-2, BCL-X(L) sequester BH3 domain-only molecules preventing BAX- and BAK-mediated mitochondrial apoptosis. *Mol Cell 8*, 705-711

[60] Chou, J. J., Li, H., Salvesen, G. S., Yuan, J. and Wagner, G. (1999) Solution structure of BID, an intracellular amplifier of apoptotic signaling. *Cell 96*, 615-624

[61] Li, L. Y., Luo, X. and Wang, X. (2001) Endonuclease G is an apoptotic DNase when released from mitochondria. *Nature 412*, 95-99

[62] Pennacchio, L. A., Bouley, D. M., Higgins, K. M., Scott, M. P., Noebels, J. L. and Myers, R. M. (1998) Progressive ataxia, myoclonic epilepsy and cerebellar apoptosis in cystatin B-deficient mice. *Nat Genet 20*, 251-258

[63] Mathiasen, I. S., Lademann, U. and Jaattela, M. (1999) Apoptosis induced by vitamin D compounds in breast cancer cells is inhibited by Bcl-2 but does not involve known caspases or p53. *Cancer Res 59*, 4848-4856

[64] Mathiasen, I. S., Sergeev, I. N., Bastholm, L., Elling, F., Norman, A. W. and Jaattela, M. (2002) Calcium and calpain as key mediators of apoptosis-like death induced by vitamin D compounds in breast cancer cells. *J Biol Chem 277*, 30738-30745

[65] Kuida, K., Zheng, T. S., Na, S., Kuan, C., Yang, D., Karasuyama, H., Rakic, P. and Flavell, R. A. (1996) Decreased apoptosis in the brain and premature lethality in CPP32-deficient mice. *Nature 384*, 368-372

[66] Vercammen, D., Beyaert, R., Denecker, G., Goossens, V., Van Loo, G., Declercq, W., Grooten, J., Fiers, W. and Vandenabeele, P. (1998) Inhibition of caspases increases the sensitivity of L929 cells to necrosis mediated by tumor necrosis factor. *J Exp Med 187*, 1477-1485

[67] Chan, F. K., Shisler, J., Bixby, J. G., Felices, M., Zheng, L., Appel, M., Orenstein, J., Moss, B. and Lenardo, M. J. (2003) A role for tumor necrosis factor receptor-2 and receptor-interacting protein in programmed necrosis and antiviral responses. *J Biol Chem 278*, 51613-51621

[68] Holler, N., Zaru, R., Micheau, O., Thome, M., Attinger, A., Valitutti, S., Bodmer, J. L., Schneider, P., Seed, B. and Tschopp, J. (2000) Fas triggers an alternative, caspase-8-independent cell death pathway using the kinase RIP as effector molecule. *Nat Immunol 1*, 489-495

[69] Lin, Y., Choksi, S., Shen, H. M., Yang, Q. F., Hur, G. M., Kim, Y. S., Tran, J. H., Nedospasov, S. A. and Liu, Z. G. (2004) Tumor necrosis factor-induced nonapoptotic cell death requires receptor-interacting protein-mediated cellular reactive oxygen species accumulation. *J Biol Chem 279*, 10822-10828

[70] Vercammen, D., Brouckaert, G., Denecker, G., Van de Craen, M., Declercq, W., Fiers, W. and Vandenabeele, P. (1998) Dual signaling of the Fas receptor: initiation of both apoptotic and necrotic cell death pathways. *J Exp Med 188*, 919-930

[71] Kawahara, A., Ohsawa, Y., Matsumura, H., Uchiyama, Y. and Nagata, S. (1998) Caspase-independent cell killing by Fas-associated protein with death domain. *J Cell Biol 143*, 1353-1360

[72] Kataoka, T., Budd, R. C., Holler, N., Thome, M., Martinon, F., Irmler, M., Burns, K., Hahne, M., Kennedy, N., Kovacsovics, M. and Tschopp, J. (2000) The caspase-8

inhibitor FLIP promotes activation of NF-kappaB and Erk signaling pathways. *Curr Biol 10*, 640-648

[73] Irisarri, M., Plumas, J., Bonnefoix, T., Jacob, M. C., Roucard, C., Pasquier, M. A., Sotto, J. J. and Lajmanovich, A. (2000) Resistance to CD95-mediated apoptosis through constitutive c-FLIP expression in a non-Hodgkin's lymphoma B cell line. *Leukemia 14*, 2149-2158

[74] Carini, R., Autelli, R., Bellomo, G., Dianzani, M. U. and Albano, E. (1995) Sodium-mediated cell swelling is associated with irreversible damage in isolated hepatocytes exposed to hypoxia or mitochondrial toxins. *Biochem Biophys Res Commun 206*, 180-185

[75] Okada, Y., Maeno, E., Shimizu, T., Dezaki, K., Wang, J. and Morishima, S. (2001) Receptor-mediated control of regulatory volume decrease (RVD) and apoptotic volume decrease (AVD). *J Physiol 532*, 3-16

[76] Barros, L. F., Stutzin, A., Calixto, A., Catalan, M., Castro, J., Hetz, C. and Hermosilla, T. (2001) Nonselective cation channels as effectors of free radical-induced rat liver cell necrosis. *Hepatology 33*, 114-122

[77] Lemasters, J. J. (1999) V. Necrapoptosis and the mitochondrial permeability transition: shared pathways to necrosis and apoptosis. *Am J Physiol 276*, G1-6

[78] Levitsky, J., Gurell, D. and Frishman, W. H. (1998) Sodium ion/hydrogen ion exchange inhibition: a new pharmacologic approach to myocardial ischemia and reperfusion injury. *J Clin Pharmacol 38*, 887-897

[79] Gunter, T. E. and Pfeiffer, D. R. (1990) Mechanisms by which mitochondria transport calcium. *Am J Physiol 258*, C755-786

[80] Kim, J. S., He, L. and Lemasters, J. J. (2003) Mitochondrial permeability transition: a common pathway to necrosis and apoptosis. *Biochem Biophys Res Commun 304*, 463-470

[81] Syntichaki, P., Xu, K., Driscoll, M. and Tavernarakis, N. (2002) Specific aspartyl and calpain proteases are required for neurodegeneration in C. elegans. *Nature 419*, 939-944

[82] Crook, N. E., Clem, R. J. and Miller, L. K. (1993) An apoptosis-inhibiting baculovirus gene with a zinc finger-like motif. *J Virol 67*, 2168-2174

[83] Birnbaum, M. J., Clem, R. J. and Miller, L. K. (1994) An apoptosis-inhibiting gene from a nuclear polyhedrosis virus encoding a polypeptide with Cys/His sequence motifs. *J Virol 68*, 2521-2528

[84] Deveraux, Q. L. and Reed, J. C. (1999) IAP family proteins--suppressors of apoptosis. *Genes Dev 13*, 239-252

[85] Uren, A. G., Coulson, E. J. and Vaux, D. L. (1998) Conservation of baculovirus inhibitor of apoptosis repeat proteins (BIRPs) in viruses, nematodes, vertebrates and yeasts. *Trends Biochem Sci 23*, 159-162

[86] Deveraux, Q. L., Takahashi, R., Salvesen, G. S. and Reed, J. C. (1997) X-linked IAP is a direct inhibitor of cell-death proteases. *Nature 388*, 300-304

[87] Deveraux, Q. L., Roy, N., Stennicke, H. R., Van Arsdale, T., Zhou, Q., Srinivasula, S. M., Alnemri, E. S., Salvesen, G. S. and Reed, J. C. (1998) IAPs block apoptotic events

induced by caspase-8 and cytochrome c by direct inhibition of distinct caspases. *Embo .J 17*, 2215-2223

[88] Costantini, P., Bruey, J. M., Castedo, M., Metivier, D., Loeffler, M., Susin, S. A., Ravagnan, L., Zamzami, N., Garrido, C. and Kroemer, G. (2002) Pre-processed caspase-9 contained in mitochondria participates in apoptosis. *Cell Death Differ 9*, 82-88

[89] Adams, J. M. and Cory, S. (1998) The Bcl-2 protein family: arbiters of cell survival. *Science 281*, 1322-1326

[90] Chao, D. T. and Korsmeyer, S. J. (1998) BCL-2 family: regulators of cell death. *Annu Rev Immunol 16*, 395-419

[91] Fu, Y. F. and Fan, T. J. (2002) Bcl-2 family proteins and apoptosis. *Sheng Wu Hua Xue Yu Sheng Wu Wu Li Xue Bao (Shanghai) 34*, 389-394

[92] Milosevic, J., Hoffarth, S., Huber, C. and Schuler, M. (2003) The DNA damage-induced decrease of Bcl-2 is secondary to the activation of apoptotic effector caspases. *Oncogene 22*, 6852-6856

[93] Gross, A., McDonnell, J. M. and Korsmeyer, S. J. (1999) BCL-2 family members and the mitochondria in apoptosis. *Genes Dev 13*, 1899-1911

[94] Kelekar, A. and Thompson, C. B. (1998) Bcl-2-family proteins: the role of the BH3 domain in apoptosis. *Trends Cell Biol 8*, 324-330

[95] Oltvai, Z. N., Milliman, C. L. and Korsmeyer, S. J. (1993) Bcl-2 heterodimerizes in vivo with a conserved homolog, Bax, that accelerates programmed cell death. *Cell 74*, 609-619

[96] Green, D. R. and Reed, J. C. (1998) Mitochondria and apoptosis. *Science 281*, 1309-1312

[97] Kluck, R. M., Bossy-Wetzel, E., Green, D. R. and Newmeyer, D. D. (1997) The release of cytochrome c from mitochondria: a primary site for Bcl-2 regulation of apoptosis. *Science 275*, 1132-1136

[98] Rosse, T., Olivier, R., Monney, L., Rager, M., Conus, S., Fellay, I., Jansen, B. and Borner, C. (1998) Bcl-2 prolongs cell survival after Bax-induced release of cytochrome c. *Nature 391*, 496-499

[99] Matsuyama, S., Llopis, J., Deveraux, Q. L., Tsien, R. Y. and Reed, J. C. (2000) Changes in intramitochondrial and cytosolic pH: early events that modulate caspase activation during apoptosis. *Nat Cell Biol 2*, 318-325

[100] Minn, A. J., Kettlun, C. S., Liang, H., Kelekar, A., Vander Heiden, M. G., Chang, B. S., Fesik, S. W., Fill, M. and Thompson, C. B. (1999) Bcl-xL regulates apoptosis by heterodimerization-dependent and -independent mechanisms. *Embo J 18*, 632-643

[101] Harris, M. H., Vander Heiden, M. G., Kron, S. J. and Thompson, C. B. (2000) Role of oxidative phosphorylation in Bax toxicity. *Mol Cell Biol 20*, 3590-3596

[102] Schendel, S. L., Azimov, R., Pawlowski, K., Godzik, A., Kagan, B. L. and Reed, J. C. (1999) Ion channel activity of the BH3 only Bcl-2 family member, BID. *J Biol Chem 274*, 21932-21936

[103] Huigsloot, M., Tijdens, I. B., Mulder, G. J. and van de Water, B. (2002) Differential regulation of doxorubicin-induced mitochondrial dysfunction and apoptosis by Bcl-2 in mammary adenocarcinoma (MTLn3) cells. *J Biol Chem 277*, 35869-35879

[104] Gross, A., Jockel, J., Wei, M. C. and Korsmeyer, S. J. (1998) Enforced dimerization of BAX results in its translocation, mitochondrial dysfunction and apoptosis. *Embo J 17*, 3878-3885

[105] Schlesinger, P. H., Gross, A., Yin, X. M., Yamamoto, K., Saito, M., Waksman, G. and Korsmeyer, S. J. (1997) Comparison of the ion channel characteristics of proapoptotic BAX and antiapoptotic BCL-2. *Proc Natl Acad Sci U S A 94*, 11357-11362

[106] Tsujimoto, Y. and Shimizu, S. (2000) VDAC regulation by the Bcl-2 family of proteins. *Cell Death Differ 7*, 1174-1181

[107] Tsujimoto, Y. (2003) Cell death regulation by the Bcl-2 protein family in the mitochondria. *J Cell Physiol 195*, 158-167

[108] Sattler, M., Liang, H., Nettesheim, D., Meadows, R. P., Harlan, J. E., Eberstadt, M., Yoon, H. S., Shuker, S. B., Chang, B. S., Minn, A. J., Thompson, C. B. and Fesik, S. W. (1997) Structure of Bcl-xL-Bak peptide complex: recognition between regulators of apoptosis. *Science 275*, 983-986

[109] Desagher, S., Osen-Sand, A., Nichols, A., Eskes, R., Montessuit, S., Lauper, S., Maundrell, K., Antonsson, B. and Martinou, J. C. (1999) Bid-induced conformational change of Bax is responsible for mitochondrial cytochrome c release during apoptosis. *J Cell Biol 144*, 891-901

[110] Yin, X. M., Wang, K., Gross, A., Zhao, Y., Zinkel, S., Klocke, B., Roth, K. A. and Korsmeyer, S. J. (1999) Bid-deficient mice are resistant to Fas-induced hepatocellular apoptosis. *Nature 400*, 886-891

[111] Goping, I. S., Gross, A., Lavoie, J. N., Nguyen, M., Jemmerson, R., Roth, K., Korsmeyer, S. J. and Shore, G. C. (1998) Regulated targeting of BAX to mitochondria. *J Cell Biol 143*, 207-215

[112] Desagher, S. and Martinou, J. C. (2000) Mitochondria as the central control point of apoptosis. *Trends Cell Biol 10*, 369-377

[113] Mancini, M., Nicholson, D. W., Roy, S., Thornberry, N. A., Peterson, E. P., Casciola-Rosen, L. A. and Rosen, A. (1998) The caspase-3 precursor has a cytosolic and mitochondrial distribution: implications for apoptotic signaling. *J Cell Biol 140*, 1485-1495

[114] Krajewski, S., Krajewska, M., Ellerby, L. M., Welsh, K., Xie, Z., Deveraux, Q. L., Salvesen, G. S., Bredesen, D. E., Rosenthal, R. E., Fiskum, G. and Reed, J. C. (1999) Release of caspase-9 from mitochondria during neuronal apoptosis and cerebral ischemia. *Proc Natl Acad Sci U S A 96*, 5752-5757

[115] Susin, S. A., Lorenzo, H. K., Zamzami, N., Marzo, I., Brenner, C., Larochette, N., Prevost, M. C., Alzari, P. M. and Kroemer, G. (1999) Mitochondrial release of caspase-2 and -9 during the apoptotic process. *J Exp Med 189*, 381-394

[116] Susin, S. A., Lorenzo, H. K., Zamzami, N., Marzo, I., Snow, B. E., Brothers, G. M., Mangion, J., Jacotot, E., Costantini, P., Loeffler, M., Larochette, N., Goodlett, D. R., Aebersold, R., Siderovski, D. P., Penninger, J. M. and Kroemer, G. (1999) Molecular characterization of mitochondrial apoptosis-inducing factor. *Nature 397*, 441-446

[117] Du, C., Fang, M., Li, Y., Li, L. and Wang, X. (2000) Smac, a mitochondrial protein that promotes cytochrome c-dependent caspase activation by eliminating IAP inhibition. *Cell 102*, 33-42

[118] Verhagen, A. M., Ekert, P. G., Pakusch, M., Silke, J., Connolly, L. M., Reid, G. E., Moritz, R. L., Simpson, R. J. and Vaux, D. L. (2000) Identification of DIABLO, a mammalian protein that promotes apoptosis by binding to and antagonizing IAP proteins. *Cell 102*, 43-53

[119] Wang, K. K. (2000) Calpain and caspase: can you tell the difference? *Trends Neurosci 23*, 20-26

[120] Blomgren, K., Zhu, C., Wang, X., Karlsson, J. O., Leverin, A. L., Bahr, B. A., Mallard, C. and Hagberg, H. (2001) Synergistic activation of caspase-3 by m-calpain after neonatal hypoxia-ischemia: a mechanism of "pathological apoptosis"? *J Biol Chem 276*, 10191-10198

[121] Mandic, A., Hansson, J., Linder, S. and Shoshan, M. C. (2003) Cisplatin induces endoplasmic reticulum stress and nucleus-independent apoptotic signaling. *J Biol Chem 278*, 9100-9106

[122] Nakagawa, T. and Yuan, J. (2000) Cross-talk between two cysteine protease families. Activation of caspase-12 by calpain in apoptosis. *J Cell Biol 150*, 887-894

[123] Hajnoczky, G., Davies, E. and Madesh, M. (2003) Calcium signaling and apoptosis. *Biochem Biophys Res Commun 304*, 445-454

[124] Schmitz, I., Kirchhoff, S. and Krammer, P. H. (2000) Regulation of death receptor-mediated apoptosis pathways. *Int J Biochem Cell Biol 32*, 1123-1136

[125] Thome, M., Schneider, P., Hofmann, K., Fickenscher, H., Meinl, E., Neipel, F., Mattmann, C., Burns, K., Bodmer, J. L., Schroter, M., Scaffidi, C., Krammer, P. H., Peter, M. E. and Tschopp, J. (1997) Viral FLICE-inhibitory proteins (FLIPs) prevent apoptosis induced by death receptors. *Nature 386*, 517-521

[126] Hu, S., Vincenz, C., Buller, M. and Dixit, V. M. (1997) A novel family of viral death effector domain-containing molecules that inhibit both CD-95- and tumor necrosis factor receptor-1-induced apoptosis. *J Biol Chem 272*, 9621-9624

[127] Irmler, M., Thome, M., Hahne, M., Schneider, P., Hofmann, K., Steiner, V., Bodmer, J. L., Schroter, M., Burns, K., Mattmann, C., Rimoldi, D., French, L. E. and Tschopp, J. (1997) Inhibition of death receptor signals by cellular FLIP. *Nature 388*, 190-195

[128] Tschopp, J., Irmler, M. and Thome, M. (1998) Inhibition of fas death signals by FLIPs. *Curr Opin Immunol 10*, 552-558

[129] Kirchhoff, S., Muller, W. W., Krueger, A., Schmitz, I. and Krammer, P. H. (2000) TCR-mediated up-regulation of c-FLIPshort correlates with resistance toward CD95-mediated apoptosis by blocking death-inducing signaling complex activity. *J Immunol 165*, 6293-6300

[130] Yoshikawa, H., Nakajima, Y. and Tasaka, K. (2000) Enhanced expression of Fas-associated death domain-like IL-1-converting enzyme (FLICE)-inhibitory protein induces resistance to Fas-mediated apoptosis in activated mast cells. *J Immunol 165*, 6262-6269

[131] Adrain, C., Murphy, B. M. and Martin, S. J. (2005) Molecular ordering of the caspase activation cascade initiated by the cytotoxic T lymphocyte/natural killer (CTL/NK) protease granzyme B. *J Biol Chem 280*, 4663-4673

[132] Barry, M., Heibein, J. A., Pinkoski, M. J., Lee, S. F., Moyer, R. W., Green, D. R. and Bleackley, R. C. (2000) Granzyme B short-circuits the need for caspase 8 activity

during granule-mediated cytotoxic T-lymphocyte killing by directly cleaving Bid. *Mol Cell Biol 20*, 3781-3794

[133] Ray, C. A., Black, R. A., Kronheim, S. R., Greenstreet, T. A., Sleath, P. R., Salvesen, G. S. and Pickup, D. J. (1992) Viral inhibition of inflammation: cowpox virus encodes an inhibitor of the interleukin-1 beta converting enzyme. *Cell 69*, 597-604

[134] Komiyama, T., Ray, C. A., Pickup, D. J., Howard, A. D., Thornberry, N. A., Peterson, E. P. and Salvesen, G. (1994) Inhibition of interleukin-1 beta converting enzyme by the cowpox virus serpin CrmA. An example of cross-class inhibition. *J Biol Chem 269*, 19331-19337

[135] Tewari, M. and Dixit, V. M. (1995) Fas- and tumor necrosis factor-induced apoptosis is inhibited by the poxvirus crmA gene product. *J Biol Chem 270*, 3255-3260

[136] Komiyama, T., Quan, L. T. and Salvesen, G. S. (1996) Inhibition of cysteine and serine proteinases by the cowpox virus serpin CRMA. *Adv Exp Med Biol 389*, 173-176

[137] Miura, M., Zhu, H., Rotello, R., Hartwieg, E. A. and Yuan, J. (1993) Induction of apoptosis in fibroblasts by IL-1 beta-converting enzyme, a mammalian homolog of the C. elegans cell death gene ced-3. *Cell 75*, 653-660

[138] Gurevich, R. M., Regula, K. M. and Kirshenbaum, L. A. (2001) Serpin protein CrmA suppresses hypoxia-mediated apoptosis of ventricular myocytes. *Circulation 103*, 1984-1991

[139] Zhou, Q. and Salvesen, G. S. (2000) Viral caspase inhibitors CrmA and p35. *Methods Enzymol 322*, 143-154

[140] Bertin, J., Armstrong, R. C., Ottilie, S., Martin, D. A., Wang, Y., Banks, S., Wang, G. H., Senkevich, T. G., Alnemri, E. S., Moss, B., Lenardo, M. J., Tomaselli, K. J. and Cohen, J. I. (1997) Death effector domain-containing herpesvirus and poxvirus proteins inhibit both Fas- and TNFR1-induced apoptosis. *Proc Natl Acad Sci U S A 94*, 1172-1176

[141] Bump, N. J., Hackett, M., Hugunin, M., Seshagiri, S., Brady, K., Chen, P., Ferenz, C., Franklin, S., Ghayur, T., Li, P. and et al. (1995) Inhibition of ICE family proteases by baculovirus antiapoptotic protein p35. *Science 269*, 1885-1888

[142] Xue, D. and Horvitz, H. R. (1995) Inhibition of the Caenorhabditis elegans cell-death protease CED-3 by a CED-3 cleavage site in baculovirus p35 protein. *Nature 377*, 248-251

[143] Miller, L. K. (1997) Baculovirus interaction with host apoptotic pathways. *J Cell Physiol 173*, 178-182

[144] Datta, R., Kojima, H., Banach, D., Bump, N. J., Talanian, R. V., Alnemri, E. S., Weichselbaum, R. R., Wong, W. W. and Kufe, D. W. (1997) Activation of a CrmA-insensitive, p35-sensitive pathway in ionizing radiation-induced apoptosis. *J Biol Chem 272*, 1965-1969

[145] Beidler, D. R., Tewari, M., Friesen, P. D., Poirier, G. and Dixit, V. M. (1995) The baculovirus p35 protein inhibits Fas- and tumor necrosis factor-induced apoptosis. *J Biol Chem 270*, 16526-16528

[146] Stennicke, H. R., Ryan, C. A. and Salvesen, G. S. (2002) Reprieval from execution: the molecular basis of caspase inhibition. *Trends Biochem Sci 27*, 94-101

[147] Xu, G., Cirilli, M., Huang, Y., Rich, R. L., Myszka, D. G. and Wu, H. (2001) Covalent inhibition revealed by the crystal structure of the caspase-8/p35 complex. *Nature 410*, 494-497

[148] Vier, J., Furmann, C. and Hacker, G. (2000) Baculovirus P35 protein does not inhibit caspase-9 in a cell-free system of apoptosis. *Biochem Biophys Res Commun 276*, 855-861

[149] Silverman, N. and Maniatis, T. (2001) NF-kappaB signaling pathways in mammalian and insect innate immunity. *Genes Dev 15*, 2321-2342

[150] Karin, M. and Lin, A. (2002) NF-kappaB at the crossroads of life and death. *Nat Immunol 3*, 221-227

[151] Li, Q. and Verma, I. M. (2002) NF-kappaB regulation in the immune system. *Nat Rev Immunol 2*, 725-734

[152] Kasibhatla, S., Brunner, T., Genestier, L., Echeverri, F., Mahboubi, A. and Green, D. R. (1998) DNA damaging agents induce expression of Fas ligand and subsequent apoptosis in T lymphocytes via the activation of NF-kappa B and AP-1. *Mol Cell 1*, 543-551

[153] Lin, B., Williams-Skipp, C., Tao, Y., Schleicher, M. S., Cano, L. L., Duke, R. C. and Scheinman, R. I. (1999) NF-kappaB functions as both a proapoptotic and antiapoptotic regulatory factor within a single cell type. *Cell Death Differ 6*, 570-582

[154] Zheng, Y., Ouaaz, F., Bruzzo, P., Singh, V., Gerondakis, S. and Beg, A. A. (2001) NF-kappa B RelA (p65) is essential for TNF-alpha-induced fas expression but dispensable for both TCR-induced expression and activation-induced cell death. *J Immunol 166*, 4949-4957

[155] Kucharczak, J., Simmons, M. J., Fan, Y. and Gelinas, C. (2003) To be, or not to be: NF-kappaB is the answer--role of Rel/NF-kappaB in the regulation of apoptosis. *Oncogene 22*, 8961-8982

[156] Jun, J. I., Chung, C. W., Lee, H. J., Pyo, J. O., Lee, K. N., Kim, N. S., Kim, Y. S., Yoo, H. S., Lee, T. H., Kim, E. and Jung, Y. K. (2005) Role of FLASH in caspase-8-mediated activation of NF-kappaB: dominant-negative function of FLASH mutant in NF-kappaB signaling pathway. *Oncogene 24*, 688-696

[157] Badrichani, A. Z., Stroka, D. M., Bilbao, G., Curiel, D. T., Bach, F. H. and Ferran, C. (1999) Bcl-2 and Bcl-XL serve an anti-inflammatory function in endothelial cells through inhibition of NF-kappaB. *J Clin Invest 103*, 543-553

[158] Kloppel, G. and Maillet, B. (1993) Pathology of acute and chronic pancreatitis. *Pancreas 8*, 659-670

[159] Nevalainen, T. J. and Aho, H. J. (1992) Standards of morphological evaluation and histological grading in experimental acute pancreatitis. *Eur Surg Res 24 Suppl 1*, 14-23

[160] Walker, N. I. (1987) Ultrastructure of the rat pancreas after experimental duct ligation. I. The role of apoptosis and intraepithelial macrophages in acinar cell deletion. *Am J Pathol 126*, 439-451

[161] Walker, N. I., Winterford, C. M. and Kerr, J. F. (1992) Ultrastructure of the rat pancreas after experimental duct ligation. II. Duct and stromal cell proliferation, differentiation, and deletion. *Pancreas 7*, 420-434

[162] Jimi, A., Kojiro, M., Miyasaka, K., Kono, A. and Funakoshi, A. (1997) Apoptosis in the pancreas of genetically diabetic rats with a disrupted cholecystokinin (CCK-A) receptor gene. *Pancreas 14*, 109-112

[163] Hahm, K. B., Kim, J. H., You, B. M., Kim, Y. S., Cho, S. W., Yim, H., Ahn, B. O. and Kim, W. B. (1998) Induction of apoptosis with an extract of Artemisia asiatica attenuates the severity of cerulein-induced pancreatitis in rats. *Pancreas 17*, 153-157

[164] Chao, K. C., Chao, K. F., Chuang, C. C. and Liu, S. H. (2006) Blockade of interleukin 6 accelerates acinar cell apoptosis and attenuates experimental acute pancreatitis in vivo. *Br J Surg 93*, 332-338

[165] Lupia, E., Goffi, A., De Giuli, P., Azzolino, O., Bosco, O., Patrucco, E., Vivaldo, M. C., Ricca, M., Wymann, M. P., Hirsch, E., Montrucchio, G. and Emanuelli, G. (2004) Ablation of phosphoinositide 3-kinase-gamma reduces the severity of acute pancreatitis. *Am J Pathol 165*, 2003-2011

[166] Frossard, J. L., Rubbia-Brandt, L., Wallig, M. A., Benathan, M., Ott, T., Morel, P., Hadengue, A., Suter, S., Willecke, K. and Chanson, M. (2003) Severe acute pancreatitis and reduced acinar cell apoptosis in the exocrine pancreas of mice deficient for the Cx32 gene. *Gastroenterology 124*, 481-493

[167] Gukovsky, I., Reyes, C. N., Vaquero, E. C., Gukovskaya, A. S. and Pandol, S. J. (2003) Curcumin ameliorates ethanol and nonethanol experimental pancreatitis. *Am J Physiol Gastrointest Liver Physiol 284*, G85-95

[168] Cao, Y., Adhikari, S., Ang, A. D., Clement, M. V., Wallig, M. and Bhatia, M. (2006) Crambene induces pancreatic acinar cell apoptosis via the activation of mitochondrial pathway. *Am J Physiol Gastrointest Liver Physiol*

[169] Cao, Y., Adhikari, S., Ang, A. D., Moore, P. K. and Bhatia, M. (2006) Mechanism of Induction of Pancreatic Acinar Cell Apoptosis by Hydrogen Sulfide. *Am J Physiol Cell Physiol*

[170] Gerasimenko, J. V., Gerasimenko, O. V., Palejwala, A., Tepikin, A. V., Petersen, O. H. and Watson, A. J. (2002) Menadione-induced apoptosis: roles of cytosolic Ca(2+) elevations and the mitochondrial permeability transition pore. *J Cell Sci 115*, 485-497

[171] Gukovskaya, A. S., Gukovsky, I., Jung, Y., Mouria, M. and Pandol, S. J. (2002) Cholecystokinin induces caspase activation and mitochondrial dysfunction in pancreatic acinar cells. Roles in cell injury processes of pancreatitis. *J Biol Chem 277*, 22595-22604

[172] Song, J. Y., Lim, J. W., Kim, H., Morio, T. and Kim, K. H. (2003) Oxidative stress induces nuclear loss of DNA repair proteins Ku70 and Ku80 and apoptosis in pancreatic acinar AR42J cells. *J Biol Chem 278*, 36676-36687

[173] Beil, M., Leser, J., Lutz, M. P., Gukovskaya, A., Seufferlein, T., Lynch, G., Pandol, S. J. and Adler, G. (2002) Caspase 8-mediated cleavage of plectin precedes F-actin breakdown in acinar cells during pancreatitis. *Am J Physiol Gastrointest Liver Physiol 282*, G450-460

[174] Mareninova, O. A., Sung, K. F., Hong, P., Lugea, A., Pandol, S. J., Gukovsky, I. and Gukovskaya, A. S. (2006) Cell death in pancreatitis: caspases protect from necrotizing pancreatitis. *J Biol Chem 281*, 3370-3381

[175] Gomez, G., Lee, H. M., He, Q., Englander, E. W., Uchida, T. and Greeley, G. H., Jr. (2001) Acute pancreatitis signals activation of apoptosis-associated and survival genes in mice. *Exp Biol Med (Maywood) 226*, 692-700

[176] Tomasini, R., Samir, A. A., Vaccaro, M. I., Pebusque, M. J., Dagorn, J. C., Iovanna, J. L. and Dusetti, N. J. (2001) Molecular and functional characterization of the stress-induced protein (SIP) gene and its two transcripts generated by alternative splicing. SIP induced by stress and promotes cell death. *J Biol Chem 276*, 44185-44192

[177] O'Neill, S., O'Neill, A. J., Conroy, E., Brady, H. R., Fitzpatrick, J. M. and Watson, R. W. (2000) Altered caspase expression results in delayed neutrophil apoptosis in acute pancreatitis. *J Leukoc Biol 68*, 15-20

[178] Salomone, T., Tosi, P., Raiti, C., Guariento, A., Tomassetti, P., Migliori, M., Saieva, C., Romboli, M. and Gullo, L. (2002) Apoptosis in the peripheral blood mononuclear cells as a self-limitation process in human acute pancreatitis. *Pancreatology 2*, 204-210

[179] Fujimoto, K., Hosotani, R., Doi, R., Wada, M., Lee, J. U., Koshiba, T., Miyamoto, Y. and Imamura, M. (1997) Role of neutrophils in cerulein-induced pancreatitis in rats: possible involvement of apoptosis. *Digestion 58*, 421-430

[180] Rau, B., Paszkowski, A., Esber, S., Gansauge, F., Poch, B., Beger, H. G. and Moller, P. (2001) Anti-ICAM-1 antibody modulates late onset of acinar cell apoptosis and early necrosis in taurocholate-induced experimental acute pancreatitis. *Pancreas 23*, 80-88

[181] Satoh, A., Gukovskaya, A. S., Edderkaoui, M., Daghighian, M. S., Reeve, J. R., Jr., Shimosegawa, T. and Pandol, S. J. (2005) Tumor necrosis factor-alpha mediates pancreatitis responses in acinar cells via protein kinase C and proline-rich tyrosine kinase 2. *Gastroenterology 129*, 639-651

[182] Monks, T. J., Hanzlik, R. P., Cohen, G. M., Ross, D. and Graham, D. G. (1992) Quinone chemistry and toxicity. *Toxicol Appl Pharmacol 112*, 2-16

[183] Niedoborski, T. E., Klein, B. P. and Wallig, M. A. (2001) Rapid isolation and purification of 1-cyano-2-hydroxy-3-butene (crambene) from Crambe abyssinica seed meal using immiscible solvent extraction and high-performance liquid chromatography. *J Agric Food Chem 49*, 3594-3599

[184] Wallig, M. A., Kingston, S., Staack, R. and Jefferey, E. H. (1998) Induction of rat pancreatic glutathione S-transferase and quinone reductase activities by a mixture of glucosinolate breakdown derivatives found in Brussels sprouts. *Food Chem Toxicol 36*, 365-373

[185] Bhatia, M., Brady, M., Kang, Y. K., Costello, E., Newton, D. J., Christmas, S. E., Neoptolemos, J. P. and Slavin, J. (2002) MCP-1 but not CINC synthesis is increased in rat pancreatic acini in response to cerulein hyperstimulation. *Am J Physiol Gastrointest Liver Physiol 282*, G77-85

[186] Bhatia, M., Slavin, J., Cao, Y., Basbaum, A. I. and Neoptolemos, J. P. (2003) Preprotachykinin-A gene deletion protects mice against acute pancreatitis and associated lung injury. *Am J Physiol Gastrointest Liver Physiol 284*, G830-836

[187] Nicholson, D. W. (2000) From bench to clinic with apoptosis-based therapeutic agents. *Nature 407*, 810-816

[188] Fischer, U. and Schulze-Osthoff, K. (2005) Apoptosis-based therapies and drug targets. *Cell Death Differ 12 Suppl 1*, 942-961

In: Cell Apoptotic Signaling Pathways
Editor: Charles O. Pickens, pp. 163-184

ISBN: 978-1-60021-507-0

Chapter VI

Osteoclast Apoptosis

Sophie Roux **[*]** ***and Estelle Chamoux***
Department of Rheumatology, Sherbrooke University, Sherbrooke, PQ, Canada.

Abstract

Introduction: Bone resorption and formation are closely linked during bone remodeling. The cessation of bone resorption precedes bone formation, and is characterized by osteoclast apoptosis. Although apoptosis has been studied in rodent models of osteoclast differentiation, little is known about osteoclast apoptosis in humans.

Osteoclast apoptosis: The two most common apoptotic pathways are the activation of one of the membrane death receptors, and that of the mitochondrion-activated pathway involving members of the Bcl-2 family. The induction of apoptosis by extracellular signals involves ligands related to TNF. Some death receptors have been described in detail, including CD95 (Fas) that binds the CD95 ligand (FasL), and the TRAIL receptors that bind TRAIL (TNF-Related Apoptosis-Inducing Ligand). Among others, caspase-8, -9, and -3 are situated at pivotal junctions in apoptotic pathways. Caspase-8 activation is initiated during receptor-induced cell death. Caspase-9 is activated in response to agents or insults that impair the internal Bcl-2 family members balance. Both caspase-8 and caspase-9 eventually lead to the activation of executor caspases, such as caspase-3 or caspase-7. When apoptosis occurs in osteoclasts, caspases are likely to be activated. Osteoclast apoptosis and its down-regulation by the survival factor M-CSF, may involve members of the Bcl-2 family, particularly Bcl-xL. The Fas-FasL and TRAIL-TRAILR systems may also be involved in osteoclast apoptosis; however TRAIL-/- or Fas-/- deficient mouse models suggest that these factors cannot be essential for bone biology under normal conditions. However, the two main apoptotic pathways, involving TRAIL receptors and Fas, could be implicated in human osteoclast apoptosis. The

[*] Correspondence concerning this article should be addressed to Sophie Roux, MD, PhD, Department of Rheumatology, Faculty of Medicine, Centre hospitalier universitaire de Sherbrooke 3001, 12e Avenue Nord, Fleurimont, Québec CANADA, J1H 5N4. TEL: (819) 564-5261; FAX: (819) 564-5265; email address: sophie.roux@usherbrooke.ca.

regulation of this process may be important in controlling bone homeostasis, and could be altered *in-vivo* under conditions characterized by a high level of osteoclast formation.

Conclusion: A better knowledge of the role of apoptosis as a major point of control of bone resorption, would fairly increase our understanding of osteoclast biology. This field is of considerable clinical importance, and can be expected to advance our therapeutic approach to diseases characterized by a high level of bone resorption.

INTRODUCTION

Bone remodeling is a continuous physiological process in which bone resorption is followed by the formation of new bone. The bone cells responsible for this coupled process include bone-resorbing cells, the osteoclasts, which are derived from hematopoietic cells of the monocyte-macrophage lineage, and bone-forming cells, the osteoblasts, which are of mesenchymal origin. These bone-resorbing and bone-forming processes are closely linked and tightly regulated in order to maintain skeletal homeostasis [1].

The survival and apoptosis (programmed cell death) of bone cells are both of major importance in bone remodeling, and in diseases such as osteoporosis or malignant osteolysis they can lead to crucial shifts in bone resorption. We have considerable understanding of the mechanisms of osteoclast apoptosis in rodents, but little is known about osteoclast apoptosis in humans [2]. There is a real need to elucidate this process, because osteoclast life span is the keystone to bone remodeling. This process could offer a useful target in the treatment of skeletal disorders, and in other human diseases the development of pharmacological inhibitors or inducers of apoptosis is already in progress [3].

BONE REMODELING AND OSTEOCLAST BONE RESORPTION

Osteoclasts are multinucleated cells formed by the fusion of mononuclear cells of hematopoietic origin, and they are responsible for bone resorption. Osteoclasts share many antigenic determinants with monocytes, tissue macrophages, and macrophage polykaryons. However, the osteoclast phenotype differs from the latter in several important respects; in particular, osteoclasts and their precursors express the calcitonin receptor (CTR), and *in-vitro* they produce bone resorption pits on the surface of cortical bone or dentine [4,5].

The discovery of new members of the TNF receptor-ligand family has highlighted the crucial role of RANKL (Receptor Activator of NK-κB Ligand) and its receptor, RANK, in osteoclast differentiation and activation [6-8]. RANKL is a membrane-bound, TNF-related factor expressed by osteoblast/stromal cells and activated lymphocytes, and may be cleaved to produce a soluble form (sRANKL). The absolute requirement for RANKL in osteoclast differentiation is now well established, and its soluble, recombinant form has been investigated in a number of *in-vitro* and *in-vivo* studies. In *in-vitro* murine or human osteoclast differentiation models, soluble RANKL enables osteoclast precursors to differentiate into mature osteoclasts in the presence of M-CSF, even in the absence of osteoblast/stromal cells [9,10]. Bone resorption activity and osteoclast survival are also increased by RANKL [10,11].

RANKL acts by binding to its membrane-bound receptor, RANK. This trimeric receptor is essential for the transduction of signals leading to osteoclast differentiation [12,13].

Osteoprotegerin (OPG) is another member of the TNF receptor family, but with no transmembrane domain and is secreted as a decoy receptor. OPG recognizes RANKL, and in doing so, blocks the interaction between RANK and RANKL, leading to the inhibition of osteoclast differentiation and activation [14,15] (Figure 1). *In-vitro* studies have demonstrated that OPG has a marked inhibitory effect on osteoclast differentiation, as well as on the bone-resorption activity of the osteoclasts [14,15]. In addition to its role in blocking RANKL, it has been suggested that OPG may also have a direct effect on osteoclasts [16]. In bone tissue OPG and RANKL are expressed by osteoblast/stromal cells, and the ratio of these substances may modulate the ability of these cells to stimulate osteoclast differentiation or activity, and thus affect the rate of bone resorption [7].

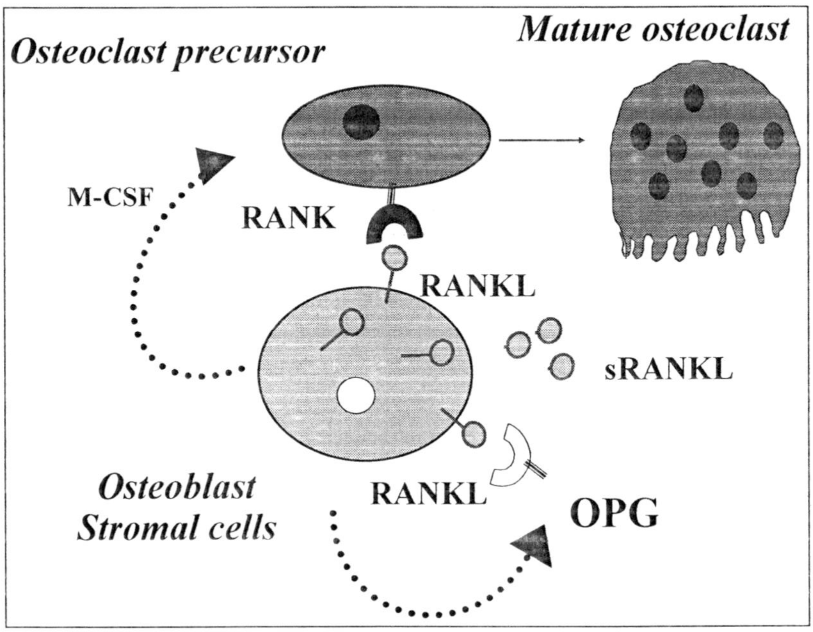

Figure 1. The role of RANKL and its receptor, RANK, in osteoclast differentiation. Osteoblast/stromal cells express RANKL, a membrane-bound TNF-related factor. Osteoclast precursors express RANK, a membrane-bound TNF-receptor that recognizes RANKL through cell-to-cell interaction with osteoblast/stromal cells. This interaction enables osteoclast precursors to differentiate in the presence of M-CSF. OPG is a member of the TNF receptor family that lacks a transmembrane domain and is a secreted TNF receptor. OPG recognizes RANKL, and this decoy receptor blocks the interaction between RANK and RANKL, leading to inhibition of osteoclast differentiation and activation.

In both pre-osteoclasts and mature osteoclasts, the interaction between RANKL and RANK results in transduction signals that activate the expression of specific genes involved in osteoclast differentiation, activation and survival. Several signaling cascades are induced, including NF-κB, JNK, p38, ERK and Src pathways [6,10]. The signaling complex that binds to the cytoplasmic domain of RANK includes several associated proteins such as the TNFR-associated factors (TRAFs), mainly TRAF6 [7,17] (Figure 2).

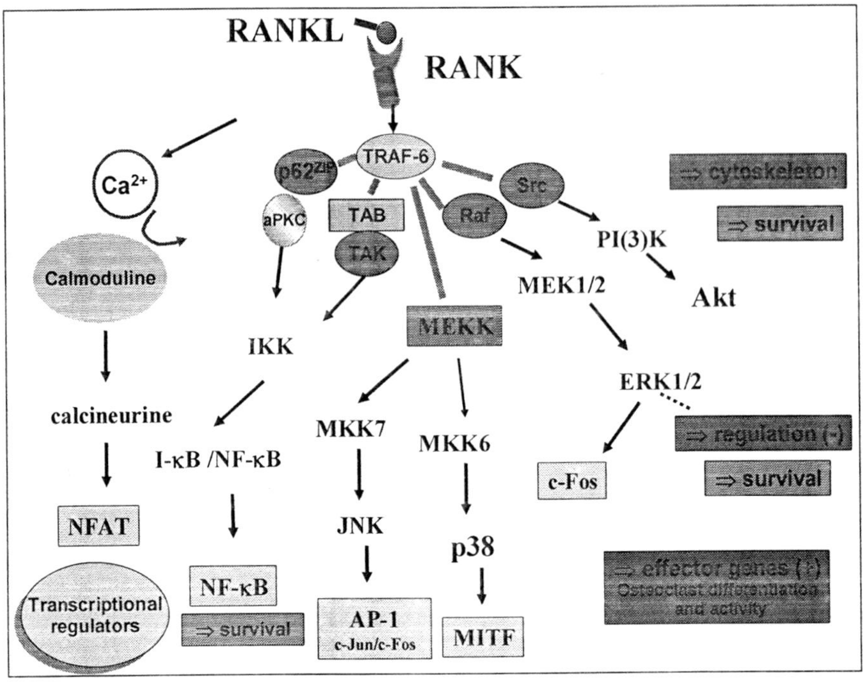

Figure 2. RANK signaling. Interactions between RANKL and RANK result in the activation of several signaling cascades, including NF-κB, JNK, p38, ERK and Src pathways. This in turn leads to the expression of specific genes involved in osteoclast differentiation, activation and survival. The signaling complex associated to the RANK cytoplasmic domain includes several proteins such as the TNFR-associated factors (TRAFs), mainly TRAF6.

Bone resorbing factors, such as PTH, 1,25$(OH)_2D_3$, IL1β, TNFα and PGE_2 increase the RANKL/OPG ratio by upregulating RANKL expression and/or downregulating OPG expression in osteoblast/stromal cells [18]. Inhibitors of bone resorption, such as IL4, IFNγ and TGFβ reduce RANKL expression or block RANKL-transduced signals [19-21]. In addition, TGFβ and estrogens increase OPG expression [18]. A new concept has emerged from these findings, which suggests that the cytokines and hormonal factors involved in bone resorption may act via a common final pathway involving RANKL and RANK [7].

Evidence from rodent models suggests that, after producing their resorption activity, the osteoclasts undergo apoptosis. Regulation of the numbers of osteoclast precursors and mature osteoclasts via the induction of cell death may therefore be a mechanism by which the number of active osteoclasts at any site is controlled, thus regulating bone resorption [22-24].

Although we know a lot about the factors that regulate osteoclast formation and activity, little is known about the factors involved in osteoclast apoptosis. Cytokines that enhance osteoclast activity do so in part by increasing their life span, as has been shown for TNFα, IL1, IL6, RANKL and M-CSF. In the same way, factors that inhibit osteoclast activity could induce osteoclast apoptosis as well as blocking osteoclast formation and bone resorption.

Such an effect has been observed in murine studies in which TGFβ and estrogens induced osteoclast apoptosis [25,26], as did OPG [27].

APOPTOSIS: GENERAL CONSIDERATIONS

Apoptosis, which is also known as programmed cell death, is characterized by morphological and biochemical changes in the cell. These include DNA fragmentation, chromatin condensation, protein degradation, and finally cell disassembly into apoptotic bodies. Caspases is the common name for members of a large family of cysteine proteases, known as the main executors of apoptotic cell death [28,29]. Caspases are synthesized as inactive precursors, and are activated by proteolytic cleavage. Among the twelve caspases encoded by the human genome, seven are known for their role in apoptosis (caspases -2,-3,-6,-7,-8,-9 and -10), the other ones being rather involved in cytokine processing [30]. There are two types of apoptotic caspases, the initiator caspases (e.g., caspases -2, -8, -9, and -10), and their targets, known as effector caspases (e.g., caspases -3, -6, and -7). The most largely studied caspases-8, -9, and -3 are located at pivotal junctions in apoptotic cascades. Apoptosis may occur via one or more pathways. Caspase-8, as part of a signaling complex of a surface receptor, is activated in response to extracellular, apoptosis-inducing ligands (extrinsic pathway). Caspase-9 is activated in response to agents or insults that trigger the release of cytochrome c from the mitochondria (intrinsic pathway). Both caspase-8 and caspase-9 can cleave and activate effector caspases like caspase-3, which will in turn cut some vital intracellular proteins and activate other caspases. These pathways may even be linked, since in some cases mitochondria-dependent caspase activation can follow caspase-8 activation [31,32].

Apoptosis Induced by Cell Surface Death Receptors (Figure 3)

The induction of apoptosis by extracellular signals involves ligands related to TNF. Conversely, the death receptors, cell surface receptors that transmit apoptosis signals, belong to the TNF receptor superfamily. All functional death receptors have an intracytoplasmic region called death domain (DD). The death receptors characterized to date include CD95 (Fas) that binds CD95 ligand (FasL), TNFR1 that binds TNF, and TRAIL receptors that bind TRAIL (TNF-Related Apoptosis-Inducing Ligand). After ligation, death receptors cluster as trimers at the plasma membrane, and promote the recruitment of specific adaptor proteins. Within the large TNF receptor superfamily, not every receptor/ligand interaction will lead to cell death and in several cases (eg RANK-RANKL), rather lead to a cascade of signalization and gene expression.

Upon ligation of TNFα, the trimerized TNFR1 recruits the adaptor molecule TRADD (TNFR-associated death domain), which subsequently recruits FADD (Fas Associated Death Domain) and the procaspase-8. As well as recruiting adaptor molecules that bind and activate caspases, TRADD also binds the serine-threonine kinase RIP (Receptor-Interacting Protein), thereby coupling the stimulation of TNFR1 to the activation of NF-κB. By engaging TNFR1,

TNFα activates transcription factors, leading to the induction of pro-inflammatory and immunomodulatory genes. In some circumstances, TNFα may also induce apoptosis. However, TNFα rarely triggers apoptosis unless protein synthesis is blocked, which suggests that there may be some cellular factors that can suppress the apoptotic stimulus generated by TNFα [33].

Triggering the Fas receptor with FasL results in receptor trimerization, clustering of the DDs, and subsequent recruitment of the dual adaptor molecule FADD. FADD contains both a DD and a DED (death effector domain), allowing it to recruit the initiator, procaspase-8. This complex of interacting proteins is known as the death-inducing signaling complex (DISC). The Fas-FasL system has been shown to play an important role in physiological apoptosis, particularly in the deletion of activated T cells at the end of the immune response, and in the ability of cytotoxic T cells and NK cells to kill virus-infected cells or cancer cells [33].

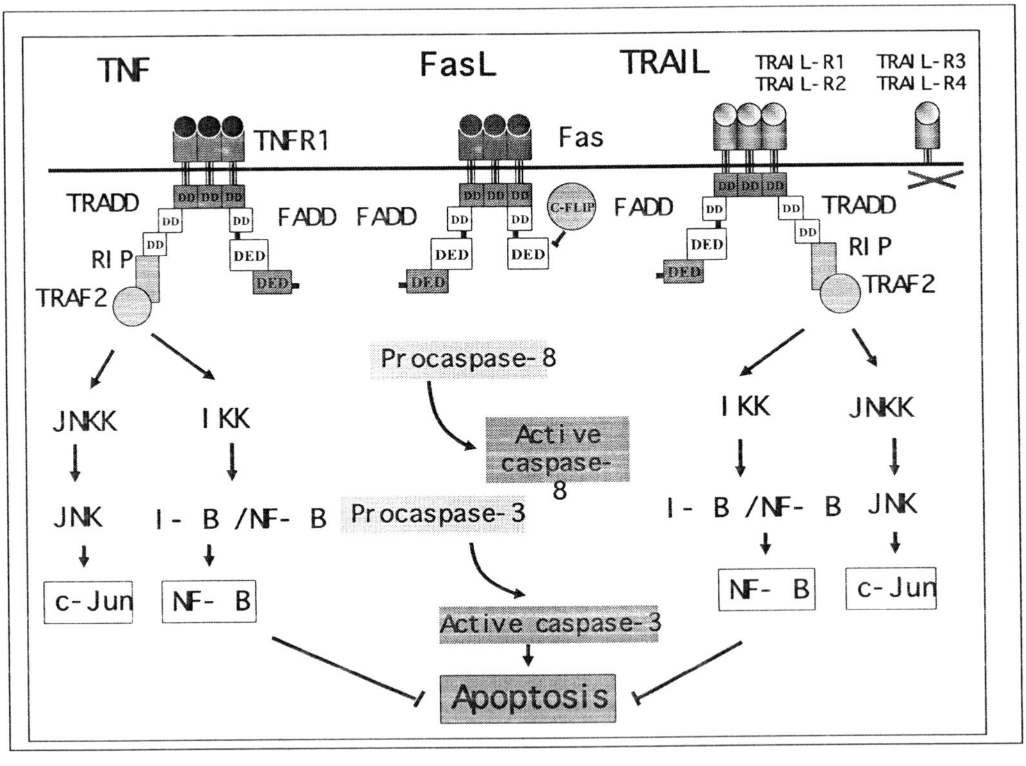

Figure 3. Extrinsic cell death signaling. Death receptors possess an intracelllular sequence motif, known as the death domain (DD). Triggering of the death receptor with its ligand results in receptor trimerisation, clustering of the receptor DD and subsequent recruitment of the dual adaptor molecule. The best characterized death receptors included CD95 (Fas), TNFR1, and the TRAIL receptors.

Unlike the expression of CD95L, which is mainly restricted to hematopoietic progenitors [34], activated T cells, NK cells, and lymphoid tissues [35], TRAIL mRNA expression occurs in many tissues [36]. To date, five receptors for TRAIL have been identified and characterized. Two of them, TRAIL-R1 and TRAIL-R2, contain a cytoplasmic DD, and are able to transduce a death signal. TRAIL-R1 and TRAIL-R2 transcripts have been found in several tissues, suggesting that there may be mechanisms that protect cells against the induction of apoptosis by TRAIL. One type of protection is based on decoy receptor expression. For instance, TRAIL-R3 and TRAIL-R4 do not have a functional DD, and are not able to mediate apoptosis. Finally, the fifth TRAIL receptor is the soluble decoy receptor, OPG, that inhibits TRAIL-mediated apoptosis [37].

TRAIL is produced by monocytes and NK cells, and appears to be an important effector molecule in defending the host against transformed cells [38-40]. In addition, TRAIL may have a role in the apoptosis of antigen-bearing dendritic cells induced by TRAIL-expressing activated T-cells [18].

The intracellular portion of both TRAIL-R1 and TRAIL-R2 contains a functional death domain that recruits the adaptor protein FADD to form a signaling complex on the cytosolic surface of the receptor. This adaptor complex then recruits caspase-8 or caspase-10. TRADD may also be recruited by TRAIL-R1 or TRAIL-R2, leading to activation of NF-κB and/or JNK [41].

The presence or absence of intracellular apoptosis inhibitors is thought to mediate resistance or sensitivity to the apoptosis mediated by TNF superfamily members. One of the mechanisms of cellular resistance to death-receptor-mediated apoptosis involves the activation of NF-κB. Modulation of NF-κB activation may be a key determinant in the sensitivity to the apoptosis induced by ligands related to TNF [42-44].

Apoptosis Induced by the Mitochondrial Pathway

In response to agents or insults that impair cell integrity (DNA damage, oxidative stresses...) the mitochondria releases several molecules including cytochrome c. The release of cytochrome c from the mitochondria into the cytosol serves as a trigger to activate caspases. Indeed, when combined to extra-mitochondrial cytochrome c, dATP and Apaf-1, the procaspase-9 is activated to form a large complex called apoptosome. This catalytic complex is responsible for the activation of the effector caspases (-3, -7). Several other factors (Apoptosis Inducing factor, Smac/diablo) are present in the intermembrane space of the mitochondria, which, once released, can further increase the activation of caspase cascade. Proteins belonging to the Bcl-2 family appear to regulate the permeability of the mitochondrial membrane to ions and to cytochrome c, and may interact with the formation of the pro-apoptotic complex. The Bcl-2 family includes about 20 proteins, each containing 1 or more Bcl-2 homology (BH) domains. Several of the Bcl2 family members display 4 BH domains and consist of a large group of anti-apoptosis proteins, which when over-expressed prevent apoptosis (e.g. Bcl-2, Bcl-xL). Other Bcl-2 proteins display a reduced number of BH domains (3 or 1), which render them pro-apoptotic (e.g. Bax, Bak, and BH3-only proteins such as Bim, Bad, Bid). The balance between anti-apoptotic and pro-apoptotic members of the Bcl-2 family may be critical in determining whether a cell undergoes apoptosis [45,46]. Other enzymes (e.g., cathepsins) or agents (e.g., calcium) can also regulate the intrinsic pathway [45,47-49].

Caspase-independent Effectors of Apoptosis

In addition to caspases, other signaling intermediates may also function as effectors of apoptosis [48]; for example non-caspase proteases such as calpain [50], cathepsins [47], the apoptosis-inducing factor (AIF) [51], or the intracellular free calcium [49].

OSTEOCLAST APOPTOSIS IN RODENTS

In rodents, osteoclast apoptosis has been studied in mature osteoclasts in both *in-vivo*, and *in-vitro* models of osteoclast differentiation [22]. Nuclear fragmentation in osteoclasts has been observed using DNA-binding fluorochrome (Hoechst). DNA strand breaks in the apoptotic cells have been confirmed by an FITC, *in-situ*, cell death detection kit (TUNEL method). This morphological approach to detecting osteoclast apoptosis is necessary both *in-vivo* in tissue sections, as well as *in-vitro,* as osteoclasts are adherent cells and their integrity could be compromised if they are detached. Such an approach has been validated in several studies [52-54]. Some evidence does indeed suggest that rodent osteoclasts undergo apoptosis, but the mechanisms involved in this process and in its regulation are not fully understood.

In vivo, targeting Simian virus 40 T antigen oncogene towards osteoclasts using the TRAP gene promoter causes functional impairment of osteoclasts *in vivo*, including a high rate of apoptosis [55].

In *in-vitro* cultures using murine bone marrow as a source of osteoclast progenitors, the presence of mRNA for caspases 2, 3, 6, 7, and 8 has been detected by RNAase protection assay [56]. In the absence of both RANKL and M-CSF, increased caspase-3 activity was also detected in cell lysates after 24 hours [56]. In a similar model, caspase expression was evaluated by Western blot after a 24-hour period. While pro-caspase 8 and its active form were not detected, procaspases -9 and -3 were expressed, and the levels of active caspase-9 and -3 were lower in the cells treated by M-CSF [57]. In addition, caspase-3 and caspase-9 inhibitors partially reversed the apoptosis of osteoclast precursors [57]. These studies therefore provide further evidence that apoptosis does occur in murine osteoclastic cells, and indeed that this process may involve caspases.

Cytokines and Survival Factors

In rodent models, some cytokines that are known to stimulate osteoclast formation, such as IL1α, M-CSF [58], TNFα [59] and RANKL [56], also stimulate osteoclast survival. The effects of IL1α, M-CSF and TNFα on osteoclast survival are partly related to the suppression of caspase activation [52,59]. In the absence of the survival factors M-CSF and RANKL, osteoclasts rapidly undergo apoptosis [56,57,60]. On the other hand, estrogens, which inhibit bone resorption, are able to induce osteoclast apoptosis [26], and may act via TGFβ [25].

Osteoclast cultures exposed to M-CSF, an osteoclast survival factor, contain increased amounts of bcl-2 and bcl-xL mRNA [56]. In this study, RANKL did not alter the expression of either bcl-2 or bcl-xL [56]. These findings suggest that osteoclast apoptosis, and its regulation by M-CSF, may involve members of the bcl-2 family, particularly bcl-xL. However the role of bcl-2 at this stage is controversial [52,57]. M-CSF could also upregulate the expression of the endogenous caspase inhibitor protein XIAP (X-linked inhibitor of apoptosis) [61]. It has been suggested that Bcl-xL plays a role in osteoclast survival in transgenic mice bearing the bcl-xL gene under the control of the TRAP gene promoter. Osteoclast formation was increased in bone marrow cultures from these transgenic mice. When both SV40 Tag and Bcl-xL were targeted to cells of the osteoclast-lineage, the pre-osteoclast survival was greatly increased, as well as the resistance of the cultures to serum deprivation [62].

In addition, M-CSF withdrawal-induced apoptosis has been associated with an increased expression of Bim, a pro-apoptotic member of the BH3-only Bcl2 family [63]. M-CSF is involved in the regulation of Bim ubiquitylation that leads to its degradation by the ubiquitin-proteasome pathways. This could explain, at least in part, the apoptosis induced by M-CSF withdrawal [63]. Osteoclasts from bim -/- transgenic mice displayed prolonged survival *in vivo*, and bim -/- osteoclasts were resistant to M-CSF withdrawal-induced apoptosis *in vitro*. However, these bim-/- deficient mice had mild osteosclerosis with reduced bone resorption, and *in vitro* the bim -/- osteoclasts had a reduced bone resorption activity [63]. These findings are not fully understood, but may reflect interactions with other pathways involved in osteoclast function.

Small G-proteins, such as Rho or Rac, regulate the cytoskeletal organization in various types of cells [64]. Other studies have shown that some members of the Rho family small G-proteins, e.g. Rac, also mediate anti-apoptotic signals [65]. Moreover, it has been demonstrated that the small G-protein Rac1 is involved in M-CSF receptor signaling, and mediated survival signaling of osteoclasts, primarily by modulating PI3K/Akt pathways [66]. Akt is a mediator of cell survival, inactivating caspase-9 and Bad, and promoting the degradation of IκB, the inhibitor of the transcription factor NF-κB [67].

Tyrosine Kinase c-Src

C-Src is a tyrosine-kinase that can be activated downstream of RANK, and it clearly plays major role in osteoclast biology, since c-src knock-out mice develop osteopetrosis characterized by the presence of a normal number of osteoclasts that are unable to resorb bone due to their inability to form the ruffled membrane [68,69]. C-Src has been implicated in vesicle transport and cytoskeleton organization, including the formation of the actin ring and ruffled border [70-72]. In addition, c-Src is involved in osteoclast resistance to apoptosis by mediating survival signals via the PI3 kinase/Akt pathway [59,73]. In osteoclasts, RANK signaling leads to c-Src induction, and via TRAF6, to PI3 kinase activation. The subsequent Akt activation induces Bad and caspase-9 phosphorylation, thus inactivating these pro-apoptotic molecules and preventing the activation of apoptosis [74]. In addition, some c-Src inhibitors disrupt osteoclast formation *in vivo* and *in vitro*, and induce osteoclast apoptosis via a mechanism involving sustained phosphorylation of ERK1/2 [75].

Transcription Factors: NF-κB and NFAT

Treating purified osteoclasts obtained from mice or rabbits with inhibitors of NF-κB induces apoptosis [76-78], although there is some controversy about this [79]. These results could be very useful in the field of bone resorption, because of the important role that NF-κB plays in osteoclast formation. Indeed, double knock-out mice for the p50 and p52 NF-κB subunits develop a form of osteopetrosis involving a defect in osteoclast differentiation [80,81]. Furthermore, NF-κB activation is stimulated by RANKL, TNFα and IL1α, which are all potent stimulators of osteoclast formation and survival. Thus, NF-κB activation, which is involved in apoptosis regulation in other cellular models, may influence osteoclast survival.

NFATc1, a member of the NFAT family transcription factors, is strongly induced and activated after RANKL stimulation in osteoclasts, and is involved in the RANKL-induced osteoclastogenesis [82]. RANKL induces the NFATc1 gene via TRAF6 and c-Fos signaling pathways. C-Fos deficient mice develop an osteopetrosis characterized by a lineage shift between osteoclasts and macrophages resulting in increased numbers of bone marrow macrophages [83]. In c-fos-/- precursor cells, overexpression of NFAT can restore the RANKL-induced osteoclastogenesis [84]. RANKL also induces and activates NFATc1 through calcium signaling involving a calcineurin-mediated activation of NFAT [85]. Cyclosporin A (CsA) and FK506, calcineurin inhibitors, inhibit osteoclastogenesis [82,86], and induce osteoclast apoptosis in mouse bone marrow cultures [87]. The caspase inhibitor z-VAD partially prevented the inhibitory effects of CsA and FK506 on the survival of osteoclast-like cells in these cultures [87], suggesting that other mechanisms may be involved in the inhibitory effects of these molecules on osteoclasts. Thus, NFATc1 appears to be a transcriptional factor involved in osteoclastogenesis and in osteoclast survival.

αvβ3 Integrin

αvβ3 is the major integrin expressed by osteoclasts [88]. The engagement of αvβ3 integrin leads to the formation of a multi-molecular complex that includes c-Src and PI3-K [67]. In addition to its role in cell-matrix attachment, αvβ3 is also involved in other aspects of osteoclast biology. The bone matrix mediates anti-apoptotic signals via integrins, and αvβ3 occupancy promotes osteoclast survival [89]. In contrast, unoccupied αvβ3 sites may induce osteoclast apoptosis via caspase-8 activation [90,91]. Furthermore, in the absence of αvβ3 ($\beta3^{-/-}$ transgenic mice), delayed cell death occurs as a result of caspase-9 activation [91]. These findings suggest that unoccupied αvβ3 sites transmit a positive death signal regulated by caspase-8. Interestingly, the S^{752} residue, essential for all previously reported integrin's functions in osteoclasts such as adhesion and bone resorption [92], does not interfere with the integrin-mediated apoptosis [91], suggesting that another component of the unligated β3 subunit is involved in this process.

Death Receptors

The TNF superfamily members have been shown to be involved in rodent osteoclast apoptosis. TNFα appears to be a survival factor during osteoclast differentiation acting through TNFR1. TNFα-enhanced survival of osteoclasts appears to be associated with a reduction in apoptosis and the suppression of caspase activation [59]. TNFα acts by activating NF-κB, through the recruitment of TRADD. Inhibition of NF-κB activation blocks osteoclast formation [93]. In a study using a dominant-negative IκB to block the NF-κB pathway, TNFα promoted osteoclast apoptosis, and this was associated with activation of caspase-3, -9, PARP degradation and an increased expression of Bax. This was also accompanied by the reduction of cIAP-1, a cellular inhibitor of apoptosis, Bcl-X_L, and TRAF6 expression. These findings suggest that TNFα may be involved in osteoclast apoptosis when NF-κB activity levels are low [94].

Recent studies suggest that the Fas-FasL system may be involved in murine osteoclast apoptosis [95,96]. Apoptosis pathways were studied using cultures from mouse bone marrow, as well as RAW 264.7 cells. These cells expressed Fas, and a Fas-activating antibody induced apoptosis in mature osteoclasts [96]. RANKL could modulate Fas-induced apoptosis, by up-regulating Fas expression in osteoclast progenitors, and down-regulating Fas expression in mature osteoclasts [97]. Apoptosis was induced by adding IL12 to murine osteoclast-like cells. When Fas blocking antibodies were added to these cultures, which were treated with M-CSF, TNFα, and IL12, apoptotic changes were inhibited [95]. Other studies have confirmed the expression of Fas by osteoclast precursors and mature osteoclasts [75,98]. Interestingly, the Fas ligand was also shown to be expressed in osteoclastic cells [98]. The role of FasL has been studied *in vitro* in murine osteoclasts derived from bone marrow cultures. Recombinant FasL upregulated the secretion of IL-1β, TNFα, and the activation of NF-κB. FasL enhanced osteoclast differentiation, and induced mature osteoclast apoptosis; but osteoclast precursors were relatively resistant to FasL-induced apoptosis [98]. Thus the FasL pathway could be involved in osteoclast apoptosis, however the effects of defects in the Fas/FasL system *in vivo* are controversial. An increase in bone mass has been observed in Fas-/- Gld mice, and the numbers of osteoclasts in bone-marrow cultures from these animals were similar to those found in cultures from wild-type mice [99]. In contrast, in Fas-/- Lpr mice, which contain a different Fas mutation, bone mineral density was significantly reduced, TRAP staining was more intense, and bone histomorphometry revealed a decrease in bone volume, although no changes were observed in the Gld mice or controls [96].

In a study to investigate the possible physiological role of TRAIL *in vivo*, TRAIL -/- deficient mice were generated. These mice were more susceptible to tumor burden, suggesting that TRAIL may play an important part in controlling tumors *in vivo*. However X-ray analysis did not indicate any bone abnormalities, and *in-vitro* osteoclast differentiation was normal in these mice, suggesting that TRAIL cannot be essential for bone biology under normal conditions [100].

Osteoclast Apoptosis in Humans

Although osteoclast apoptosis and its regulation are still not fully understood in animal models, considerable progress has been made in the last decade, and this is of great importance for our understanding of the behavior of osteoclast cells in bone resorption and bone remodeling. Indeed, bone resorption, which is the first stage of bone renewal, is the keystone in the bone remodeling process, and its regulation is of major importance in controlling bone homeostasis. This highlights the importance of investigating osteoclast life span. The data obtained from rodent models will contribute to the general understanding of osteoclast apoptosis, but there is a real need to focus research on human cells, in the hope of identifying potential targets that could be used to treat skeletal disorders. However few human data are yet available.

Using cord blood monocytes (CBMs) as the source of human osteoclast precursors, we showed that osteoclast-like cells from CBM cultures expressed TRAIL-receptors, both TRAIL-R1 and TRAIL-R2 that are able to transduce apoptosis, and TRAIL-R3 and TRAIL-R4 that are decoy receptors [101]. This is in line with findings of other authors showing that human osteoclast precursors expressed TRAIL receptors [102]. It has also been shown that Fas is expressed by human osteoclasts obtained from peripheral blood monocytes (PBMC) [96]. In addition, we have shown that these pathways are indeed functional as, in human osteoclasts obtained from long-term cultures of CBMs, apoptosis occurred and was significantly greater in the presence of TRAIL, and of a Fas-activating antibody [101]. These findings suggest that the TRAIL and FasL pathways could be involved in osteoclast apoptosis in humans.

The potential role of Fas in human osteoclast physiology has been highlighted by recent data showing that DcR3, a secreted member of the TNF receptor superfamily, induces osteoclast formation by promoting TNFα production [103]. It is interesting to note that DcR3 has been shown to be a decoy receptor for several members of the TNF family, including the Fas ligand. It is not clear, however, whether DcR3 acts by neutralizing FasL. Nevertheless, our findings show clearly that Fas stimulation did induce osteoclast apoptosis in human osteoclasts.

In the study by Zauli *et al.*, osteoclast precursors obtained from human PBMC expressed TRAIL receptors, and TRAIL inhibited osteoclast differentiation and blocked osteoclast bone resorption, by inhibiting the p38/MAPK pathway without producing any cytotoxic effect on osteoclast precursors [102]. Unlike Zauli *et al.*, we found that TRAIL induced osteoclast apoptosis [101]. This discrepancy may have been due to differences in the experimental procedures.

These findings are of considerable interest with regard to bone physiology. In fact, the RANKL-RANK osteoclast survival pathway and the TRAIL-TRAILR apoptotic pathway are both regulated by the same inhibitor, OPG, the importance of which has been amply demonstrated in the physiology of osteoclasts. The effect of OPG may depend on the cellular state and responsiveness, and the concentrations of RANKL and TRAIL in the bone micro-environment.

Osteoclast Apoptosis in Human Diseases

Multiple Myeloma

One prominent feature in multiple myeloma is the occurrence of skeletal events. The excessive bone resorption may be due to the production of several osteoclast-activating factors either by the myeloma cells themselves or by the bone marrow microenvironment, including MIP-1α, IL6 and RANKL. The expression of RANKL by stromal cells is considerably increased in myeloma, and is associated with a concomitant decrease in OPG expression. The RANKL-OPG imbalance could play a decisive role in the lytic bone lesions of myeloma [104]. On the other hand, deregulation of the apoptotic pathways could be involved and lead to abnormal survival of the osteoclasts.

In a previous study, we found that TRAIL-R3 and TRAIL-R4 were strongly expressed in osteoclasts from myeloma bone lesions, but that the expression of TRAIL-R1 and TRAIL–R2 was weak or absent [101]. It is of interest to note that none of these receptors was detected in normal bone, suggesting that the TRAIL pathway could be induced in osteoclasts *in-vivo* under conditions characterized by a high level of osteoclast formation. The over-expression of the decoy receptors (TRAIL-R3 and TRAIL-R4) could decrease osteoclast apoptosis, and contribute to the high level of osteoclast activity observed in myeloma. These findings are in agreement with data showing that osteoclasts from PBMC generated in the presence of myeloma T cells over-expressed TRAIL-R4 and had a decreased expression of TRAIL-R1 [105], suggesting induced-resistance to TRAIL apoptosis.

Paget's Disease of Bone

Paget's disease of bone (PDB) is a common disorder of the elderly, and is characterized by excessive resorption and formation of bone, and phenotypically abnormal osteoclasts with numerous nuclei and increased survival [106]. Mutations in the gene encoding p62 have been shown to be involved in the pathogenesis of PDB [107-110], however its relationship to the Paget phenotype is still unclear. Abnormal regulation of osteoclast apoptosis could be involved, as suggested by the increased expression of Bcl-2 in pagetic osteoclasts. Mutations of the bcl-2 promoter have been identified in some patients, suggesting that at least in some cases, this could account for the overproduction of Bcl-2 that results in delayed apoptosis [111].

Bisphosphonate-induced Apoptosis

Bisphosphonates inhibit bone resorption, and are widely used in the treatment of diseases associated with excessive osteoclast-mediated bone resorption, such as osteoporosis, malignant osteolysis or Paget's disease of bone. Bisphosphonates directly suppress osteoclast activity. Numerous studies have shown that one of the principle mechanisms of bisphosphonate action is the induction of osteoclast apoptosis both *in vitro* and *in vivo*

[53,112]. Small G-proteins, such as those of the Ras, Rho and Rab families, could be targets of nitrogen-containing bisphosphonates. These bisphosphonates inhibit the post-translational prenylation of small G-proteins, and this may be a mechanism of their induction of osteoclast apoptosis [113]. Nitrogen-containing bisphosphonates inhibit farnesyl diphosphate synthase (FPP synthase), an enzyme that belongs to the mevalonate pathway, thus depleting cells of FPP and GGPP (geranylgeranyl diphosphate) [114,115]. These two molecules are required for the prenylation of small G-proteins, which is essential for their localization and function. Small G-proteins are crucial signaling proteins that regulate a number of processes necessary for osteoclast function, such as cytoskeletal organization and cell survival [116]. On the other hand, bisphosphonates that do not contain nitrogen are metabolized in the cell to form methylene-containing analogues of ATP, the accumulation of which in the osteoclasts could induce cell apoptosis [117]

Conclusion

Osteoclast apoptosis is an important point of control of bone resorption, although little is known about the mechanisms involved, and its possible role in bone physiology and pathology has been largely ignored. Understanding the mechanisms involved in human osteoclast apoptosis should provide fresh information about the impact of factors known to modulate osteoclast bone resorption. There is now increasing evidence suggesting that changes in the regulation of osteoclast death may contribute to clinically important bone diseases, such as osteoporosis, malignant osteolysis and Paget's disease of bone, and that the induction of osteoclast apoptosis is a potential therapeutic tool for treating these diseases.

References

[1] Parfitt AM. The cellular basis of bone remodeling: the quantum concept reexamined in light of recent advances in the cell biology of bone. Calcif Tissue Int 1984, 36: S37-S45.

[2] Xing L, Boyce BF. Regulation of apoptosis in osteoclasts and osteoblastic cells. *Biochem Biophys Res Commun* 2005, 328: 709-20.

[3] Green DR, Kroemer G. Pharmacological manipulation of cell death: clinical applications in sight? *J Clin Invest* 2005, 115: 2610-7.

[4] Hattersley G, Chambers TJ. Calcitonin receptors as markers for osteoclastic differentiation: correlation between generation of bone-resorptive cells and cells that express calcitonin receptors in mouse bone marrow cultures. *Endocrinology* 1989, 125: 1606 - 1612.

[5] Quinn JM, Morfis M, Lam MH, Elliott J, Kartsogiannis V, Williams ED, Gillespie MT, Martin TJ, Sexton PM. Calcitonin receptor antibodies in the identification of osteoclasts. *Bone* 1999, 25: 1-8.

[6] Boyle WJ, Simonet WS, Lacey DL. Osteoclast differentiation and activation. *Nature* 2003, 423: 337-42.

[7] Hofbauer LC, Khosla S, Dunstan CR, Lacey DL, Boyle WJ, Riggs BL. The roles of osteoprotegerin and osteoprotegerin ligand in the paracrine regulation of bone resorption. *J Bone Miner Res* 2000, 15: 2-12.

[8] Suda T, Takahashi N, Udagawa N, Jimi E, Gillespie MT, Martin TJ. Modulation of osteoclast differentiation and function by the new members of the tumor necrosis factor receptor and ligand families. *Endocr Rev* 1999, 20: 345-57.

[9] Quinn JM, Elliott J, Gillespie MT, Martin TJ. A combination of osteoclast differentiation factor and macrophage- colony stimulating factor is sufficient for both human and mouse osteoclast formation in vitro. *Endocrinology* 1998, 139: 4424-7.

[10] Jimi E, Akiyama S, Tsurukai T, Okahashi N, Kobayashi K, Udagawa N, Nishihara T, Takahashi N, Suda T. Osteoclast differentiation factor acts as a multifunctional regulator in murine osteoclast differentiation and function. *J Immunol* 1999, 163: 434-42.

[11] Lacey DL, Timms E, Tan HL, Kelley MJ, Dunstan CR, Burgess T, Elliott R, Colombero A, Elliott G, Scully S, Hsu H, Sullivan J, Hawkins N, Davy E, Capparelli C, Eli A, Qian YX, Kaufman S, Sarosi I, Shalhoub V, Senaldi G, Guo J, Delaney J, Boyle WJ. Osteoprotegerin ligand is a cytokine that regulates osteoclast differentiation and activation. *Cell* 1998, 93: 165-76.

[12] Nakagawa N, Kinosaki M, Yamaguchi K, Shima N, Yasuda H, Yano K, Morinaga T, Higashio K. RANK is the essential signaling receptor for osteoclast differentiation factor in osteoclastogenesis. *Biochem Biophys Res Commun* 1998, 253: 395-400.

[13] Hsu H, Lacey DL, Dunstan CR, Solovyev I, Colombero A, Timms E, Tan HL, Elliott G, Kelley MJ, Sarosi I, Wang L, Xia XZ, Elliott R, Chiu L, Black T, Scully S, Capparelli C, Morony S, Shimamoto G, Bass MB, Boyle WJ. Tumor necrosis factor receptor family member RANK mediates osteoclast differentiation and activation induced by osteoprotegerin ligand. *Proc Natl Acad Sci U S A* 1999, 96: 3540-5.

[14] Tsuda E, Goto M, Mochizuki S, Yano K, Kobayashi F, Morinaga T, Higashio K. Isolation of a novel cytokine from human fibroblasts that specifically inhibits osteoclastogenesis. *Biochem Biophys Res Commun* 1997, 234: 137-42.

[15] Simonet WS, Lacey DL, Dunstan CR, Kelley M, Chang MS, Luthy R, Nguyen HQ, Wooden S, Bennett L, Boone T, Shimamoto G, DeRose M, Elliott R, Colombero A, Tan HL, Trail G, Sullivan J, Davy E, Bucay N, Renshaw-Gegg L, Hughes TM, Hill D, Pattison W, Campbell P, Boyle WJ, et al. Osteoprotegerin: a novel secreted protein involved in the regulation of bone density. *Cell* 1997, 89: 309-19.

[16] Wittrant Y, Couillaud S, Theoleyre S, Dunstan C, Heymann D, Redini F. Osteoprotegerin differentially regulates protease expression in osteoclast cultures. *Biochem Biophys Res Commun* 2002, 293: 38-44.

[17] Wada T, Nakashima T, Hiroshi N, Penninger JM. RANKL-RANK signaling in osteoclastogenesis and bone disease. *Trends Mol Med* 2006, 12: 17-25.

[18] Kong YY, Boyle WJ, Penninger JM. Osteoprotegerin ligand: a regulator of immune responses and bone physiology. *Immunol Today* 2000, 21: 495-502.

[19] Takayanagi H, Ogasawara K, Hida S, Chiba T, Murata S, Sato K, Takaoka A, Yokochi T, Oda H, Tanaka K, Nakamura K, Taniguchi T. T-cell-mediated regulation of

osteoclastogenesis by signalling cross-talk between RANKL and IFN-gamma. *Nature* 2000, 408: 600-5.

[20] Lubberts E, Joosten LA, Chabaud M, van Den Bersselaar L, Oppers B, Coenen-De Roo CJ, Richards CD, Miossec P, van Den Berg WB. IL-4 gene therapy for collagen arthritis suppresses synovial IL-17 and osteoprotegerin ligand and prevents bone erosion. *J Clin Invest* 2000, 105: 1697-710.

[21] Wei S, Wang MW, Teitelbaum SL, Ross FP. Interleukin-4 reversibly inhibits osteoclastogenesis via inhibition of NF-kappa B and mitogen-activated protein kinase signaling. *J Biol Chem* 2002, 277: 6622-30.

[22] Hughes DE, Boyce BF. Apoptosis in bone physiology and disease. *Mol Pathol* 1997, 50: 132-7.

[23] Parfitt AM, Mundy GR, Roodman GD, Hughes DE, Boyce BF. A new model for the regulation of bone resorption, with particular reference to the effects of bisphosphonates. *J Bone Miner Res* 1996, 11: 150- 159.

[24] Roodman GD. Advances in bone biology: the osteoclast. Endocr Rev 1996, 17: 308-332.

[25] Hughes DE, Dai A, Tiffee JC, Li HH, Mundy GR, Boyce BF. Estrogen promotes apoptosis of murine osteoclasts mediated by TGF-β. *Nature Medicine* 1996, 2: 1132-1136.

[26] Saintier D, Khanine V, Uzan B, Ea HK, de Vernejoul MC, Cohen-Solal ME. Estradiol inhibits adhesion and promotes apoptosis in murine osteoclasts in vitro. *J Steroid Biochem Mol Biol* 2006, 99: 165-73.

[27] Shiotani A, Takami M, Itoh K, Shibasaki Y, Sasaki T. Regulation of osteoclast differentiation and function by receptor activator of NFkB ligand and osteoprotegerin. *Anat Rec* 2002, 268: 137-46.

[28] Lavrik IN, Golks A, Krammer PH. Caspases: pharmacological manipulation of cell death. *J Clin Invest* 2005, 115: 2665-72.

[29] Thornberry NA, Lazebnik Y. Caspases: enemies within. *Science* 1998, 281: 1312-6.

[30] Kroemer G, Martin SJ. Caspase-independent cell death. *Nat Med* 2005, 11: 725-30.

[31] Luo X, Budihardjo I, Zou H, Slaughter C, Wang X. Bid, a Bcl2 interacting protein, mediates cytochrome c release from mitochondria in response to activation of cell surface death receptors. *Cell* 1998, 94: 481-90.

[32] Suliman A, Lam A, Datta R, Srivastava RK. Intracellular mechanisms of TRAIL: apoptosis through mitochondrial-dependent and -independent pathways. *Oncogene* 2001, 20: 2122-33.

[33] Ashkenazi A, Dixit VM. Death receptors: signaling and modulation. *Science* 1998, 281: 1305-8.

[34] Nagafuji K, Shibuya T, Harada M, Mizuno S, Takenaka K, Miyamoto T, Okamura T, Gondo H, Niho Y. Functional expression of Fas antigen (CD95) on hematopoietic progenitor cells. *Blood* 1995, 86: 883-9.

[35] Nagata S. Apoptosis by death factor. *Cell* 1997, 88: 355-65.

[36] Wiley SR, Schooley K, Smolak PJ, Din WS, Huang CP, Nicholl JK, Sutherland GR, Smith TD, Rauch C, Smith CA, et al. Identification and characterization of a new member of the TNF family that induces apoptosis. *Immunity* 1995, 3: 673-82.

[37] Emery JG, McDonnell P, Burke MB, Deen KC, Lyn S, Silverman C, Dul E, Appelbaum ER, Eichman C, DiPrinzio R, Dodds RA, James IE, Rosenberg M, Lee JC, Young PR. Osteoprotegerin is a receptor for the cytotoxic ligand TRAIL. *J Biol Chem* 1998, 273: 14363-7.

[38] Takeda K, Hayakawa Y, Smyth MJ, Kayagaki N, Yamaguchi N, Kakuta S, Iwakura Y, Yagita H, Okumura K. Involvement of tumor necrosis factor-related apoptosis-inducing ligand in surveillance of tumor metastasis by liver natural killer cells. *Nat Med* 2001, 7: 94-100.

[39] Smyth MJ, Cretney E, Takeda K, Wiltrout RH, Sedger LM, Kayagaki N, Yagita H, Okumura K. Tumor necrosis factor-related apoptosis-inducing ligand (TRAIL) contributes to interferon gamma-dependent natural killer cell protection from tumor metastasis. *J Exp Med* 2001, 193: 661-70.

[40] Cretney E, Takeda K, Yagita H, Glaccum M, Peschon JJ, Smyth MJ. Increased susceptibility to tumor initiation and metastasis in TNF-related apoptosis-inducing ligand-deficient mice. *J Immunol* 2002, 168: 1356-61.

[41] MacFarlane M. TRAIL-induced signalling and apoptosis. Toxicol Lett 2003, 139: 89-97.

[42] Harper N, Farrow SN, Kaptein A, Cohen GM, MacFarlane M. Modulation of tumor necrosis factor apoptosis-inducing ligand- induced NF-kappa B activation by inhibition of apical caspases. *J Biol Chem 2001*, 276: 34743-52.

[43] Liu ZG, Hsu H, Goeddel DV, Karin M. Dissection of TNF receptor 1 effector functions: JNK activation is not linked to apoptosis while NF-kappaB activation prevents cell death. *Cell* 1996, 87: 565-76.

[44] Rivera-Walsh I, Cvijic ME, Xiao G, Sun SC. The NF-kappa B signaling pathway is not required for Fas ligand gene induction but mediates protection from activation-induced cell death. *J Biol Chem* 2000, 275: 25222-30.

[45] Adams JM, Cory S. The Bcl-2 protein family: arbiters of cell survival. *Science* 1998, 281: 1322-6.

[46] Bouchier-Hayes L, Lartigue L, Newmeyer DD. Mitochondria: pharmacological manipulation of cell death. *J Clin Invest* 2005, 115: 2640-7.

[47] Chwieralski CE, Welte T, Buhling F. Cathepsin-regulated apoptosis. *Apoptosis* 2006, 11: 143-9.

[48] Hail N, Jr., Carter BZ, Konopleva M, Andreeff M. Apoptosis effector mechanisms: A requiem performed in different keys. *Apoptosis* 2006, 11: 889-904.

[49] Hajnoczky G, Davies E, Madesh M. Calcium signaling and apoptosis. *Biochem Biophys Res Commun* 2003, 304: 445-54.

[50] Goll DE, Thompson VF, Li H, Wei W, Cong J. The calpain system. *Physiol Rev* 2003, 83: 731-801.

[51] Cregan SP, Dawson VL, Slack RS. Role of AIF in caspase-dependent and caspase-independent cell death. *Oncogene* 2004, 23: 2785-96.

[52] Okahashi N, Koide M, Jimi E, Suda T, Nishihara T. Caspases (interleukin-1beta-converting enzyme family proteases) are involved in the regulation of the survival of osteoclasts. *Bone* 1998, 23: 33-41.

[53] Hughes DE, Wright KR, Uy HL, Sasaki A, Yoneda T, Roodman GR, Mundy GR, Boyce FB. Bisphosphonates promote apoptosis in murine osteoclasts in vitro and in vivo. *J Bone Mineral Res* 1995, 10: 1478- 1487.

[54] Kameda T, Ishikawa H, Tsutsui T. Detection and characterization of apoptosis in osteoclasts in vitro. *Biochem Biophys Res Commun* 1995, 207: 753-60.

[55] Boyce BF, Wright K, Reddy SV, Koop BA, Story B, Devlin R, Leach RJ, Roodman GD, Windle JJ. Targeting simian virus 40 T antigen to the osteoclast in transgenic mice causes osteoclast tumors and transformation and apoptosis of osteoclasts. *Endocrinology* 1995, 136: 5751- 5759.

[56] Lacey DL, Tan HL, Lu J, Kaufman S, Van G, Qiu W, Rattan A, Scully S, Fletcher F, Juan T, Kelley M, Burgess TL, Boyle WJ, Polverino AJ. Osteoprotegerin ligand modulates murine osteoclast survival in vitro and in vivo. *Am J Pathol* 2000, 157: 435-48.

[57] Woo KM, Kim HM, Ko JS. Macrophage colony-stimulating factor promotes the survival of osteoclast precursors by up-regulating Bcl-X(L). *Exp Mol Med* 2002, 34: 340-6.

[58] Jimi E, Shuto T, Koga T. Macrophage colony-stimulating factor and interleukin-1a maintain the survival of osteoclast-like cells. *Endocrinology* 1995, 136: 808- 811.

[59] Lee SE, Chung WJ, Kwak HB, Chung CH, Kwack KB, Lee ZH, Kim HH. Tumor necrosis factor-alpha supports the survival of osteoclasts through the activation of Akt and ERK. *J Biol Chem* 2001, 276: 49343-9.

[60] Fuller K, Owens JM, Jagger CJ, Wilson A, Moss R, Chambers TJ. Macrophage colony-stimulating factor stimulates survival and chemotactic behavior in isolated osteoclasts. *J Exp Med* 1993, 178: 1733-44.

[61] Kanaoka K, Kobayashi Y, Hashimoto F, Nakashima T, Shibata M, Kobayashi K, Kato Y, Sakai H. A common downstream signaling activity of osteoclast survival factors that prevent nitric oxide-promoted osteoclast apoptosis. *Endocrinology* 2000, 141: 2995-3005.

[62] Hentunen TA, Reddy SV, Boyce BF, Devlin R, Park HR, Chung H, Selander KS, Dallas M, Kurihara N, Galson DL, Goldring SR, Koop BA, Windle JJ, Roodman GD. Immortalization of osteoclast precursors by targeting Bcl -XL and Simian virus 40 large T antigen to the osteoclast lineage in transgenic mice. *J Clin Invest* 1998, 102: 88-97.

[63] Akiyama T, Bouillet P, Miyazaki T, Kadono Y, Chikuda H, Chung UI, Fukuda A, Hikita A, Seto H, Okada T, Inaba T, Sanjay A, Baron R, Kawaguchi H, Oda H, Nakamura K, Strasser A, Tanaka S. Regulation of osteoclast apoptosis by ubiquitylation of proapoptotic BH3-only Bcl-2 family member Bim. *Embo J* 2003, 22: 6653-64.

[64] Burridge K, Wennerberg K. Rho and Rac take center stage. Cell 2004, 116: 167-79.

[65] Nishida K, Kaziro Y, Satoh T. Anti-apoptotic function of Rac in hematopoietic cells. *Oncogene* 1999, 18: 407-15.

[66] Fukuda A, Hikita A, Wakeyama H, Akiyama T, Oda H, Nakamura K, Tanaka S. Regulation of osteoclast apoptosis and motility by small GTPase binding protein Rac1. *J Bone Miner Res* 2005, 20: 2245-53.

[67] Golden LH, Insogna KL. The expanding role of PI3-kinase in bone. *Bone* 2004, 34: 3-12.

[68] Boyce BF, Yoneda T, Lowe C, Soriano P, Mundy GR. Requirement of pp60c-src expression for osteoclasts to form ruffled borders and resorb bone in mice. *J Clin Invest* 1992, 90: 1622- 1627.

[69] Soriano P, Montgomery C, Geske R, Bradley A. Targeted disruption of the c-src proto-oncogene leads to osteopetrosis in mice. *Cell* 1991, 64: 693- 702.

[70] Abu-Amer Y, Ross FP, Schlesinger P, Tondravi MM, Teitelbaum SL. Substrate recognition by osteoclast precursors induces C-src/microtubule association. *J Cell Biol* 1997, 137: 247-58.

[71] Nakamura I, Jimi E, Duong LT, Sasaki T, Takahashi N, Rodan GA, Suda T. Tyrosine phosphorylation of p130Cas is involved in actin organization in osteoclasts. *J Biol Chem* 1998, 273: 11144-9.

[72] Teti A, Taranta A, Migliaccio S, Degiorgi A, Santandrea E, Villanova I, Faraggiana T, Chellaiah M, Hruska KA. Colony stimulating factor-1-induced osteoclast spreading depends on substrate and requires the vitronectin receptor and the c-src proto-oncogene. *J Bone Miner Res* 1998, 13: 50-8.

[73] Xing L, Venegas AM, Chen A, Garrett-Beal L, Boyce BF, Varmus HE, Schwartzberg PL. Genetic evidence for a role for Src family kinases in TNF family receptor signaling and cell survival. *Genes Dev* 2001, 15: 241-53.

[74] Wong BR, Besser D, Kim N, Arron JR, Vologodskaia M, Hanafusa H, Choi Y. TRANCE, a TNF family member, activates Akt/PKB through a signaling complex involving TRAF6 and c-Src. *Mol Cell* 1999, 4: 1041-9.

[75] Recchia I, Rucci N, Funari A, Migliaccio S, Taranta A, Longo M, Kneissel M, Susa M, Fabbro D, Teti A. Reduction of c-Src activity by substituted 5,7-diphenyl-pyrrolo[2,3-d]-pyrimidines induces osteoclast apoptosis in vivo and in vitro. Involvement of ERK1/2 pathway. *Bone* 2004, 34: 65-79.

[76] Jimi E, Nakamura I, Ikebe T, Akiyama S, Takahashi N, Suda T. Activation of NF-kappaB is involved in the survival of osteoclasts promoted by interleukin-1. *J Biol Chem* 1998, 273: 8799-805.

[77] Ozaki K, Takeda H, Iwahashi H, Kitano S, Hanazawa S. NF-kappaB inhibitors stimulate apoptosis of rabbit mature osteoclasts and inhibit bone resorption by these cells. *FEBS Lett* 1997, 410: 297-300.

[78] Gingery A, Bradley E, Shaw A, Oursler MJ. Phosphatidylinositol 3-kinase coordinately activates the MEK/ERK and AKT/NFkappaB pathways to maintain osteoclast survival. *J Cell Biochem* 2003, 89: 165-79.

[79] Miyazaki T, Katagiri H, Kanegae Y, Takayanagi H, Sawada Y, Yamamoto A, Pando MP, Asano T, Verma IM, Oda H, Nakamura K, Tanaka S. Reciprocal role of ERK and NF-kappaB pathways in survival and activation of osteoclasts. *J Cell Biol* 2000, 148: 333-42.

[80] Franzoso G, Carlson L, Xing L, Poljak L, Shores EW, Brown KD, Leonardi A, Tran T, Boyce BF, Siebenlist U. Requirement for NF-kappaB in osteoclast and B-cell development. *Genes Dev* 1997, 11: 3482-96.

[81] Iotsova V, Caamano J, Loy J, Yang Y, Lewin A, Bravo R. Osteopetrosis in mice lacking NF-kappaB1 and NF-kappaB2. *Nat Med* 1997, 3: 1285-9.

[82] Hirotani H, Tuohy NA, Woo JT, Stern PH, Clipstone NA. The calcineurin/nuclear factor of activated T cells signaling pathway regulates osteoclastogenesis in RAW264.7 cells. *J Biol Chem* 2004, 279: 13984-92.

[83] Grigoriadis AE, Wang ZQ, Cecchini MG, Hofstetter W, Felix R, Fleisch HA, Wagner EF. c-Fos: a key regulator of osteoclast-macrophage lineage determination and bone remodeling. *Science* 1994, 266: 443-8.

[84] Matsuo K, Galson DL, Zhao C, Peng L, Laplace C, Wang KZ, Bachler MA, Amano H, Aburatani H, Ishikawa H, Wagner EF. Nuclear factor of activated T-cells (NFAT) rescues osteoclastogenesis in precursors lacking c-Fos. *J Biol Chem* 2004, 279: 26475-80.

[85] Takayanagi H. Mechanistic insight into osteoclast differentiation in osteoimmunology. *J Mol Med* 2005, 83: 170-9.

[86] Ishida N, Hayashi K, Hoshijima M, Ogawa T, Koga S, Miyatake Y, Kumegawa M, Kimura T, Takeya T. Large scale gene expression analysis of osteoclastogenesis in vitro and elucidation of NFAT2 as a key regulator. *J Biol Chem* 2002, 277: 41147-56.

[87] Igarashi K, Hirotani H, Woo JT, Stern PH. Cyclosporine A and FK506 induce osteoclast apoptosis in mouse bone marrow cell cultures. *Bone* 2004, 35: 47-56.

[88] Horton MA. The alpha v beta 3 integrin "vitronectin receptor". *Int J Biochem Cell Biol* 1997, 29: 721-5.

[89] Frisch SM, Screaton RA. Anoikis mechanisms. *Curr Opin Cell Biol* 2001, 13: 555-62.

[90] Stupack DG, Puente XS, Boutsaboualoy S, Storgard CM, Cheresh DA. Apoptosis of adherent cells by recruitment of caspase-8 to unligated integrins. *J Cell Biol* 2001, 155: 459-70.

[91] Zhao H, Ross FP, Teitelbaum SL. Unoccupied alpha(v)beta3 integrin regulates osteoclast apoptosis by transmitting a positive death signal. *Mol Endocrinol* 2005, 19: 771-80.

[92] Feng X, Novack DV, Faccio R, Ory DS, Aya K, Boyer MI, McHugh KP, Ross FP, Teitelbaum SL. A Glanzmann's mutation in beta 3 integrin specifically impairs osteoclast function. *J Clin Invest* 2001, 107: 1137-44.

[93] Abu-Amer Y, Dowdy SF, Ross FP, Clohisy JC, Teitelbaum SL. TAT fusion proteins containing tyrosine 42-deleted IkappaBalpha arrest osteoclastogenesis. *J Biol Chem* 2001, 276: 30499-503.

[94] Abbas S, Abu-Amer Y. Dominant-negative IkappaB facilitates apoptosis of osteoclasts by tumor necrosis factor-alpha. *J Biol Chem* 2003, 278: 20077-82.

[95] Kitaura H, Nagata N, Fujimura Y, Hotokezaka H, Yoshida N, Nakayama K. Effect of IL-12 on TNF-alpha-mediated osteoclast formation in bone marrow cells: apoptosis mediated by Fas/Fas ligand interaction. *J Immunol* 2002, 169: 4732-8.

[96] Wu X, McKenna MA, Feng X, Nagy TR, McDonald JM. Osteoclast apoptosis: the role of Fas in vivo and in vitro. *Endocrinology* 2003, 144: 5545-55.

[97] Wu X, Pan G, McKenna MA, Zayzafoon M, Xiong WC, McDonald JM. RANKL Regulates Fas Expression and Fas-Mediated Apoptosis in Osteoclasts. *J Bone Miner Res* 2005, 20: 107-16.

[98] Park H, Jung YK, Park OJ, Lee YJ, Choi JY, Choi Y. Interaction of Fas ligand and Fas expressed on osteoclast precursors increases osteoclastogenesis. *J Immunol* 2005, 175: 7193-201.

[99] Katavic VV, Lukic IK, Kovacic N, Grcevic D, Lorenzo JA, Marusic A. Increased Bone Mass Is a Part of the Generalized Lymphoproliferative Disorder Phenotype in the Mouse. *J Immunol* 2003, 170: 1540-1547.

[100] Sedger LM, Glaccum MB, Schuh JC, Kanaly ST, Williamson E, Kayagaki N, Yun T, Smolak P, Le T, Goodwin R, Gliniak B. Characterization of the in vivo function of TNF-alpha-related apoptosis-inducing ligand, TRAIL/Apo2L, using TRAIL/Apo2L gene-deficient mice. *Eur J Immunol* 2002, 32: 2246-54.

[101] Roux S, Lambert-Comeau P, Saint-Pierre C, Lepine M, Sawan B, Parent JL. Death receptors, Fas and TRAIL receptors, are involved in human osteoclast apoptosis. *Biochem Biophys Res Commun* 2005, 333: 42-50.

[102] Zauli G, Rimondi E, Nicolin V, Melloni E, Celeghini C, Secchiero P. TNF-related apoptosis-inducing ligand (TRAIL) blocks osteoclastic differentiation induced by RANKL plus M-CSF. *Blood* 2004, 104: 2044-50.

[103] Yang CR, Wang JH, Hsieh SL, Wang SM, Hsu TL, Lin WW. Decoy receptor 3 (DcR3) induces osteoclast formation from monocyte/macrophage lineage precursor cells. *Cell Death Differ* 2004, 11 Suppl 1: S97-107.

[104] Roux S, Mariette X. The high rate of bone resorption in multiple myeloma is due to RANK (receptor activator of nuclear factor-kappaB) and RANK Ligand expression. *Leuk Lymphoma* 2004, 45: 1111-8.

[105] Colucci S, Brunetti G, Rizzi R, Zonno A, Mori G, Colaianni G, Del Prete D, Faccio R, Liso A, Capalbo S, Liso V, Zallone A, Grano M. T cells support osteoclastogenesis in an in vitro model derived from human multiple myeloma bone disease: the role of the OPG/TRAIL interaction. *Blood* 2004, 104: 3722-30.

[106] Roodman GD, Windle JJ. Paget disease of bone. *J Clin Invest* 2005, 115: 200-8.

[107] Good DA, Busfield F, Fletcher BH, Lovelock PK, Duffy DL, Kesting JB, Andersen J, Shaw JT. Identification of SQSTM1 mutations in familial Paget's disease in Australian pedigrees. *Bone* 2004, 35: 277-82.

[108] Hocking LJ, Lucas GJ, Daroszewska A, Cundy T, Nicholson GC, Donath J, Walsh JP, Finlayson C, Cavey JR, Ciani B, Sheppard PW, Searle MS, Layfield R, Ralston SH. Novel UBA domain mutations of SQSTM1 in Paget's disease of bone: genotype phenotype correlation, functional analysis, and structural consequences. *J Bone Miner Res* 2004, 19: 1122-7.

[109] Johnson-Pais TL, Wisdom JH, Weldon KS, Cody JD, Hansen MF, Singer FR, Leach RJ. Three novel mutations in SQSTM1 identified in familial Paget's disease of bone. *J Bone Miner Res* 2003, 18: 1748-53.

[110] Laurin N, Brown JP, Morissette J, Raymond V. Recurrent mutation of the gene encoding sequestosome 1 (SQSTM1/p62) in Paget disease of bone. *Am J Hum Genet* 2002, 70: 1582-8.

[111] Brandwood CP, Hoyland JA, Hillarby MC, Berry JL, Davies M, Selby PL, Mee AP. Apoptotic gene expression in Paget's disease: a possible role for Bcl-2. *J Pathol* 2003, 201: 504-12.

[112] Hiroi-Furuya E, Kameda T, Hiura K, Mano H, Miyazawa K, Nakamaru Y, Watanabe-Mano M, Okuda N, Shimada J, Yamamoto Y, Hakeda Y, Kumegawa M. Etidronate (EHDP) Inhibits Osteoclastic-Bone Resorption, Promotes Apoptosis and Disrupts Actin Rings in Isolate-Mature Osteoclasts. *Calcif Tissue Int* 1999, 64: 219-223.

[113] Luckman SP, Hughes DE, Coxon FP, Graham R, Russell G, Rogers MJ. Nitrogen-containing bisphosphonates inhibit the mevalonate pathway and prevent post-translational prenylation of GTP-binding proteins, including Ras. *J Bone Miner Res* 1998, 13: 581-9.

[114] Dunford JE, Thompson K, Coxon FP, Luckman SP, Hahn FM, Poulter CD, Ebetino FH, Rogers MJ. Structure-activity relationships for inhibition of farnesyl diphosphate synthase in vitro and inhibition of bone resorption in vivo by nitrogen-containing bisphosphonates. *J Pharmacol Exp Ther* 2001, 296: 235-42.

[115] van Beek E, Pieterman E, Cohen L, Lowik C, Papapoulos S. Nitrogen-containing bisphosphonates inhibit isopentenyl pyrophosphate isomerase/farnesyl pyrophosphate synthase activity with relative potencies corresponding to their antiresorptive potencies in vitro and in vivo. *Biochem Biophys Res Commun* 1999, 255: 491-4.

[116] Coxon FP, Thompson K, Rogers MJ. Recent advances in understanding the mechanism of action of bisphosphonates. *Curr Opin Pharmacol* 2006, 6: 307-12.

[117] Frith JC, Monkkonen J, Auriola S, Monkkonen H, Rogers MJ. The molecular mechanism of action of the antiresorptive and antiinflammatory drug clodronate: evidence for the formation in vivo of a metabolite that inhibits bone resorption and causes osteoclast and macrophage apoptosis. *Arthritis Rheum* 2001, 44: 2201-10.

In: Cell Apoptotic Signaling Pathways
Editor: Charles O. Pickens, pp. 185-209

ISBN: 978-1-60021-507-0

Chapter VII

MOLECULAR CHARACTERIZATION AND EVOLUTION OF LIPOPOLYSACCHARIDE-INDUCED APOPTOTIC PATHWAYS IN SPONGES (DEMOSPONGIAE, PORIFERA)

Matthias Wiens**, Stefanie Klein, Petra Wrede, David Brandt and Werner E. G. Müller
Institut für Physiologische Chemie, Abteilung Angewandte Molekularbiologie, Universität, Duesbergweg 6, D-55099 Mainz; Germany.

ABSTRACT

Apoptosis, or programmed cell death, is a physiological process that not only plays an essential role during development, morphogenesis, and tissue homeostasis but is also part of various defense mechanisms of multicellular animals. Following exposure to countless stimuli, physiological and non-physiological, apoptotic cells actively participate in their own demise through activation of an intrinsic program that consists of a series of signal cascades. Dysregulation of the mechanisms controlling these apoptotic cascades may result in various disorders, such as auto-immune or neurodegenerative diseases, tumorigenesis, and infections.

Sponges (phylum Porifera) represent the next closest relatives to the common ancestral phylum Urmetazoa, ancestor of all metazoan phyla. Thus, the discovery of core components of the poriferan apoptotic cell death machinery contributes to advancing our understanding of both molecular evolution and basic control of apoptotic mechanisms. The apoptotic molecules that have been preserved from Porifera to man include members of the Bcl-2 family (apoptotic regulators), death domain proteins (adapters), caspases

* Correspondence concerning this article should be addressed to Dr. M. Wiens; Institut für Physiologische Chemie, Abteilung Angewandte Molekularbiologie, Universität, Duesbergweg 6, D-55099 Mainz; Germany. e-mail: wiens@uni-mainz.de.

(effectors), and various proteins, whose part in apoptosis has not been completely elucidated yet.

Since sponges are filter feeders, they are abundantly exposed to hazardous microorganisms. Consequently, they are provided with efficient defense mechanisms. However, breach of those defensive barriers inevitably requires the disposal of infected and of damaged cells through apoptotic processes. Here, we describe the identification of apoptotic and defensive molecules and their implication in the poriferan response to bacterial lipopolysaccharide (LPS), an endotoxic component of the outer envelope of Gram-negative bacteria. These poriferan proteins comprise the transcription factor LITAF_SUBDO (responsible for the regulated expression of apoptotic molecules), AIF_SUBDO (a mitochondrial pro-apoptotic effector), and VLP_SUBDO (mediator of the innate immune response). Characterization of the respective sponge molecules revealed both conserved as well as sponge specific features. In combination with functional and phylogenetic analyses we attempt to substantiate the overlapping pathways of poriferan apoptosis and innate immunity.

INTRODUCTION

Programmed cell death (PCD) and apoptosis, its morphological manifestation, is an innate mechanism by which multicellular organisms eliminate unwanted, infected, damaged, and superfluous cells, following exposure to countless stimuli. Dysregulation of PCD – either too little or not enough – inevitably results in disease (Thompson 1995; Yuan and Yankner 2000; Green and Evans 2002; Kirkin et al. 2004). The morphological hallmarks of PCD (cytoplasmic blebbing and condensation, cell shrinkage, chromatin condensation, and cell fragmentation) are controlled by the activation of nucleases and restricted-target proteases (caspases). Caspases (cysteine dependant aspartyl-specific proteases) are highly selective cysteine proteases that cleave their substrate after an aspartic acid residue that is part of a conserved tetrapeptide motif. Every caspase is expressed as a very weakly active proenzyme (procaspase), comprising a large and a small subunit, as well as a prodomain. In order to be fully activated, caspases have to be processed proteolytically. During this processing, carried out autocatalytically or mediated in a cascade-like reaction by other caspases, the subunits are separated from each other and from the prodomain. According to their prodomain structure caspases can be grouped into two general categories (Salvesen and Dixit 1997; Thornberry and Lazebnik 1998). Effector caspases carry only a short prodomain whereas initiator caspases bear a large prodomain that contains a protein-protein interaction domain.

Two general pathways can be distinguished that lead to the activation of these effector molecules. The extrinsic pathway is initiated through extracellular signals that are recognized by transmembrane receptors of the cell surface. The death receptors involved are members of the tumor necrosis factor receptor family (TNFR family) that trimerize upon binding to their ligands (Zheng et al. 1995; Nagata 1997; Ashkenazi and Dixit 1998). In conjunction with adapter proteins several procaspase-8 molecules are subsequently tethered to this multimeric protein aggregate via their prodomain to form the death inducing signaling complex (DISC; Krammer 2000). In this close proximity the low intrinsic enzymatic activity of aggregated initiator procaspases is sufficient to facilitate intermolecular processing (Muzio et al. 1998).

The resulting catalytically active and mature forms in their turn process effector caspases, resulting in a positive feedback loop and an amplification of the apoptotic signal.

In contrast, the intrinsic pathway is initiated through mitochondrial changes. In the course of these changes the mitochondria become permeable and release pro-apoptotic molecules into the cytoplasm, in particular cytochrome c, AIF (apoptosis inducing factor, also termed programmed cell death protein 8 [PCD8]), and procaspases (Liu et al. 1996; Susin et al. 1999a, b; Qin et al. 2001). Cytoplasmic cytochrome c induces oligomerization of the adapter Apaf-1 (apoptotic protease activating factor-1) and subsequent recruitment of a cytosolic initiator caspase (procaspase-9) and of dATP. The resulting high molecular weight caspase-activating complex (apoptosome; Li et al 1997) once more promotes procaspase self-processing and consequently triggers the caspase cascade. Furthermore, in response to apoptogenic stimuli mitochondrial AIF translocates to the nucleus where it causes large-scale DNA fragmentation. Since AIF does not contain intrinsic nuclease activity the underlying molecular mechanisms remain unclear. However, AIF effects are caspase-independent (Lorenzo et al. 1999).

The mitochondrial and the death receptor pathway are intimately connected; the main link between both is a member of the Bcl-2 family (tBid). The Bcl-2 family comprises both pro-apoptotic and anti-apoptotic members, classified into three subfamilies according to their function and domain architecture: (1) a Bcl-2 subfamily with anti-apoptotic function, (2) a Bax subfamily and (3) a BH3 subfamily, both with pro-apoptotic activity. The functional mechanisms of Bcl-2 proteins are still controversial; however, it appears that they are involved in regulating the mitochondrial membrane permeabilization. Whereas anti-apoptotic Bcl-2 proteins favor the closure of the mitochondrial permeability transition pore (PTP), pro-apoptotic Bcl-2 proteins promote the opening of this protein complex that is located at contact sites between outer and inner mitochondrial membranes (Shimizu et al. 1999). An alternative model suggests that some pro-apoptotic Bcl-2 proteins possess the ability to form homomeric or heteromeric mitochondrial channels that mediate the release of apoptogenic factors into the cytosol (Muchmore et al. 1996; Schendel et al. 1997). Thus, upon death receptor mediated activation, caspase-8 cleaves Bid (a member of the BH3 subfamily) at its amino-terminus. Subsequently, truncated Bid (tBid) is translocated to the mitochondria and induces the release of pro-apoptotic molecules (Li et al. 1998).

Innate immunity represents an evolutionary ancient defense mechanism that is based on germ-cell encoded receptors. These structurally and functionally diverse proteins were termed pattern recognition receptors (PRR) because they recognize conserved microbial structures, pathogen-associated molecular patterns (PAMPs), that are exclusively produced by invading microorganisms but not the host (Janeway 1989). Different prototypic PRRs are know, including mannose receptor and further C type lectins, macrophage scavanger receptors, peptidoglycan recognition proteins, RNA helicases, NOD-LRR proteins, and Toll-like receptors (Gordon 2002; Iwanaga and Lee 2005; Akira et al. 2006; Werts et al. 2006). The signal cascades thus activated are often interwoven.

Toll-like receptors (TLRs) are among the best characterized PRRs. Their ligands comprise a number of microbial products, e.g., lipoproteins and lipopolysaccharides (LPS), glycolipids, flagellin, and CpG oligonucleotides (Medzhitov and Janeway 2000, Janeway and Medzhitov 2002; O'Neill 2004; Kaisho and Akira 2006). Upon ligation of PAMPs TLRs recruit intracellular adapter proteins, in particular MyD88, Mal, TRAM, and TRIF. Subsequently, these multi-protein complexes trigger intricate signal cascades, ultimately resulting in the activation of various protein kinases (IRAK, p38, MAPK) and transcription factors (NF-κB, IRF) (Takeda 2005; McGettrick and O'Neill 2004; Kaisho and Akira 2006). Furthermore, several PAMPs have been shown to trigger apoptosis through TLR-mediated signaling (Lopez et al. 2003). This pathway bifurcates at the level of TLR-binding adapters, either via MyD88 and subsequent activation of the caspase-cascade or via TRIF and subsequent stimulation of the intrinsic pathway (Aliprantis et al. 1999, 2000; Hsu et al. 2004).

According to both palaeontological and molecular biological data sponges (taxon Porifera) evolved 600 - 800 Mya. They represent the evolutionary oldest living relicts of a successful transition from the unicellular to the multicellular state. Consequently, they received the status of "living fossil" (Müller 1998, 2001). It was during this transition that PCD and apoptosis arose, involved in a network of mechanisms that are characteristic of every multicellular animal: immunity, morphogenesis, development, etc. Thus, sponges are highly suitable model organisms to trace the evolutionary origin of metazoan-specific processes.

In the course of the last years many core-components of the poriferan apoptotic cell death machinery were identified and functionally analysed (Wiens and Müller 2006). At the same time, it was demonstrated that the bacterial model compound LPS triggers signal cascades of the poriferan innate immune system, involving the pattern recognition receptor SLIP (sponge LPS interacting protein), the adapter MyD88 and the anti-bacterial effector MPEG (macrophage-expressed protein) (Wiens et al. 2005). Furthermore, the stress-response kinases p38 and JNK were phosphorylated upon LPS challenge and the expression of bioactive, low-molecular weight lipid derivatives was upregulated (Böhm et al. 2001; Müller et al. 2004).

This chapter will focus on ancient molecules whose identification, molecular biological characterization, and phylogentic analysis promise to advance our understanding of the evolutionary conserved link between apoptotic cell death and innate immune response.

MATERIAL AND METHODS

Sponges

Live specimens of *Suberites domuncula* (Porifera, Demospongiae, Hadromerida; Olivi, 1792) were collected in the Adriatic Sea near Rovinj (Croatia). Prior to their use the sponges were kept in aquaria in Mainz (Germany) at 17°C for at least three months.

Differential Display Analyses and Cloning of *SDLITAF*

Tissue samples of *S. domuncula* were incubated with 1 μg LPS (from *Escherichia coli* (L2880); Sigma, Deisenhofen, Germany) per ml seawater (Sigma) for the indicated time and then immediately pulverized in liquid nitrogen. Subsequently, total RNA was prepared with

TRIzol, as recommended by the manufacturer (Invitrogen, Karlsruhe, Germany). Following the procedures previously described (Lohmann et al. 1995; McClelland et al. 1995), with modifications according to Müller et al. (2004), RNA was submitted to differential display analyses. In short, 5 µg RNA of LPS-stimulated and unstimulated samples was reverse transcribed (Superscript RT; Invitrogen), using anchored $dT_{23}X$ primers (X =G, C, or A). The cDNA thus prepared was employed as template during a polymerase chain reaction (PCR), applying the anchored primer in combination with an arbitrary primer (12-mer) that was 5'-infrared tagged. PCR was carried out at an initial denaturation for 3 min at 95°C, followed by 40 amplification cycles at 95°C for 30 sec, 40°C for 120 sec and 74°C for 30 sec, and terminated with a final extension at 74°C for 10 min. Subsequently, the samples were size-separated, using an automated DNA sequencer (LI-COR 4200). Differentially expressed PCR-products detected (Odyssey infrared imaging system, LI-COR) in the LPS-treated samples were isolated, reamplified, T/A-cloned (pGEMT; Promega, Madison, WI) and sequenced. Finally, the complete cDNAs were obtained through primer walking (Ausubel et al. 1995; Wiens et al. 1998), using a *S. domuncula* lambda Zapp Express cDNA library (Kruse et al. 1997).

Northern Blotting Analyses

5 µg total RNA of LPS-treated and untreated *S. domuncula* tissue was size-separated on denaturating 1.1% formaldehyde/agarose gels and subsequently blotted on Hybond-N^+ membranes (GE Healthcare Europe GmbH, Munich, Germany). The blots were probed for transcripts of *SDLITAF*, *SDPCD8*, *SDVLP*, and *SDCASPR* with probes that have been DIG-11-dUTP labeled through the PCR-DIG-probe synthesis kit, according to the manufacturer's instructions (Roche Applied Science, Mannheim, Germany). The *SDLITAF* probe included the LITAF domain (nt_{191}-$_{507}$), the *SDPCD8* probe comprised a large part of the oxidoreductase domain (nt_{428}-$_{1584}$), the *SDVLP* probe detected a region that contained the SAM-domain (nt_{196}-$_{786}$), and the *SDCASPR* probe targeted the p20/p10 subunit (nt_{790}-$_{1526}$; GenBank accession number AJ426651). A further probe was designed to monitor the expression of the housekeeping gene β-tubulin (*SDTUB*, nt_{83}-$_{483}$; AJ550806). Hybridized probes were detected by anti-DIG Fab fragments (conjugated to alkaline phosphatase; Roche Applied Science) and visualized, using a chemiluminescence substrate ([disodium 2-chloro-5-(4-methoxyspiro {1,2-dioxetane-3,2'-(5'-chloro)-tricyclo [3.3.1.1.3,7]decan}-4-yl)phenyl phosphate], CDP-Star; Roche Applied Science).

Isolation of the cDNAs for Poriferan Programmed Cell Death 8/Apoptosis Inducing Factor (*SDPCD8*) and Viperin (*SDVLP*)

The cDNA encoding the putative poriferan Programmed Cell Death 8/Apoptosis Inducing Factor (*SDPCD8*) was isolated from the *S. domuncula* cDNA library by applying the degenerate forward primer 5'-CAT/C CAA/G A/TC/GI ATG TTC/T TGG-3' (where I = inosine). This primer targeted a highly invariant region within the oxidoreductase domain of PCD8, H-Q-S-M-F-W. Likewise, *SDVLP* was isolated by means of a degenerate forward primer, 5'-TGG AAG/A GTI TTC/T CAA/G TGT/C T/CT-3', that was directed against the conserved sequence W-K-V-F-Q-C-L within the radical_SAM domain (Pfam accession number PF04055) of viperin. The PCRs consisted of an initial denaturation for 5 min at 95°C, followed by 35 amplification cycles (three steps) at 95°C for 30 sec (denaturation), 54°C to 58°C for 45 sec (primer annealing), 72°C for 50 sec (elongation), and a final extension at 74°C for 10 min. The resulting PCR fragments (≈550 bp for *SDPCD8* and ≈580 bp for *SDVLP*) were cloned, sequenced, and identified through BLAST and FASTA homology searches. Afterwards, they were completed by using a combination of sequence specific and library specific primers. Ultimately, northern blot analyses confirmed that the size of each transcript was consistent with that of the cDNA isolated.

Sequence Analyses

Homology analyses were performed searching protein data bases through the programs FASTA and BLAST available at the European Bioinformatics Institute, Hinxton, UK (http://www.ebi.ac.uk/fasta33/), and at the National Center for Biotechnology Information, Bethesda, USA (http://www.ncbi.nlm.nih.gov/BLAST/). Multiple alignments were performed using ClustalW Version 1.6 sequence alignment program (Thompson et al. 1994) and by manual adjustment. Phylogenetic trees were based on the distances calculated between all pairs of sequence from a multiple alignment and subsequent application of the neighbor-joining method to the resulting distance matrix (Saitou and Nei 1987; Dayhoff et al. 1978). To derive confidence values for their internal branches, the trees were bootstrapped according to Felsenstein (1993). The graphical output of the bootstrap figures was prepared with Treeview (R.D.M. Page, University of Glasgow, http://taxonomy.zoology.gla.ac.uk/rod/treeview.html). Further graphic presentations were generated with the GeneDoc program (Nicholas and Nicholas 1996). Domains, transmembrane regions, and signal sequences were predicted after searching the databases of Pfam (Finn et al. 2006; http://www.sanger.ac.uk/Software/Pfam/), SMART (Letunic et al. 2006; http://smart.embl-heidelberg.de/), and WoLF PSORT (Horton et al. 2006; http://wolfpsort.seq.cbrc.jp/).

Western Blotting Analyses

In order to analyze the LPS-stimulated release of PCD8_SUBDO, mitochondria were isolated from homogenized LPS-treated or untreated *S. domuncula* tissue, using the mitochondria isolation kit (Pierce, Rockford, IL). Protein concentrations were measured with the 2-D Quant kit (GE Healthcare Europe GmbH). Then, 5 μg mitochondrial and cytosolic proteins were size-separated on 10% sodium dodecyl sulfate-polyacrylamide gels (SDS-PAGE), transferred on polyvinylidene difluoride membranes (PVDF; Millipore, Billerica, MA) and probed for with a polyclonal anti-AIF antibody (pAb), raised against a conserved epitope of human AIF (Santa Cruz Biotechnology, Santa Cruz, CA). The resulting immune complexes were visualized by successive incubation with a horseradish peroxidase-coupled secondary antibody and the "BM Chemiluminescence Western Blotting Substrate" (Roche Applied Science), according to the manufacturer's instructions.

RESULTS AND DISCUSSION

The *S. domuncula* LPS-Induced TNF-α Factor

Characterization

In the course of differential display analyses of sponge RNA, application of $dT_{23}A$ and the arbitrary primer 5'-TGC TCC AGG ACC-3' revealed two transcripts whose expression was strongly induced in LPS-treated tissue (Figure 1). Subsequently, the respective DNA was eluted, cloned, and sequenced. The first sample was identified as the previously published caspase-related polypeptide *SDCASPR* (GenBank accession number AJ426651; Wiens et al. 2003). The second sample revealed significant sequence homology to the vertebrate LPS-induced TNF-α factor (LITAF). Ultimately, the complete LITAF cDNA was isolated from the *S. domuncula* cDNA library through primer walking. *SDLITAF* consists of 639 nt (excluding the poly(A) tail) with an open reading frame (ORF) contained within that ranges from nt_{91}-$_{93}$ (Met_{start}) to nt_{508}-$_{510}$ (stop codon). The putative protein LITAF_SUBDO consists of 139 aa with a calculated Mw of 14,991 and a pI of 4.70. It revealed significant sequence similarity to the *Danio rerio* LITAF homologues with an expect value (*E*-value; Coligan et al. 2000) of $2e^{-04}$, and of $5e^{-04}$ to chicken TNF-α factor. The homology to the predicted proteins of invertebrates, e.g., *Caenorhabditis elegans* (F36G3.3) and *Drosophila melanogaster* (RH52355p), was found to be considerably less (*E*-value = $2e^{-03}$ and $3.8e^{-02}$, respectively).

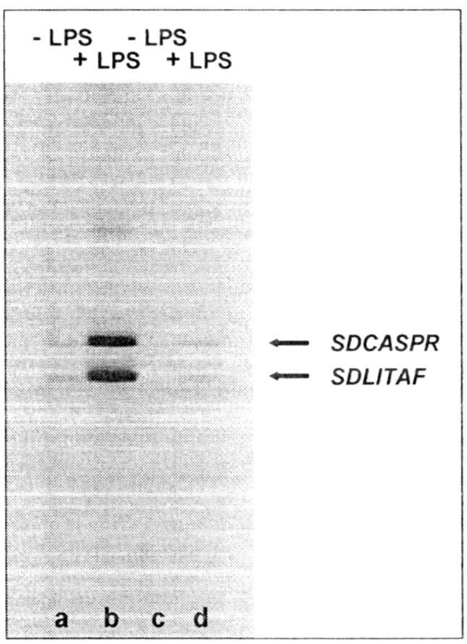

Figure 1. Differential display of LPS-stimulated and unstimulated transcription in *S. domuncula*. Transcripts of LPS-treated (16 h, 1 μg LPS) and untreated sponge tissue were isolated, reverse transcribed, PCR-amplified, and size-separated on polyacrylamide gels. Lane a and c contain the

control sample (– LPS), Lane b and d the LPS-treated sample (+ LPS). Samples were applied either diluted (1:5, Lane c, d) or undiluted (lane a, b). Differentially expressed transcripts were isolated, cloned and identified (arrows).

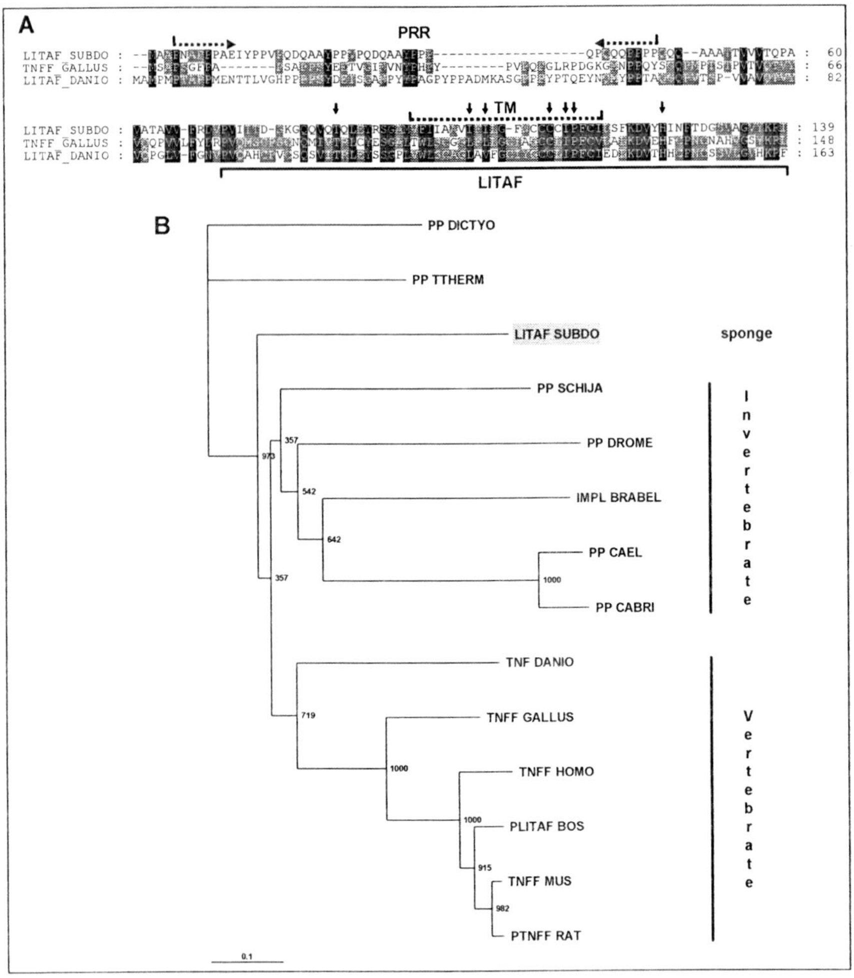

Figure 2. *S. domuncula* LITAF. **A** The aa sequence of the deduced poriferan LPS-induced TNF-α factor (LITAF_SUBDO) was aligned with related sequences of *Gallus gallus* (TNFF_GALLUS; GenBank accession number AAU95774), and *Danio rerio* (LITAF_DANIO; AAH74085). The LITAF domain and the putative transmembrane stretch (TM) within are marked as is the amino-terminal proline-rich region (PRR). The arrows indicate the seven aa conserved in all LITAF sequences analyzed (see "Results"). Residues conserved (identical or similar with respect to their physico-chemical properties, BLOSUM 62 Matrix) in all sequences are shown in white letters on black; those conserved in 60% are in white letters on grey. **B** Phylogenetic tree constructed after alignment of the above-mentioned sequences, furthermore including LITAF domain containing proteins (predicted) of *Schistosoma japonicum* (PP_SCHIJA; AAX26919), *Drosophila melanogaster* (PP_DROME, AAM29605), *Branchiostoma belcheri tsingtaunese* (IMPL_BRABEL; AAT09132), *Caenorhabditis elegans* (PP_CAEL; CAE17781), *Caenorhabditis briggsae* (PP_CABRI; CAE58714), and *Tetrahymena thermophila* (PP_TTHERM; EAS01693). Also included were LITAFs (predicted) of *Homo sapiens* (TNFF_HOMO; AAH46154), *Bos taurus* (PLITAF_BOS; XP_598811), mouse (TNFF_MUS; AAH18559), and rat (PTNFF_RAT; XP_343857). The clusters of vertebrate and invertebrate sequences are indicated as is the poriferan sequence. The tree was rooted with the distantly related LITAF-bearing

sequence of *Dictyostelium discoideum* (PP_DICTYO; XP_636385) and computed by neighbor-joining and distance matrix determinations. The scale bar indicates an evolutionary distance of 0.1 aa substitutions per position in the sequence. The numbers at the nodes are an indication of the level of confidence for branching (1000 bootstrap replicates).

LITAFs are characterized by a large domain, the LITAF domain (SMART accession number SM00714), which is probably membrane-associated. LITAF_SUBDO displays the LITAF domain ($1.7e^{-07}$), bearing a putative transmembrane stretch within (Kyte and Doolittle 1982) (Figure 2A). Intriguingly, the poriferan domain is missing a second characteristic feature, two C-X-X-C knuckles that are thought to bind Zn^{2+} (Collet et al, 2003). Of the seven Cys residues that had been considered to be highly conserved, only three are present in sponge LITAF. However, by including the newly identified poriferan LITAF into an alignment of various vertebrate, invertebrate, and protist LITAF and LITAF-similar sequences, three highly conserved aa could be identified, Cys_{110}, Pro_{113}, His_{124} (numbering based on the poriferan LITAF sequence). Three further conserved positions display Val, Leu, Ile, or Met (Ile_{101}, Ile_{103}, and Ile_{112}). Finally, one conserved position consists of Ser or Thr (Thr_{84}). By integrating the poriferan sequence data it was thus possible to extract a conserved pattern within LITAF domains, [TS]-x(6)-[GS]-x(3)-[WHFY]-x(5)-[LI]-x(1)-[LIVM]-[LF]-x(2,5)-[CL]-C-x(1)-[ILV]-P-[FCY]-[CG]-x(5)-[DS]-x(2)-H-x(1)-[CN]-[PTS]-x(1)-[CD], that will facilitate the future identification of novel metazoan LITAF molecules.

Furthermore, the poriferan sequence bears another feature generally found in LITAFs, an amino-terminal proline-rich region (aa_{4}-$_{45}$), consisting of close to 43% proline residues.

Phylogenetic Analysis

LITAF was initially identified as one of the transcription factors that bind to the promoter of TNF-α, following stimulation with bacterial lipopolysaccharides (Myokai et al. 1999). Since then LITAF-like proteins have been identified in several vertebrates, and GenBank database research led to the identification of a number of predicted invertebrate proteins, also bearing the LITAF-domain. This includes proteins of the cephalochordate *Branchiostoma belcheri*, the fly *D. melanogaster*, the nematodes *Caenorhabditis briggsae* and *C. elegans,* as well as the platyhelminth *Schistosoma japonicum*. In this context it is interesting to note that LITAF_SUBDO revealed a greater sequence homology to the vertebrate LITAF-like proteins than to those of invertebrates. As a general rule, sponge proteins are significantly more similar to vertebrate (human) proteins than to those of classic invertebrate model organisms, such as *C. elegans* (Gamulin et al. 2000). Thus, poriferan LITAF shares 31% identical and 43% similar aa with chicken LITAF, and 30%/40% with *D. rerio* LITAF (Figure 2A), but only 17%/39% with the *D. melanogaster* protein, or 15%/27% with the *B. belcheri* LITAF. The homology of LITAF_SUBDO to these sequences was even more pronounced when only the respective LITAF domain was considered for homology analysis. Accordingly, the poriferan domain shares 43% identical and 59% similar aa with its *D. rerio* counterpart, 40%/60% with the chicken LITAF domain and 31%/54% with the *B. belcheri* domain. The rooted phylogenetic tree generated from these sequence data revealed two distinct clusters (Figure 2B). The first cluster incorporates the LITAF-bearing invertebrate sequences of platyhelminth, nematode, insect, and cephalochordate. The second one comprises the vertebrate sequences of fish, bird, and mammal. The poriferan LITAF forms the base of those two clusters, indicating an ancient origin of the transcription factor that has distant relatives within fungi and ciliates; however, no related sequence was found in the kingdom of bacteria. Furthermore, no resolution was attained between the predicted proteins of *Tetrahymena thermophila* and the facultative multicellular fungus *Dictyostelium discoideum* so that the branching order of the deep divergences cannot be ascertained.

The *S. domuncula* Programmed Cell Death 8/Apoptosis Inducing Factor

Characterization

The poriferan cDNA putatively coding for a PCD8-like protein was identified through screening of the *S. domuncula* cDNA library with a degenerate primer that was directed against a conserved sequence within the oxidoreductase domain. By means of a combination of sequence and vector specific oligonucleotides the complete cDNA (*SDPCD8*) was finally isolated. It consisted of 1961 nt (not including the poly(A) tail) that display an ORF of 590 aa (nt_{12-14} [Met_{start}] - $nt_{1782-1784}$ [stop]). The predicted sponge protein PCD8_SUBDO has a calculated Mw of 63,364 and a pI of 6.43. PCD8_SUBDO revealed significant sequence homology to chicken programmed cell death 8 (*E*-value = $6e^{-118}$) and to the human molecule ($1e^{-115}$). The homology to invertebrate sequences was considerably lower (e.g., sea urchin predicted PCD8 [$9e^{-81}$], and *D. melanogaster* predicted protein SD03428p [$2e^{-102}$]).

Consistent with the domain architecture of other known PCD8/AIF molecules, PCD8_SUBDO bears a significant homology to the NADH oxidoreductase domain ($aa_{140-590}$; $2e^{-30}$). This COOH-terminal part is linked via a spacer to an amino-terminal mitochondrial localization signal (MLS, aa_{1-65}; WoLF PSORT) that is putatively cleaved off upon entering of the proprotein into mitochondria. Thus, the mature protein retains the spacer and the oxidoreductase domain, furthermore including a nuclear localization signal (NLS, $aa_{451-457}$), a HSP70 binding domain (HSP, $aa_{156-235}$), and a chromatin condensation domain (CCD, $aa_{536-584}$) (Figure 3A).

Phylogenetic Analyses

In the course of the apoptotic cascade the programmed cell death 8 protein/AIF is translocated from the mitochondrial intermembrane space into the nucleus, where it is involved in causing large scale DNA fragmentation. PCD8 is a phylogenetically ancient FAD binding flavoprotein whose homologues have been found (predicted) in vertebrates (human, mouse, chicken, amphibia [*Xenopus laevis*], fish [*Tetraodon nigroviridis*]) and invertebrates (echinoderms [*Strongylocentrotus purpuratus*], insects [*D. melanogaster*, *Tribolium castaneum*, *Aedes aegypti*, *Apis mellifera*], hydrozoan [*Hydra vulgaris*]). The overall domain architecture is similarly conserved, consisting of an amino-terminal mitochondrial localization signal attached via a spacer to an oxidoreductase domain that comprises binding sites for NADH and HSP, a nuclear localization signal, and a chromatin condensation domain. Moreover, important homology to oxidoreductases has been discovered in kingdoms that lack "classic" apoptotic s, i. pathway, plants (*Arabidopsis thaliana*, *Oryza sativa*) and bacteria (*Nitrosospira multiformis*).

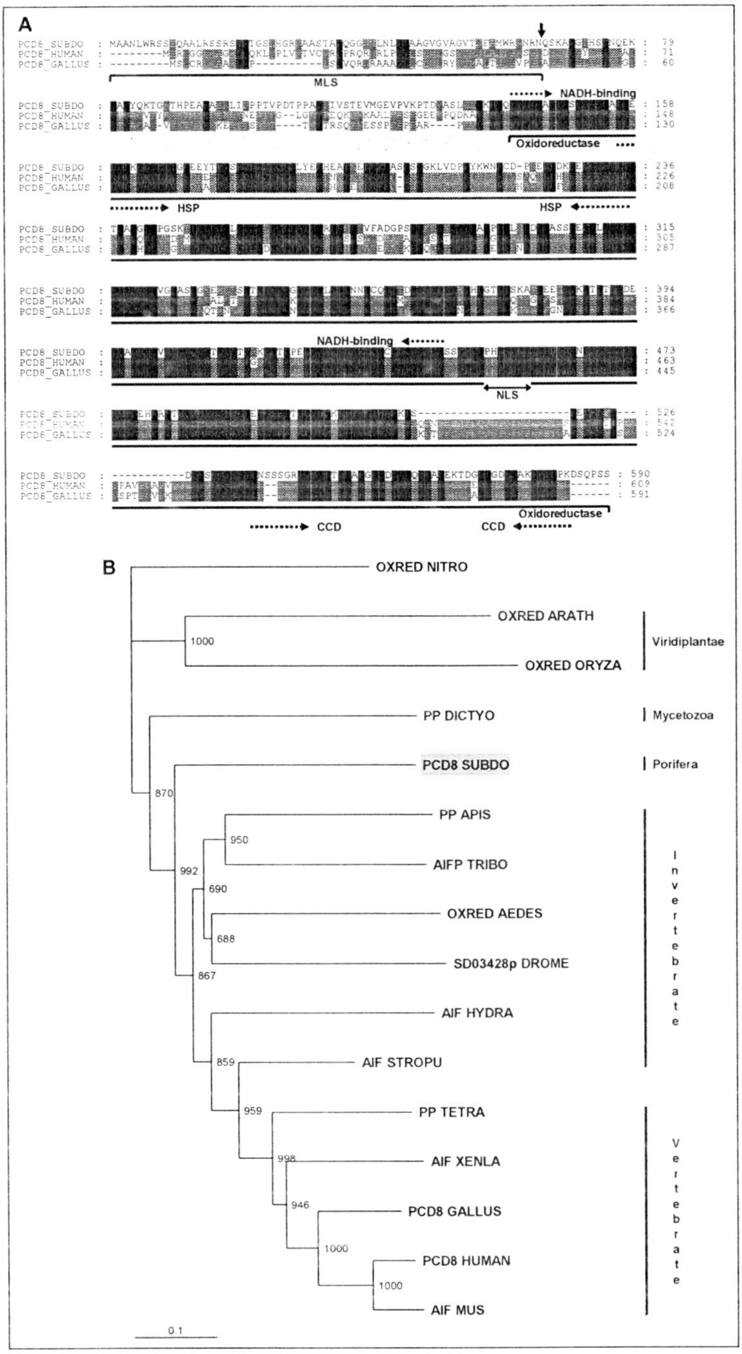

Figure 3. *S. domuncula* PCD8/AIF **A** Alignment of the deduced programmed cell death 8 protein sequence of *S. domuncula* (PCD8_SUBDO) with related sequences, human PCD8 isoform 2 (PCD8_HUMAN; GenBank accession number NP_665811) and chicken PCD8 (PCD8_GALLUS; NP_001007491). The putative nuclear localization (NLS) sequence is underlined (the arrow indicates the cleavage site), as well as the mitochondrial localization sequence (MLS). Also, the NADH binding domain, the HSP70 binding site (HSP), and the chromatin condensation domain (CCD) are marked. All domains mentioned are part of a larger FAD binding oxidoreductase domain. **B** Rooted phylogenetic tree generated after alignment of the above-mentioned sequences, further including PCD8 (apoptosis-

inducing factor AIF) of *Xenopus laevis* (AIF_XENLA; NP_001017244), PCD8 of *Mus musculus* (AIF_MUS; AAH03292), unnamed protein CAG02984 of *Tetraodon nigroviridis* (PP_TETRA; CAG02984), predicted AIF of *Strongylocentrotus purpuratus* (AIF_STROPU; XP_783530), predicted proteins SD03428p of *Drosophila melanogaster* (SD03428p_DROME; AAK93507) and of *Apis mellifera* (PP_APIS; XP_395674), AIF of *Tribolium castaneum* (AIFP_TRIBO; XP_972831), disulfide oxidoreductase of *Aedes aegypti* (OXRED_AEDES; EAT46873), AIF of *Hydra vulgaris* (AIF_HYDRA; AAX13996), predicted protein DDB0191137 of *D. discoideum* (PP_DICTYO; XP_636815), disulfide oxidoreductase of *Arabidopsis thaliana* (OXRED_ARATH; NP_189420), putative monodehydroascorbate reductase of *Oryza sativa* (OXRED_ORYZA; XP_467388), and FAD-dependent pyridine nucleotide-disulphide oxidoreductase of *Nitrosospira multiformis* (OXRED_NITRO; YP_411055). The tree was rooted with the latter sequence.

The poriferan PCD8-like protein revealed highest homology to vertebrate PCD8/AIF, especially to the chicken (38% identical and 56% similar aa), human and *X. laevis* (38%/55%, both) proteins. The homology to PCD8 of insect or sea urchin was less pronounced (31%/46% and 26%/35%, respectively), though still higher than compared to oxidoreductases of plants (16%/28% [*A. thaliana*]). Based on these sequences a phylogenetic tree was generated and rooted with a bacterial oxidoreductase sequence (Figure 3B). This tree revealed the poriferan PCD8 at the basis of two clades containing the predicted PCD8/AIF/oxidoreductases of protostomes and deuterostomes, preceded only by the mycetozoa and plants. Bootstrapping indicated stable branching points. However, it is interesting to note the hydrozoan AIF branch at the basis of the deuterostomes cluster. Thus, a larger data set of basal metazoans is required to verify this positioning.

The *S. domuncula* Viperine-Like Protein

Characterization

Similarly to *SDPCD8* cDNA of the viperin-like protein *SDVLP* was obtained through PCR-screening, applying a degenerate primer that was directed against a stretch of conserved aa within the radical_SAM domain of viperins. The isolated cDNA has a size of 1283 nt, excluding the poly(A) tail, with an ORF encoding the putative *S. domuncula* viperin-like protein VLP_SUBDO. The coding sequence ranges from nt_{99-101} (Met_{start}) to the stop codon at $nt_{1074-1076}$, comprising a 325 aa long protein (Figure 4A). The calculated Mw of VLP_SUBDO is 37,164 with a theoretical pI of 8.20. Database research revealed closest similarity of VLP_SUBDO to viperin of human (E-value = $2e^{-131}$) and other mammals, of mandarin fish *Siniperca chuatsi* ($4e^{-127}$) and chicken ($3e^{-119}$).

Radical_SAM domain containing proteins are able to generate a radical species by reductive cleavage of S-adenosylmethionine (SAM) through an unusual Fe-S centre (Sofia et al. 2001). The SAM domain of VLP_SUBDO is located at aa_{41-204}, including the characteristic motifs (aa_{41-57} [motif 1], aa_{84-95} [motif 2] and $aa_{166-179}$ [motif 3]) (Sofia et al. 2001). With two exceptions (marked in bold) the poriferan protein follows the published consensus pattern of motif 1 [LIVF]-x(3)-C-[NDP]-[LIVMF**Y**]-[DNQRS]-C-x-[FYM]-C (prosite entry PDOC01009). The three closely spaced Cys residues are important for iron-sulfur cluster coordination.

The amino-terminal 40 aa of VLP_SUBDO contain a putative transmembrane stretch (aa_{4-24}; SMART and WoLF PSORT prediction), however, the overall sequence homology concerning this region is low among viperins. It is assumed that the amino-terminus functions as an anchor to the ER, whereas the remaining conserved region is exposed to the cytoplasm (Sun and Nie 2004).

Phylogenetic Analysis

Viperins are radical_SAM domain containing proteins that are part of a large superfamily. The radical_SAM superfamily further includes MoaA, MiaB, NifB, LipA, and several other families (Sofia et al. 2001). The poriferan VLP shares highest sequence similarity with members of the viperin subfamily, e.g., human (60% identical and 72% similar amino acids) and chicken viperin (54%/70%). With the exception of VLP_SUBDO no other invertebrate viperin homologues have been identified until now. In contrast, VLP_SUBDO is considerably less related to the other radical_SAM subfamilies; the next closest relatives are members of the molybdenum cofactor biosynthesis protein (MoaA) family. The MoaA family not only includes human (11%/27%; $1.1e^{-01}$) and chicken (10%/26%; $3.4e^{-02}$) molybdenum cofactor biosynthesis protein 1 A but also invertebrate proteins such as the *D. melanogaster* molybdenum cofactor synthesis-step 1 protein A splice type I (11%/28%; $3e^{-04}$).

A phylogenetic tree with branch confidence values based on 1000 bootstrap replicates was constructed after alignment of viperin or viperin-like protein sequences of sponge and vertebrates. The alignment also comprised sequences of the molybdenum cofactor biosynthesis enzyme (MOCS) of vertebrates and invertebrates. The tree was rooted with the bacterial predicted MOCS of *Vibrio vulnificus* and revealed two major branches (Figure 4B). The first one exclusively comprised MOCS sequences, whereas the second branch was made up of viperin or viperin-like sequences with the poriferan protein at its base. This second cluster showed a further branching point between fish and mammalian sequences.

Gene Expression in Response to LPS-Challenge

Bacterial lipopolysaccharide (LPS) is an endotoxic component of the outer envelope of Gram-negative bacteria that is released from the surface of dying or replicating cells. It represents an important class of PAMPs that are known to induce a broad spectrum of host-defense mechanisms. Thus, on one hand LPS triggers the apoptotic cell death machinery via activation of caspases and mitochondrial permeabilization (Hull et al. 2002). On the other hand, recognition of bacterial lipopolysaccharide through TLRs elicits the activation of transcription factors (in particular NF-κB and LITAF), which gives rise to the expression of pro-inflammatory and anti-microbial proteins (Janeway and Medzhitov 2002; Kaisho and Akira 2006).

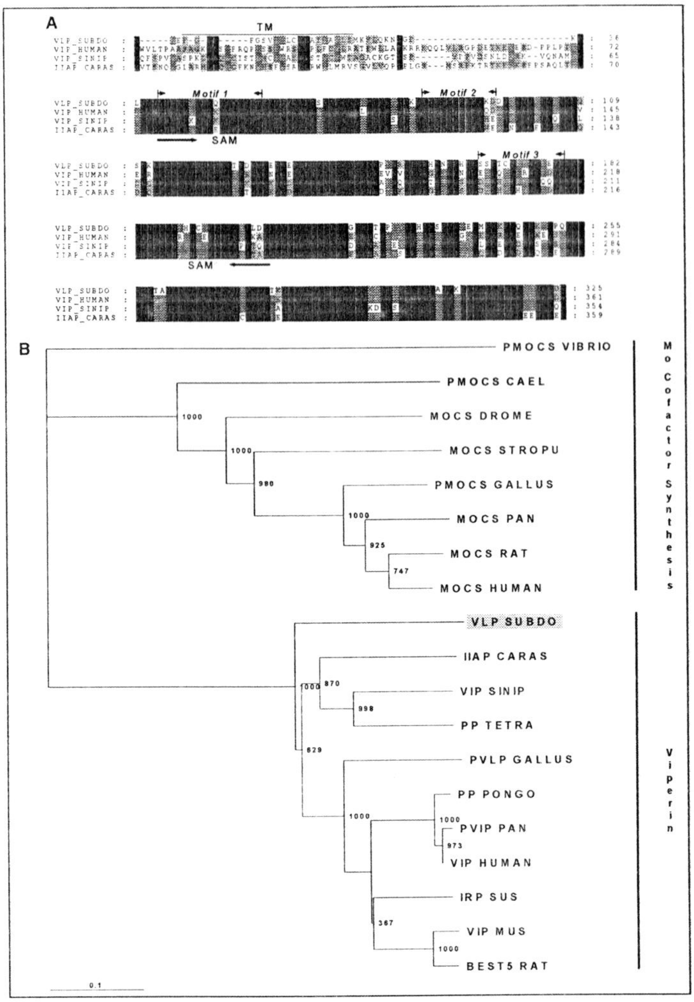

Figure 4. *Suberites domuncula* VLP. **A** Alignment of the deduced viperin-like protein sequence from *S. domuncula* (VLP_SUBDO) with the related viperin sequences of the mandarin fish *Siniperca chuatsi* (VIP_SINIP; GenBank accession number AAS77258) and of human (VIP_HUMAN; AAL50053), also including the *Carassius auratus* IFN-inducible and antiviral protein sequence (IIAP_CARAS; AAP68824). The radical_SAM domain (SAM) with its three characteristic motifs is marked as is the putative transmembrane region (TM). **B** Rooted phylogenetic tree constructed from the above-mentioned sequences as well as from the molybdenum cofactor synthesis enzyme (MOCS, predicted) sequences of *C. elegans* (PMOCS_CAEL; NP_509612), *D. melanogaster* (MOCS_DROME; AAF67855), *S. purpuratus* (MOCS_STROPU; XP_783948), chicken (PMOCS_GALLUS; XP_419475), rat (MOCS_RAT; XP_236911), human (MOCS_HUMAN; O14940), *Pan troglodytes* (MOCS_PAN; XP_527381), the viperin or viperin-like protein sequences (predicted) of chicken (PVLP_GALLUS; XP_426208), *P. troglodytes* (PVIP_PAN; XP_515283), mouse (VIP_MUS; AAL50054), the rat Best5 protein (BEST5_RAT; NP_620236), the *Sus scrofa* inflammatory response protein 6 (IRP_SUS; NP_998982), the predicted protein of *Pongo pygmaeus* (PP_PONGO; CAH90389), and the unnamed protein of the fish *T. nigroviridis* (PP_TETRA; CAG10497). The tree was rooted with the predicted MOCS sequence of *Vibrio vulnificus* as outgroup (PMOCS_VIBRIO; AE016811). The clusters of molybdenum cofactor synthesis enzymes and viperins are marked.

In order to analyze the LPS-induced gene expression in sponges, northern blot analyses were performed. For this purpose, sponge tissue was incubated with LPS for 8 h and 16 h. Subsequently, RNA was extracted, electrophoresed, blotted, and then identified through hybridization with DIG-labeled probes. The probes were generated to detect transcripts of the poriferan LITAF (*SDLITAF*), PCD8/AIF (*SDPCD8*), caspase (*SDCASPR*), viperin (*SDVLP*), and tubulin (*SDTUB*). The expression of the housekeeping gene tubulin was used as an internal control to verify that the same amounts of RNA were loaded on the gels.

In untreated tissue the hybridized probes revealed a relatively low steady-state level of transcription of *SDLITAF*, *SDPCD8*, *SDVLP*, and *SDCASPR*. However, in the course of 16 h LPS significantly increased the expression of the four genes (Figure 5). In particular the expression of *SDLITAF* was rapidly upregulated. In contrast, the signal-intensities of the tubulin transcripts remained at the same level, indicating the unchanged and constitutive expression of this housekeeping gene.

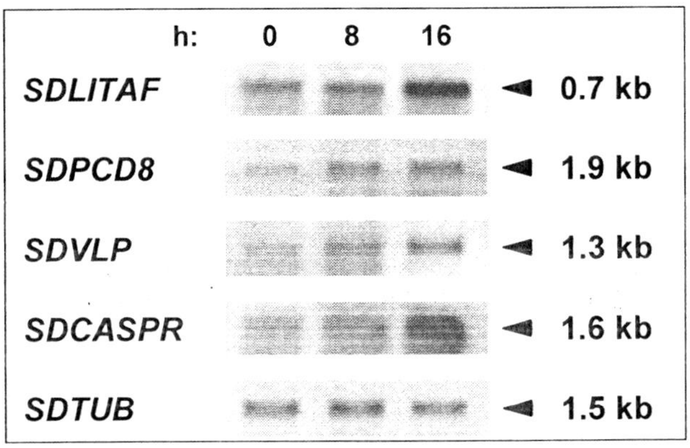

Figure 5. LPS-induced gene expression in *S. domuncula*. RNA of untreated and LPS-treated sponge tissue (0 h, 8 h or 16 h incubation, 1 µg) was isolated, size-separated (equal amounts of RNA were loaded), and blotted. The blots were probed for transcripts of poriferan LITAF (*SDLITAF*), PCD8 (*SDPCD8*), VLP (*SDVLP*), caspase (*SDCASPR*), and tubulin (*SDTUB*). The transcript sizes are indicated.

LPS-Induced PCD8-Translocation

PCD8/AIF is an evolutionary ancient molecule that is released from mitochondria into the cytosol upon activation of the intrinsic apoptotic pathway (Susin et al. 1999a; Lorenzo et al. 1999). To examine the PCD8 redistribution in sponges in the course of LPS-challenge, western blotting analyses were performed. Following the incubation of sponge tissue with LPS for 8 h and 16 h, equal concentrations of mitochondrial and cytosolic proteins were submitted to SDS-PAGE, blotted on PVDF-membranes and then probed for with a pAb that cross-reacted with a conserved aa stretch of poriferan PCD8.

This western blot data clearly demonstrated a reduction of the PCD8-signal in mitochondrial extracts that correlated inversely with an increased intensity of the PCD8-signal in cytosolic extracts, occurring already after exposure to LPS for 8 h (Figure 6). The effect was even more pronounced after 16 h of LPS-challenge, clearly indicating a mitochondrial permeabilization. Interestingly, a low level of PCD8 was even detected in the cytosolic fraction of unstimulated samples. Since PCD8 is nuclear encoded, the signals detected probably represent the newly synthesized protein prior to its transport into mitochondria.

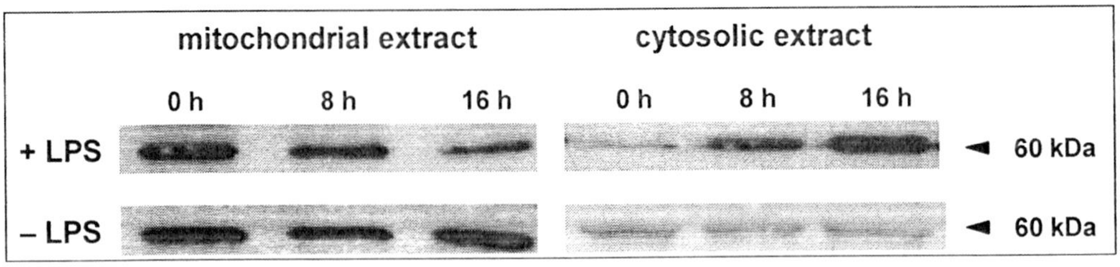

Figure 6. LPS-induced release of mitochondrial PCD8. Following LPS-treatment (for the indicated time), mitochondrial and cytosolic *S. domuncula* proteins were isolated and submitted to SDS-PAGE (equal loading). Subsequently, anti-AIF/PCD8 immuno complexes were detected on western blots, confirming the time-dependant and LPS-stimulated release of poriferan PCD8 into the cytosol. Untreated tissue revealed no such PCD8 translocation.

Discussion

In the human system bacterial lipopolysaccharide is involved in many live-threatening pathological conditions, including sepsis and multiorgan failure. Upon ligation of LPS, the Toll-like receptor 4 (TLR4) homodimerizes and recruits the adapter protein MyD88 (myeloid differentiation factor 88; Medzhitov et al. 1998). MyD88 comprises two homophilic protein interaction domains, the TIR domain and the death domain (DD). TIR facilitates the interaction between TLRs and adaper proteins. In contrast, DD was originally identified as a region of sequence similarity between the tumor necrosis factor (TNF) receptors, required for the induction of apoptosis. However, DD-DD interaction was also found to mediate TLR signaling. Thus, after binding to TLRs, MyD88 associates with a member of the IRAK family, IRAK-4 (IL-1R-associated kinase-4), also bearing a DD. Ultimately, the LPS-stimulated pathway elicits an anti-microbial response via activation of transcription factors. In contrast, pro-apoptotic pathways, also inducible through LPS and other bacterial compounds, remain more elusive. However, since MyD88 also interacts with FADD (Fas-associated death domain), a DD adapter protein of the extrinsic apoptotic pathway, TLRs are coupled to the activation of the caspase cascade (Aliprantis et al. 2000; Lopez et al. 2003). Furthermore, apoptotic pathways often involve activation and increased expression of transcription. Hence, in several human cell types the expression of LITAF was found to increase after LPS-stimulation, subsequently promoting the expression of TNF-α (Myokai et al. 1999; Bolcato-Bellemin et al. 2004), a pleiotropic cytokine that leads to apoptosis and inflammation (Cross et al. 1989; Zheng et al. 1995; Ashkenazi and Dixit 1998). Because of the fundamental connection between the transcriptional activator LITAF and the cytokine TNF-α, the dysregulated expression or activity of either one often bears fatal consequences (Matsumura et al. 2004; Zhu et al. 2005; Street et al. 2003).

In the poriferan model organism *S. domuncula*, LPS triggers a recently described MyD88-dependent signaling cascade that upregulates the expression of MyD88 and of the anti-bacterial perforin-like molecule MPEG (Wiens et al. 2005). This chapter described additional differential display and northern blot analyses of LPS-treated sponge tissue. These analyses revealed further molecules involved in LPS-induced signaling, thus substantiating the link between innate immune response and apoptosis. Consequently, it is proposed that initiation of the poriferan MyD88-dependent pathway causes activation of the transcription factor LITAF, resulting in an augmented expression of various molecules, anti-microbial (viperin-like protein VLP, MPEG) or pro-apoptotic (CASPR, PCD8). Since LITAF was also expressed in untreated sponge tissue, though at a considerable lower level, it might additionally regulate its own expression in a positive feedback loop.

Viperin (virus inhibitory protein, endoplasmic reticulum-associated, interferon-inducible) was initially identified through differential display analysis as a virus-induced gene in vertebrates (Boudinot et al. 1999, 2000). Subsequent functional analyses demonstrated its induction not only in response to a diversity of viruses but also to LPS, dsRNA, and interferon, implicating a general involvement of viperin in pathological conditions such as infections (Olofsson et al. 2005; Sun and Nie 2004; Chin and Creswell 2001; Boudinot et al. 2000). In very recent studies the LPS-induced viperin expression was related to the TLR4-meditated pathway (Severa et al. 2006). Viperin contains a radical_SAM domain and displays significant homology to members of the SAM superfamily. SAM proteins are required for biosynthetic pathways during which highly reactive radical species are generated by reductive cleavage of the cofactor S-adenoylmethionine (SAM). The *modus operandi* of viperin is still speculative. However, through protein radical formation and biosynthesis of cofactors viperin might participate in non-specific responses to pathological insults. The putative anti-microbial VLP_SUBDO was identified through PCR-screening of a *S. domuncula* cDNA library, and similar to the recently discovered poriferan MPEG its expression was found to be LPS-inducible. To the best of our knowledge, the poriferan VLP represents the first viperin identified in invertebrates. It comprises the characterisitc signatures of viperns and the phylogenetic tree generated upon the sequence data revealed unambiguously a clustering of VLP with all vertebrate viperins known so far. Therefore, it is proposed to define viperins as a new class of evolutionary ancient mediators of the innate immune system.

Continuous LPS-challenge of *S. domuncula* not only induced the expression of anti-microbial molecules but also led to an elevated expression of the caspase CASPR and the flavoprotein PCD8. The expression of CASPR was already recently shown to be stimulated following exposure to ocadaic acid, a toxic polyether fatty-acid derivative of bacteria and dinoflagellates (Wiens et al. 2003). Therefore, these results suggest that bacterial products not only elicit an innate immune response but also trigger cell death. Intriguingly, concurrent transcriptional activation of CASPR and PCD8 coupled the proteolytic cascade to the caspase-independent DNA fragmentation (see below).

In the course of the apoptotic cascade mitochondrial proteins, normally confined to the intermembrane space, are released into the cytosol due to a permeabilization of the outer mitochondrial membrane. Among the molecules released cytochrome c was found to associate in an ATP-dependent way with cytosolic apoptotic protease activating factor-1

(Apaf-1) and procaspase-9. The resulting apoptosome induces a cascade-like activation of caspases (Li et al. 1997; Baliga and Kumar 2003). Other mitochondrial proteins released in response to death stimuli include HSP10, SMAC, procaspases, and AIF (Martinou and Green 2001; Susin et al. 1999). Cytosolic AIF (apoptosis-inducing factor, synonymous with PCD8) is very rapidly translocated into the nucleus where it causes peripheral chromatin condensation and large-scale DNA fragmentation, preceding oligonucleosomal DNA fragmentation (Susin et al. 1999a; 2000; Yu et al. 2002). Yet, so far it remains elusive how AIF exerts its lethal functions.

The poriferan protein PCD8/AIF, described in this chapter, displays a significant homology to the vertebrate AIF. Based on sequence comparisons a domain structure characteristic of AIF was predicted. Furthermore, phylogenetic analysis underlined the evolutionary ancient character of PCD8 that was placed at the root of the metazoan phylogenetic tree. Nevertheless, since upon exposure to apoptogenic LPS poriferan PCD8 was released into the cytosol, a similar pro-apoptotic function as that of its vertebrate counterpart was suggested. Nonetheless, several important differences in sequence (namely the omission of two stretches of 23 and 14 aa at the amino-terminus; Figure 3B) imply considerable functional or regulatory modifications that await further characterization.

CONCLUSION

The findings of this study taken with previous results indicate that upon ligation of LPS to the poriferan pattern recognition receptor SLIP a heteromeric protein complex is formed. This complex comprises a SLIP-dimer and the DD-bearing adapter MyD88. In an early response, subsequent signaling gives rise to the expression of anti-bacterial proteins via activation of the transcription factor LITAF. However, upon prolonged exposure to LPS the signal cascade bifurcates at a yet undetermined point, escalating in apoptotic cell death. Thus, increased expression of CASPR and PCD8 correlates with mitochondrial permeabilization (i.e., release of mitochondrial PCD8), until the death pathway overwhelms the survival pathway (Figure 7). Therefore, both signal cascades are intimately connected and *S. domuncula*, a member of the evolutionary oldest metazoan taxon, represents a highly suitable model to investigate the ancient link.

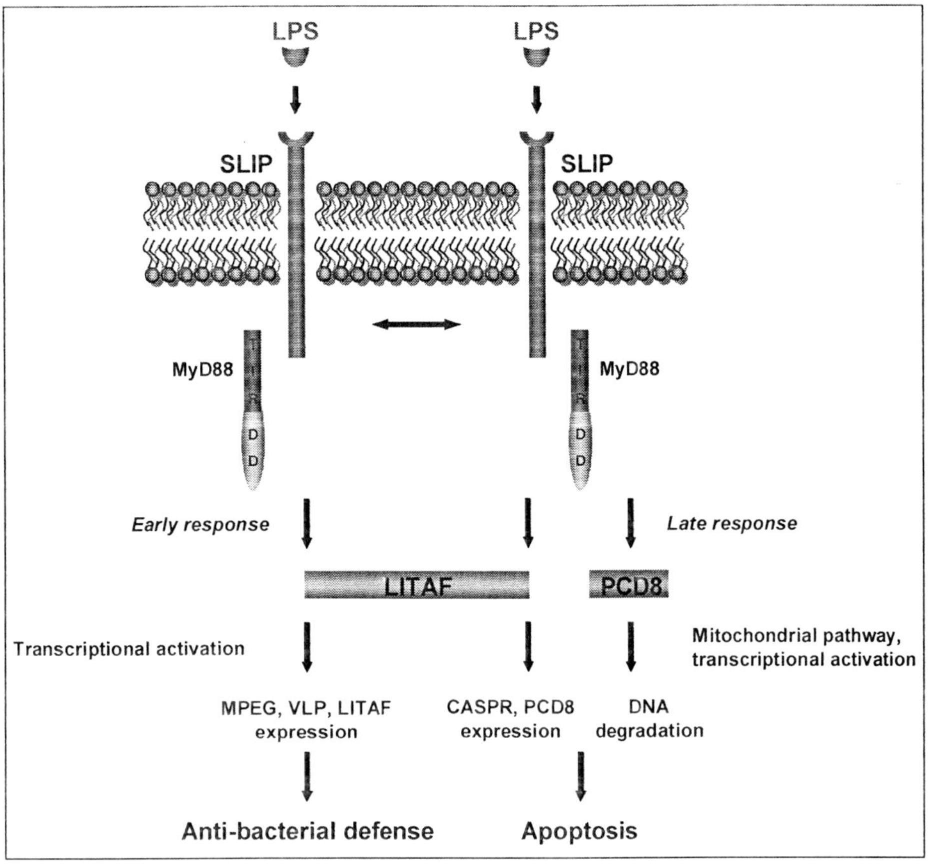

Figure 7. Poriferan LPS-induced signaling - a schematic model. Recognition of LPS through the transmembrane receptor SLIP induces dimerization of SLIP and subsequent recruitment of MyD88, an adapter molecule bearing two protein interaction domains (DD and TIR). Intracellular signaling elicits activation of the transcription factor LITAF. In an early response to LPS the transcription of anti-bacterial factors (MPEG, VLP) and of LITAF is upregulated. Continuous exposure to LPS results in an expression of pro-apoptotic proteins (CASPR, PCD8) and the translocation of mitochondrial PCD8 into the nucleus where the degradation of genomic DNA is initiated (late response).

Acknowlegments

This work was supported by grants from the Deutsche Forschungsgemeinschaft Mü/14-4 and the Bundesministerium für Bildung und Forschung, Germany [project: Center of Excellence; BIOTECmarin].

References

Akira, S., Uematsu, S. & Takeuchi, O. (2006). Pathogen recognition and innate immunity. *Cell, 124*, 783-801.

Aliprantis, A. O., Yang, R. B., Mark, M. R., Suggett, S., Devaux, B., Radolf, J. D., Klimpel, G. R., Godowski, P. & Zychlinsky, A. (1999). Cell activation and apoptosis by bacterial lipoproteins through Toll-like receptor-2. *Science, 285*, 736-739.

Aliprantis, A. O., Yang, R. B., Weiss, D. S., Godowski, P. & Zychlinsky, A. (2000). The apoptotic signaling pathway activated by Toll-like receptor-2. *EMBO J, 19*, 3325-3336.

Ashkenazi, A. & Dixit, V. M. (1998). Death receptors: signaling and modulation, *Science, 281*, 1305-1308.

Ausubel, F. M., Brentc, R., Kingston, R. E., Moore, D. D., Smith, J. A., Seidmann, J. G. & Struhl, K. (1995). *Current protocols in molecular biology*. New York, John Wiley & Sons, Inc.

Baliga, B. & Kumar, S. (2003). Apaf-1/cytochrome c apoptosome: an essential initiator of caspase activation or just a sideshow? *Cell Death Differ, 10*, 16-18.

Böhm, M., Hentschel, U., Friedrich, A., Fieseler, L., Steffen, R., Gamulin, V., Müller, I. M. & Müller W. E. G. (2001). Molecular response of the sponge *Suberites domuncula* to bacterial infection. *Mar Biol, 139*, 1037-1045.

Bolcato-Bellemin, A. L., Mattei, M. G., Fenton, M., & Amar, S. (2004). Molecular cloning and characterization of mouse LITAF cDNA: role in the regulation of tumor necrosis factor-α (TNF-α) gene expression. *J Endotoxin Res, 10*, 15-23.

Boudinot, P., Massin, P., Blanco, M., Riffault, S. & Benmansour, A. (1999). Vig-1, a new fish gene induced by the rhabdovirus glycoprotein, has a virus-induced homologue in humans and shares conserved motifs with the MoaA family. *J Virol, 73*, 1846-1852.

Boudinot, P., Riffault, S., Salhi, S., Carrat, C., Sedlik, C., Mahmoudi, N., Charley, B. & Benmansour, A. (2000). Vesicular stomatitis virus and pseudorabies virus induce a vig1/cig5 homologue in mouse dendritic cells via different pathways. *J Gen Virol, 81*, 2675-2682.

Chin, K.C. & Cresswell, P. (2001). Viperin (cig5), an IFN-inducible antiviral protein directly induced by human cytomegalovirus. *Proc Natl Acad Sci USA, 98*, 15125-15130.

Coligan, J. E., Dunn, B. M., Ploegh, H. L., Speicher, D. W. & Wingfield, P. T. (2000). *Current Protocols in Protein Science*. New York: John Wiley & Sons, Inc.

Collet, J. F., D'Souza, J. C., Jakob, U. & Bardwell, J. C. (2003). Thioredoxin 2, an oxidative stress-induced protein, contains a high affinity zinc binding site. *J Biol Chem, 278*, 45325-45332.

Cross, A. S., Sadoff, J. C., Kelly, N., Bernton, E. & Gemski, P. (1989). Pretreatment with recombinant murine tumor necrosis factor alpha/cachectin and murine interleukin 1 alpha protects mice from lethal bacterial infection. *J Exp Med, 169*, 2021-2027.

Dayhoff, M. O., Schwartz, R. M. & Orcutt, B. C. (1978). A Model of Evolutionary Change in Proteins. In M. O. Dayhoff (Ed.), *Atlas of Protein Sequence and Structure* (pp. 345-352). Washington, D.C.: National Biomedical Research Foundation.

Felsenstein, J. (1993). PHYLIP-Phylogeny inference package. *Cladistics, 5*, 164-166.

Finn, R. D., Mistry, J., Schuster-Böckler, B., Griffiths-Jones, S., Hollich, V., Lassmann, T., Moxon, S., Marshall, M., Khanna, A., Durbin, R., Eddy, S. R., Sonnhammer, E. L. L. & Bateman, A. (2006). Pfam: clans, web tools and services. *Nucleic Acids Res, 34*, D247-D251.

Gamulin, V., Müller, I. M. & Müller, W. E. G. (2000). Sponge proteins are more similar to those of *Homo sapiens* than to *Caenorhabditis elegans*. *Biol J Linnnean Soc, 71*, 821-828.

Gordon, S. (2002). Pattern recognition receptors: doubling up for the innate immune response. *Cell, 111*, 927-930.

Green, D. R. & Evans, G. I. (2002). A matter of life and death. *Cancer Cell, 1*, 19-30.

Horton, P., Park, K. J., Obayashi, T. & Nakai, K. (2006). Protein Subcellular Localization Prediction with WoLF PSORT. Proceedings of the 4th Annual Asia Pacific Bioinformatics Conference APBC06, Taipei, Taiwan, 39-48.

Hsu, L. C., Park, J. M., Zhang, K., Luo, J. L., Maeda, S., Kaufman, R. J., Eckmann, L., Guiney, D. G. & Karin, M. (2004). The protein kinase PKR is required for macrophage apoptosis after activation of Toll-like receptor 4. *Nature, 428*, 341-345.

Hull, C., McLean, G., Wong, F., Duriez, P. J. & Karsan, A. (2002). Lipopolysaccharide signals an endothelial apoptosis pathway through TNF receptor-associated factor 6-mediated activation of c-Jun NH_2-terminal kinase. *J Immunol, 169*, 2611-2618.

Iwanaga, S. & Lee, B. L. (2005). Recent advances in the innate immunity of invertebrate animals. *J Biochem Mol Biol, 38*, 128-150.

Janeway C. A. Jr. (1989). Approaching the asymptote? Evolution and revolution in immunology. *Cold Spring Harb Symp Quant Biol, 54*, 1-13.

Janeway, C. A. Jr. & Medzhitov, R. (2002). Innate immune recognition. *Annu Rev Immunol, 20*, 197-216.

Kaisho, T. & Akira, S. (2006). Toll-like receptor function and signaling. *J Allergy Clin Immunol, 117*, 979-987.

Kirkin, V., Joos, S. & Zornig, M. (2004). The role of Bcl-2 family members in tumorigenesis. *Biochem Biophys Acta, 1644*, 229-249.

Krammer, P. H. (2000). CD95's deadly mission in the immune system. *Nature, 407*, 789-795.

Kruse, M., Müller, I. M. & Müller, W. E. G. (1997). Early evolution of metazoan serine/threonine- and tyrosine kinases: identification of selected kinases in marine sponges. *Mol Biol Evol, 14*, 1326-1334.

Kyte, J. & Doolittle, R. F. (1982). A simple method for displaying the hydropathic character of a protein. *J Mol Biol, 157*, 105-132.

Letunic, I., Copley, R. R., Pils, B., Pinkert, S., Schultz, J. & Bor, P. (2006). SMART 5: domains in the context of genomes and networks. *Nucleic Acids Res, 34*, D257-D260.

Li, H., Zhu, H., Xu, C. J. & Yuan, J. (1998). Cleavage of BID by caspase 8 mediates the mitochondrial damage in the Fas pathway of apoptosis. *Cell, 94*, 491-501.

Li, P., Nijhawan, D., Budihardjo, I., Srinivasula, S. M., Ahmad, M., Alnemri, E. S. & Wang, X. (1997). Cytochrome c and dATP-dependent formation of Apaf-1/caspase-9 complex initiates an apoptotic protease cascade. *Cell, 91*, 479-489.

Liu, X., Kim, C. N., Yang, J., Jemmerson, R. & Wang, X. (1996). Induction of apoptotic program in cell-free extracts: requirement for dATP and cytochrome c. *Cell, 86*, 147-150.

Lohmann, J., Schickel, H. & Bosch, T. C. G (1995). REN display, a rapid and efficient method for nonradioactive differential display and mRNA isolation. *BioTechniques, 18*, 200-201.

Lopez, M., Sly, L. M., Luu, Y., Young, D., Cooper, H. & Reiner, N. E. (2003). The 19-kDa *Mycobacterium tuberculosis* protein induces macrophage apoptosis through Toll-like receptor-2. *J Immunol, 170,* 2409-2416.

Lorenzo, H. K., Susin, S. A., Penninger, J. & Kroemer, G. (1999). Apoptosis inducing factor (AIF): a phylogenetically old, caspase-independent effector of cell death. *Cell Death Differ, 6,* 516-524.

Martinou, J. C. & Green, D. R. (2001). Breaking the mitochondrial barrier. *Nat Rev Mol Cell Biol, 2,* 63-67.

Matsumura, Y., Matsumura, Y., Nishigori, C., Horio, T. & Miyachi, Y. (2004). *PIG7/LIT*AF gene mutation and overexpression of its gene product in extramammary Paget's disease. *Int J Cancer, 111,* 218-223.

McClelland, M., Mathieu-Daude, F. & Welsh, J. (1995). RNA finger-printing and differential display using arbitrarily primed PCR. *Trends Genet, 11,* 242-246.

McGettrick, A. F. & O'Neill, L. A. J. (2004). The expanding family of MyD88-like adaptors in Toll-like receptor signal transduction. *Mol Immunol 41,* 577-582.

Medzhitov, R. & Janeway, C. (2000). The Toll receptor family and microbial recognition. *Trends Microbiol, 8,* 452-456.

Medzhitov, R., Preston-Hurlburt, P., Kopp, E., Stadlen, A., Chen, C., Ghosh, S. & Janeway, C. A. Jr. (1998). MyD88 is an adaptor protein in the hToll/IL-1 receptor family signaling pathways. *Mol Cell, 2,* 253-258.

Muchmore, S. W., Sattler, M., Liang, H., Meadows, R. P., Harlan, J. E., Yoon, H. S., Nettesheim, D., Chang, B. S., Thompson, C. B., Wong, S. L., Ng, S. L. & Fesik, S. W. (1996). X-ray and NMR structure of human Bcl-xL, an inhibitor of programmed cell death. *Nature, 381,* 335-341.

Müller, W. E. G. (1998). Origin of Metazoa: sponges as living fossils. Naturwiss, 85, 11-25.

Müller, W. E. G. (2001). How was metazoan threshold crossed: the hypothetical Urmetazoa. *Comp Biochem Physiol [A], 129,* 433-460.

Müller, W. E. G., Klemt, M., Thakur, N. L., Schröder, H. C., Aiello, A., D'Esposito, M., Menna, M. & Fattorusso, E. (2004). Molecular/chemical ecology in sponges: evidence for an adaptive antibacterial response in *Suberites domuncula. Mar Biol, 144,* 19-29.

Muzio, M., Stockwell, B. R., Stennicke, H. R., Salvesen, G. S. & Dixit, V. M. (1998). An induced proximity model for caspase-8 activation. *J Biol Chem, 273,* 2926-2930.

Myokai, F., Takashiba, S., Lebo, R. & Amar, S. (1999). A novel liposaccharide-induced transcription factor regulating tumor necrosis factor alpha gene expression: molecular cloning, sequencing, characterization, and chromosomal assignment. *Proc Natl Acad Sci USA, 96,* 4518-4523.

Nagata, S. (1997). Apoptosis by death factor. *Cell, 88,* 355-365.

Nicholas, K. B. & Nicholas, H. B. Jr. (1997). GeneDoc: a tool for editing and annotating multiple sequence alignments. Version 2.6.002. Available from: URL: http://www.psc.edu/biomed/genedoc/

Olofsson, P. S., Jatta, K., Wägsäter, D., Gredmark, S., Hedin, U., Paulsson-Berne, G., Söderberg-Naucler, C., Hansson, G. K. & Sirsjö, A. (2005). The antiviral cytomegalovirus inducible gene 5/viperin is expressed in atherosclerosis and regulated by proinflammatory agents. *Arterioscler Thromb Vasc Biol, 25,* e113-e116.

O'Neill, L. A. (2004). TLRs: Professor Mechnikov, sit on your hat. *Trends Immunol, 25*, 687-693.

Qin, Z. H., Wang, Y., Kikly, K. K., Sapp, E., Kegel, K. B., Aronin, N. & DiFiglia, M. (2001). Pro-caspase-8 is predominantly localized in mitochondria and released into cytoplasm upon apoptotic stimulation. *J Biol Chem, 276*, 8079-8086.

Saitou, N. & Nei, M. (1987). The neighbor-joining method: a new method for reconstructing phylogenetic trees. *Mol Biol Evol, 4*, 406-425.

Salvesen, G. S. & Dixit V. M. (1997). Caspases: intracellular signaling by proteolysis. *Cell, 91*, 443-446.

Schendel, S. L., Xie, Z., Montal, M. O., Matsuyama, S., Montal, M. & Reed, J.C. (1997). Channel formation by antiapoptotic protein Bcl-2. *Proc Natl Acad Sci USA, 94*, 5113-5118.

Severa, M., Coccia, E. M. & Fitzgerald, K. A. (2006). Toll-like receptor-dependent and -independent viperin gene expression and counter-regulation by PRD-I-binding factor-1/BLIMP-1. *J Biol Chem*, in press.

Shimizu, S., Narita, M. & Tsujimoto, Y. (1999). Bcl-2 family proteins regulate the release of apoptogenic cytochrome c by the mitochondrial channel VDAC. *Nature, 399*, 483-487.

Sofia, H. J., Chen, G., Hetzler, B. G., Reyes-Spindola, J. F. & Miller, N. E. (2001). Radical SAM, a novel protein superfamily linking unresolved steps in familiar biosynthetic pathways with radical mechanisms: functional characterization using new analysis and information visualization methods. *Nucleic Acids Res, 29*, 1097-1106.

Street, V. A., Bennett, C. L., Goldy, J. D., Shirk, A. J., Kleopa, K. A., Tempel, B. L., Lipe, H. P., Scherer, S. S., Bird, T. D. & Chance, P. F. (2003). Mutation of a putative protein degradation gene LITAF/SIMPLE in Charcot-Marie-Tooth disease 1C. *Neurology, 60*, 22-26.

Sun, B. J. & Nie, P. (2004). Molecular cloning of the viperin gene and its promoter region from the mandarin fish *Siniperca chuatsi*. *Vet Immunol Immunopathol, 101*, 161-170.

Susin, S. A., Lorenzo, H. K., Zamzami, N., Marzo, I., Snow, B. E., Brothers, G. M., Mangion, J., Jacotot, E., Costantini, P., Loeffler, M., Larochette, N., Goodlett, D. R., Aebersold, R., Siderovski, D. P., Penninger, J. M. & Kroemer, G. (1999a). Molecular characterization of mitochondrial apoptosis-inducing factor. *Nature, 397*, 441-446.

Susin, S. A., Lorenzo, H. K., Zamzami, N., Marzo, I., Brenner, C., Larochette, N., Prevost, M. C., Alzari, P. M. & Kroemer, G. (1999b). Mitochondrial release of caspase-2 and -9 during the apoptotic process. *J Exp Med, 189*, 381-394.

Susin, S. A., Daugas, E., Ravagnan, L., Samejima, K., Zamzami, N., Loeffler, M., Costantini, P., Ferri, K. F., Irinopoulou, T., Prevost, M. C., Brothers, G., Mak, T. W., Penninger, J., Earnshaw, W. C. & Kroemer, G. (2000). Two distinct pathways leading to nuclear apoptosis. *J Exp Med 192*, 571-580.

Takeda, K. 2005. Evolution and integration of innate immune recognition systems: the Toll-like receptors. *J Endotoxin Res, 11*, 51-55.

Thompson, C. B. (1995). Apoptosis in the pathogenesis and treatment of disease. Science, 267, 1456-1462.

Thompson, J. D., Higgins, D. G. & Gibson, T. J. (1994). CLUSTAL W: improving the sensitivity of progressive multiple sequence alignment through sequence weighting, position-specific gap penalties and weight matrix choice. *Nucleic Acids Res, 22*, 4673-4680.

Thornberry, N. A. & Lazebnik, Y. (1998). Caspases: Enemies within. *Science, 281*, 1312-1316.

Werts, C., Girardin, S. E. & Philpott, D. J. (2006). TIR, CARD and PYRIN: three domains for an antimicrobial triad. *Cell Death Differ, 13*, 798-815.

Wiens, M. & Müller W. E. G. (2006). Cell death in Porifera: molecular players in the game of apoptotic cell death in living fossils. *Can J Zool, 84*, 307–321.

Wiens, M., Koziol, C., Hassanein, H. M. A., Batel, R. & Müller, W. E. G. (1998). Expression of the chaperones 14-3-3 and HSP70 induced by PCB 118 (2,3',4,4',5-pentachlorobiphenyl) in the marine sponge *Geodia cydonium*. *Mar Ecol Prog Ser, 165*, 247-257.

Wiens, M., Luckas, B., Brümmer, F., Shokry, M., Ammar, A., Steffen, R., Batel, R., Diehl-Seifert, B., Schröder, H. C. & Müller W. E. G. (2003). Okadaic acid: a potential defense molecule for the sponge *Suberites domuncula*. *Mar Biol, 142*, 213-223.

Wiens, M., Korzhev, M., Krasko, A., Thakur, N. L., Perović-Ottstadt, S., Breter, H. J., Ushijima, H., Diehl-Seifert, B., Müller, I. M. & Müller, W. E. G. (2005). Innate immune defense of the sponge *Suberites domuncula* against bacteria involves a MyD88-dependent signaling pathway. Induction of a perforin-like molecule. *J Biol Chem, 280*, 27949-27959.

Yu, S. W., Wang, H., Poitras, M. F., Coombs, C., Bowers, W. J., Federoff, H. J., Poirier, G. G., Dawson, T. M. & Dawson, V. L. (2002). Mediation of poly(ADP-ribose) polymerase-1-dependent cell death by apoptosis-inducing factor. *Science, 297*, 259-263.

Yuan, J. & Yanker, B. A. (2000). Apoptosis in the nervous system. *Nature, 407*, 802-809.

Zheng, L., Fisher, G., Miller, R. E., Peschon, J., Lynch, D. H. & Lenardo, M. J. 1995. Induction of apoptosis in mature T-cells by tumour necrosis factor. *Nature, 377*, 348-351.

Zhu, Y., Mahon, B. D., Froicu, M. & Cantorna, M. T. (2005). Calcium and 1α,25-dihydroxyvitamin D3 target the TNF-α pathway to supress experimental inflammatory bowel disease. *Eur J Immunol, 35*, 217-224.

INDEX

A

B

C

D

E

F

G

H

I

K

L

M

N

O

P

T

U

V

W

X

Y

Z